AF356384

Water Resource Management: Hydrological Modelling, Hydrological Cycles, and Hydrological Prediction

Water Resource Management: Hydrological Modelling, Hydrological Cycles, and Hydrological Prediction

Guest Editors

Agnieszka Rutkowska
Katarzyna Baran-Gurgul

Basel • Beijing • Wuhan • Barcelona • Belgrade • Novi Sad • Cluj • Manchester

Guest Editors

Agnieszka Rutkowska
Department of Applied
Mathematics
University of Agriculture
Cracow
Kraków
Poland

Katarzyna Baran-Gurgul
Department of Geoengineering
and Water Resources Management
Cracow University
of Technology
Kraków
Poland

Editorial Office
MDPI AG
Grosspeteranlage 5
4052 Basel, Switzerland

This is a reprint of the Special Issue, published open access by the journal *Water* (ISSN 2073-4441), freely accessible at: www.mdpi.com/journal/water/special_issues/UF438HK5CP.

For citation purposes, cite each article independently as indicated on the article page online and using the guide below:

Lastname, A.A.; Lastname, B.B. Article Title. *Journal Name* **Year**, *Volume Number*, Page Range.

ISBN 978-3-7258-3462-4 (Hbk)
ISBN 978-3-7258-3461-7 (PDF)
https://doi.org/10.3390/books978-3-7258-3461-7

Contents

About the Editors

Agnieszka Rutkowska

Agnieszka Rutkowska, PhD, is an assoc. prof. at the Department of Applied Mathematics at the University of Agriculture in Cracow, Poland, where she teaches mathematics, data analysis, information technology, and statistics applied to water management, meteorology, and environmental sciences. Her research interests cover applied statistics in hydrology and water management, particularly time series analysis; nonstationarity detection; hydro-climatic extreme events; temporal and spatial variability; the impact of climate change on hydrological characteristics; anthropogenic impacts on extreme events; precipitation indices, and drought indices. She has authored or co-authored over 40 journal and conference papers.

Katarzyna Baran-Gurgul

Katarzyna Baran-Gurgul, PhD Eng., is an assistant professor at the Cracow University of Technology, Faculty of Environmental Engineering and Energy. She graduated from the same Faculty and holds a PhD in Environmental Engineering Sciences (specialization: hydrology and applied statistics). At the Department of Geoengineering and Water Management, she deals with issues related to extreme hydrological phenomena (mainly droughts), the temporal and spatial variability of flows in rivers, and the application of statistics in environmental engineering, including the analysis of time series stationarity and the identification of the probability distributions of the hydrological processes' characteristics. She has authored or co-authored over 30 journal articles, conference papers, and academic textbooks. She is also a lecturer, teaching mainly statistics and riverbed hydraulics.

Preface

The reprint contains the papers that were published in the Special Issue of the journal Water, titled "Water Resource Management: Hydrological Modelling, Hydrological Cycles, and Hydrological Prediction". The issue was open between February 2023 and August 2024. The main objective was to design a platform for researchers to share their advances in various topics of water resource management, such as the assessment of river flow and precipitation variability, extreme events like floods and droughts, links between river flows and natural or human-induced catchment alterations, the assessment of uncertainty in hydrological predictions, and the impact of climate change on water resources. Ten original papers and one editorial were published. These papers cover the following research areas: water balance, hydrological modeling, seasonal and long-term variability, hydrological prediction, and the impact of anthropogenic activities on river flow changes.

We wish to express our gratitude to the authors who contributed their publications to the Special Issue. We sincerely appreciate your willingness to share your extensive knowledge and experience with our readers.

Our sincere gratitude also goes to the referees of the papers. We deeply appreciate your time and consideration. Your valuable and constructive comments made a significant contribution to improving the papers.

We also gratefully acknowledge the editorial team for their effort in making the Special Issue a success. Finally, our heartfelt thanks go to Ms. Helen Jing, Managing Editor. We deeply appreciate your generous support, inspiration, and the welcoming atmosphere you created.

Agnieszka Rutkowska and Katarzyna Baran-Gurgul
Guest Editors

Editorial

Water Resource Management: Hydrological Modelling, Hydrological Cycles, and Hydrological Prediction

Katarzyna Baran-Gurgul [1,*] and Agnieszka Rutkowska [2]

1   Department of Geoengineering and Water Resources Management, Faculty of Environmental Engineering and Energy, Cracow University of Technology, 31-155 Kraków, Poland
2   Department of Applied Mathematics, Faculty of Environmental Engineering and Land Surveying, University of Agriculture in Krakow, 30-198 Kraków, Poland; rmrutkow@cyf-kr.edu.pl
*   Correspondence: katarzyna.baran-gurgul@pk.edu.pl

Citation: Baran-Gurgul, K.; Rutkowska, A. Water Resource Management: Hydrological Modelling, Hydrological Cycles, and Hydrological Prediction. *Water* **2024**, *16*, 3689. https://doi.org/10.3390/w16243689

Received: 10 December 2024
Accepted: 13 December 2024
Published: 21 December 2024

1. Introduction

This editorial provides a definitive review of ten articles published in the Special Issue of the journal *Water*, entitled "Water Resource Management: Hydrological Modelling, Hydrological Cycles, and Hydrological Prediction". Each article makes an original and significant contribution to the field of water resource management. The ten articles address three core research areas: (a) water balance, hydrological modelling, and seasonal and multi-year variability, (b) hydrological prediction, and (c) the impact of anthropogenic activity on river flow changes. The papers present high-quality research, drawing practical conclusions and insights for implementing new water resources management methods, supported by case studies from different world regions. The proper assessment of river flow and precipitation variability, the frequency and severity of extreme events, and the correct development of hydrological forecasts become challenging tasks due to climate change and land-use and land cover changes, which strongly affect water resources. We are thankful to all the authors for their substantial contributions to the topics of this Special Issue. We also appreciate the efforts of the reviewers of the papers. Their valuable comments and suggestions helped the authors to improve the quality of their papers.

Water resources are impacted by external factors resulting from climate variability and human activity. Due to intensive development and industrialisation, the level of pollution in water resources is expected to rise, making water less available for various types of consumption. Different approaches, such as hydrological modelling and water allocation, can help forecast water availability under different climatic and institutional scenarios. Thus, the field of water management is constantly changing and developing, resulting in an increasing number of scientific papers. Using the Scopus database search engine, a bibliographic survey was carried out using the terms included in the Special Issue title, i.e., (Water Resource Management) AND ((Hydrological Modelling) OR (Hydrological Cycles) OR (Hydrological Prediction)). The number of documents containing these keywords constantly increased over the entire period under study, i.e., 1970–2024, but was small in the first several years—the number of documents recorded in Scopus per year between 1970 and 1987 did not exceed 100, and by 2002, it exceeded 1000 (Figure 1).

In the following years, a rapid increase in the number of published papers (up to a multi-year maximum of 22,020 papers in 2023) was observed, indicating the growing interest of the scientific community in issues related to water resource management. The largest number of documents found by these keywords is expected to come from the largest countries—the USA (with 52,143 documents published between 1970 and 2024) and China (51,637 documents). During the study period, more than 10,000 documents were also published by authors coming from the United Kingdom, Germany, India, Australia, and Canada. The editors of this Special Issue are from Poland, which, with 2988 documents published, is ranked 22nd in the number of documents found by Scopus according to keywords.

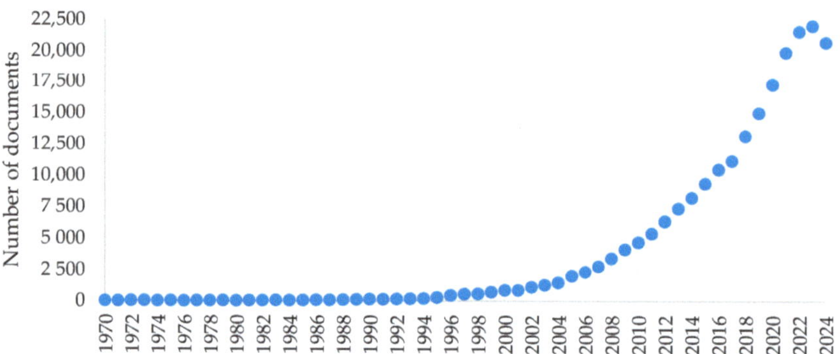

Figure 1. The number of documents in the period from 1970 to 2024 for the searched keywords (Water Resource Management) AND ((Hydrological Modelling) OR (Hydrological Cycles) OR (Hydrological Prediction)).

Water resource management encompasses the planning and control of water resources to ensure sufficient quantity and quality for people and ecosystems. Contemporary water management emphasises an Integrated Water Resources Management (IWRM) approach, which takes into account social, economic, and environmental needs, as well as climate change. IWRM is a process that promotes the coordinated development and management of water, land, and related resources to resultant economic and social welfare in an equitable manner without compromising the sustainability of vital ecosystems [1]. Legislation, such as the European Union's Water Framework Directive, plays a crucial role in supporting water quality protection and efficient water management at the international level [2].

Water resource management requires an integrated approach that includes understanding hydrological cycles and hydrological forecasting. Hydrological models allow for the simulation of flows and water use scenarios, which helps with retention planning and drought and flood protection. Research by Bisrat et al. [3] indicates that models such as SWAT-MODFLOW effectively contribute to assessing the impacts of climate change on water resources, enabling the development of crisis management strategies. These tools facilitate more sustainable water management, which not only protects against extreme weather events but also supports stable water supply, essential for the economy and ecosystems. Hydrological modelling and forecasting, combined with an understanding of hydrological cycles, provide the basis for effective water management, contributing to environmental protection and meeting the needs of the population.

Hydrological modelling, hydrological cycles, and hydrological prediction are integral components of hydrology, focusing on the understanding and management of water resources.

Hydrological modelling is the process of simulating water flows and hydrological responses in the natural environment. These models help analyse the impact of land use changes, climate change, and extreme weather events on water systems and play an important role in decision-making for water resource management, flood control, and environmental protection. The history of hydrological modelling, from the 1950s to the present, can be traced in Singh's work [4]. From the 1960s, when computers began to be used, it became possible to simulate the entire hydrological cycle; optimisation or operational study techniques were also developed, which formed the basis for reservoir management and operation and the simulation of entire river basins. Two- and three-dimensional models of groundwater were developed, as well as of infiltration and soil water flow or simultaneous simulation of water flow and sediment and pollutant transport. With the development of computational capabilities, the models used user-friendly software and tools for acquisition, storage, retrieval, processing, and dissemination. Remote sensing tools such as radar and satellites have been introduced to acquire spatial data for large areas, and geographic information systems (GIS) have been developed to process vast amounts of raster and vector data. Over the past two decades, there has been significant development in artificial neural networks, fuzzy logic, genetic programming, and wave models [4]. The recent studies proposed advanced decomposition techniques such as empirical mode decomposition (EMD), wavelet transform (WT), and variational mode decomposition (VMD) [5]. Currently, models such as SWAT (Soil and Water Assessment Tool) and MODFLOW are popular tools in hydrology. SWAT makes it possible to analyse processes related to land use, climate, and water management in catchments [3]. The long-term prediction of the impact of various factors on surface water and groundwater is possible.

The hydrological cycle, or water cycle, encompasses all processes through which water moves from land and ocean surfaces to the atmosphere and then returns to the Earth's surface as precipitation [6]. The hydrological cycle includes the processes, such as evaporation, condensation, precipitation, infiltration, and surface runoff, that regulate the circulation of water in nature. This is key for understanding the water resource availability in a given region. Understanding the hydrological cycles in a catchment and knowing the water yield potential of a catchment are essential for proper water planning and management in the area and for assessing water productivity [7]. Various human activities and interventions (e.g., groundwater and surface water exploitation, land-use changes, deforestation, reservoir construction, and inter–intra water transfer projects) directly and indirectly alter hydrological processes (e.g., evapotranspiration, infiltration, and runoff), with changes in these processes potentially influencing the development of droughts as specific hydrological events. Some of these changes may affect the hydrological drought characteristics, such as the deficit or severity [8]. Therefore, both climatic factors and human activities influence the hydrological cycle, making it essential for sustainable water management practices. Hydrological forecasting, based on advanced hydrological models and historical data, enables the prediction of future hydrological conditions and natural disaster risks. Regardless of the spatial detail of inputs, hydrological models inherently contain various uncertainties due to parameters, model structure, input data, and initial and boundary conditions [9]. Calibration and uncertainty estimation techniques, initially developed for lumped models with simpler structures, have been modified and applied to reduce the gap between distributed modelling and reality [10]. As shown by Hasan et al. [10], this approach leads to an improved forecast accuracy, provided high-resolution data are available.

2. An Overview of the Contributions to This Special Issue

In the Special Issue of *Water*, "Water Resource Management: Hydrological Modelling, Hydrological Cycles, and Hydrological Prediction", there are 10 scientific articles. It should be emphasised that the success of this Special Issue is due to the excellent work conducted by the 45 authors of the articles, as well as the wide range of issues discussed in them and the high quality of the research, the results of which are presented in the articles.

The submitted papers are strongly diverse, both in terms of the subject matter, methodology, and research area. As mentioned earlier, the subject of "water resource management" is universal and addressed in many countries; therefore, the research area in this Special Issue encompassed China in three cases and Serbia, Poland, Egypt, USA, South Korea, Romania, and Canada in the other cases.

2.1. Water Balance, Hydrological Modelling, and Seasonal and Long-Term Variability

The water balance of a catchment is a numerical summary of the individual components of the water cycle, distinguishing between inflows, i.e., water reaching the catchment in the form of precipitation, and outflows, including surface runoff, field evaporation, irrigation, and groundwater recharge. An understanding of the hydrological cycle helps in the sustainable management of water resources, especially during dry periods, as well as in making informed decisions regarding spatial planning.

In their work, Li et al. (Contribution 1) studied water balance components. They assessed the temporal and spatial distribution of surface runoff, actual evapotranspiration, and groundwater recharge in an arid area above the New Asyut Barrage (NAB) region in the Nile Valley in Upper Egypt. The authors used the WetSpass-M model for the period 2012–2020 in their calculations. The model input data were presented as raster maps, with such variables as the monthly meteorological data (temperature, precipitation, and wind speed), potential evapotranspiration, and leaf area index, as well as an elevation model, slope, land cover, irrigated area, soil map, and depth to ground water. The annual, seasonal (four seasons), and monthly scales were considered. The results showed that, at the annual scale, the groundwater recharge varied from 0 to 384 mm year^{-1}, the interception from 0 to 300 mm year^{-1}, surface runoff from 0 to 1198 mm year^{-1}, and evapotranspiration from 0.6 to 2910 mm year^{-1} in the study area. The months with the highest/lowest groundwater recharge were , September/January for evapotranspiration, and October/September for runoff. The northeast and southwest parts of the NAB experienced lower recharge than the other parts. The variability in recharge and evapotranspiration across different soil types was also shown.

The relationships between the monthly precipitation and irrigation water and the amount of real evapotranspiration, groundwater recharge, and surface runoff were found. A significant influence of the change in land use from agricultural to built-up areas with a decrease in the groundwater recharge, an increase in evapotranspiration, and an increase in surface runoff was identified.

Other authors of the articles in this section studied selected components of hydrological balance. Thus, Yang et al. (Contribution 2) dealt with groundwater, Augas et al. (Contribution 3)—snow, and Blagojević et al. (Contribution 4)—river flows.

Yang et al. (Contribution 2) studied the characteristics of the groundwater distribution of the middle and lower reaches of the Songhua River, an important agricultural region in northeastern China. The SWAT model, which was calibrated on the data from Tongjiang station, was used to simulate the runoff. Data from the period 2008–2016 were used in modelling. The MODFLOW model was introduced to compare and assess the simulation results of the SWAT model, which ascertained the characteristics of the shallow groundwater distribution and its renewal, recoverable volume, and groundwater levels in the Songhua River basin, divided into 32 sub-basins. The long-term changes in shallow aquifer water storage were identified; the rises in several sub-basins, with water resources predominantly derived from surface water, correspond with an increasing trend in precipitation from 2008 to 2016. The strength of the increase in the water storage in shallow aquifers is spatially diverse. Using the groundwater balance equation, it was found that the region is in a healthy extraction condition. When considering water storage and annual runoff, the driest year was 2011, while the wettest years were 2014 and 2016. Precipitation was identified as the primary source of groundwater recharge. Water loss can be assigned to absorption by plant roots from the superficial water layer. As regards the potential evapotranspiration, an increase in the 21st century was forecasted in the study area. The effectiveness of the SWAT and MODFLOW models in the study region was found.

In their paper, Augas et al. (Contribution 3) presented an extension of the monolayer model of a semi-distributed hydrological snow model (HYDROTEL) to a multilayer model that considers snow to be a combination of ice and air, while accounting for freezing rain. In the new model, some fundamental equations were modified, while the overall computational structure was preserved. The inclusion of new parameters was limited. The study was conducted at three stations in Canada using data from the period 2006–2011 (one station) and 2014–2017 (two stations). The performance measures (KGE, RMSE, and NSE) of the multilayer snow model for various parameter sets showed an improvement in comparison to the monolayer model. It was found that the multilayer model provides more precise estimates of the maximum snow water equivalent and total spring snowmelt dates, which is a result of the model's greater sensitivity to atmospheric thermal conditions. The authors emphasised that although the multilayer model generally improves the estimation of snow height, it shows excessive snow densities during the spring snowmelt. However, it seems that this model has the potential to improve the simulation of spring snowmelt, solving a common problem of the monolayer model.

The work of Blagojević et al. (Contribution 4) was devoted to detecting annual and seasonal changes in river flows using marginal distributions of daily flows. The analysis was carried out on daily flow sequences from ten river cross sections in Serbia, taking into account changes between two 30-year periods: 1961–1990 and 1991–2020. The probabilistic seasonal runoff pattern was constructed by determining quantiles from the marginal distributions of daily flows for each day of the year. By applying Fourier transform to the statistics of the daily flows' partial series, smooth periodical functions of the distribution parameters throughout the year and, consequently, of the quantiles were obtained. The decrease in runoff volume in most catchments was detected at the annual and seasonal scales (four seasons) with the strongest change of −37%. The change in runoff timing shown as the shift in the centroid date was also noted at the annual scale indicating an earlier occurrence of dry and average conditions. The main results are based on comparisons between the dry, average, and wet zones of hydrological conditions defined by quantiles of daily flows for selected probabilities. The authors noted that the relative change in runoff volume is most pronounced during extremely dry winter conditions, and the annual time shift is the largest in the dry and average zones.

2.2. Hydrological Forecasting

In this Special Issue, four papers were dedicated to hydrological forecasting. Hydrological forecasting, based on advanced hydrological models and historical data, enables the prediction of future hydrological conditions and the risk of natural disasters, which is important for effective water resources management.

In their work, Wang et al. (Contribution 5) introduced a hybrid machine learning model for river discharge forecasting. The authors used the LSTM (long short-term memory) model combined with variational mode decomposition (VMD) and principal component analysis (PCA). The study was conducted at the Waizhou station in the Ganjiang River Basin, China. Using the PCA method, stationary components were identified from 130 circulation indexes. The results confirm that the VMD-LSTM-LSTM model effectively addresses the issue of the low prediction accuracy at high flows caused by a limited number of samples. Compared to the single LSTM and VMD-LSTM models, this comprehensive approach significantly enhances the model's predictive accuracy during the flood season.

Xu et al. (Contribution 6) developed a hydrological real-time prediction model for runoff calculation that combines the Shuffled Complex Evolution-University of Arizona optimisation algorithm with the general unit hydrograph method. The main objective was to investigate the applicability of the general unit hydrograph in runoff calculations and its effectiveness in predicting flash flood events. In addition, the authors investigated the influence of changing the parameters of the general unit hydrograph on flood simulations and conducted a comparative analysis with the traditional Nash unit hydrograph. Data from 53 flood events were used. The results confirm that the general unit hydrograph method significantly reduced the calculation errors, improved the forecast accuracy, and significantly reduced the time difference between the peak-to-current time difference, thereby enhancing the simulation accuracy. The new model also showed robustness across various flood scenarios.

Kim et al. (Contribution 7) investigated the impact of model spatial resolution on streamflow prediction using high-resolution scenarios and parameter estimation. The study was conducted for the Geumho River catchment in South Korea using weather research and the forecasting hydrological modelling system with spatial resolutions of 100 m, 250 m, and 500 m. An automatic calibration tool based on the model-independent parameter estimation and uncertainty analysis method was developed. Two flood events with distinct origins were selected: the first flood resulted from a brief intense rainfall that occurred after long-term precipitation, while the second flood was caused by sudden heavy rainfall triggered by the impact of Typhoon Hinnamnor during very dry conditions. For both rainfall events, a significant improvement was observed after event-specific calibration at all resolutions. The highest quality of the model was shown for the spatial resolution of 250 m.

In their paper, Simeone et al. (Contribution 8) investigated the performance of two widely used hydrological models: the national water model and the national hydrologic model in characterising drought. The models' ability to classify periods of drought and non-drought was assessed, the error components were quantified, and the models' simulations of drought intensity, duration, and severity were evaluated. The study used data from over 4500 stations in the United States from 1985 to 2016. The analysis showed that the former model better simulated the timing of the drought flows, while the latter better predicted the drought flows. The authors noted that both models performed better in the wetter eastern regions than in the drier western regions of the country. Both models performed worse in regions that were most susceptible to drought.

2.3. The Influence of Anthropopressure on the Changes in River Flow

Anthropogenic pressure, including the construction of dams and barrages, urbanisation, river channel regulation, and agriculture affect the natural flow regime in the river, changing the timing, magnitude, and frequency of low and high flows.

In their study, Bărbulescu and Mohammed (Contribution 9) analysed changes in water discharge in the Buzău River after building one of Romania's largest dams, the Siru dam. The modifications in hydrological patterns were emphasised by a complex technique that involves the decomposition of time series into trends, seasonal indices, random components, and intrinsic mode functions. The hypothesis of stationarity in the flow series before and after construction of the dam was rejected for all series, with positive trends confirmed by the Mann–Kendall test. The multifractal analysis showed two distinct data series patterns. The authors observed a decline in seasonality indices, indicating a reduction in extreme events (high flows and floods). Empirical mode decomposition revealed different short-term and long-term patterns in the series. The main result, that there was a significant alteration in the river discharge after the dam's inception, provides a scientific background for flood modelling in the study area.

Mańko (Contribution 10) studied the impact of Międzyodrze revitalisation, including channel clearance and hydraulic structure repairs, on the flow patterns in the estuary of the Odra River in Poland. He used three computational scenarios: (1) treating Międzyodrze as an uncontrolled floodplain, (2) excluding Międzyodrze from the flow (as with past models), and (3) incorporating the hydraulic capacity of selected channels with hypothetical restoration. The computations were supported by the Hec-Ras software. The analyses and deductions validated the thesis proposed in this study that the potential process of channel dredging and renovation of the hydraulic infrastructure in Międzyodrze will significantly influence the flow distribution within the lower Odra River network. The curvilinear relationship between the global roughness coefficient and the prevailing flow was identified. The impact of the Międzyodrze area on water distribution in the lower course of the Odra River was successfully demonstrated.

3. Conclusions

The future of water resources management research depends on the understanding of the hydrological cycle and the interrelation of meteorological, hydrological, environmental, socioeconomic, and technological factors. The editors of this Special Issue guarantee that the work presented herein constitutes a substantial contribution to research and that the methodologies proposed by the authors will facilitate more precise modelling and forecasting of phenomena, thereby advancing understanding of water resources management.

Author Contributions: Conceptualization, K.B.-G.; writing–original draft preparation, K.B.-G. and A.R.; writing–review and editing, K.B.-G. and A.R.; supervision, K.B.-G. and A.R. All authors have read and agreed to the published version of the manuscript.

Funding: This research received no external funding.

Conflicts of Interest: The authors declare no conflicts of interest.

List of Contributions:

1. Li, Z.; Eladly, A.S.; Amen, E.M.; Salem, A.; Hassanien, M.M.; Yahya, K.E.; Liang, J. Spatial–Temporal Water Balance Evaluation in the Nile Valley Upstream of the New Assiut Barrage, Egypt, Using WetSpass-M. *Water* **2024**, *16*, 543. https://doi.org/10.3390/w16040543.
2. Yang, X.; Dai, C.; Liu, G.; Meng, X.; Li, C. Evaluation of Groundwater Resources in the Middle and Lower Reaches of Songhua River Based on SWAT Model. *Water* **2024**, *16*, 2839. https://doi.org/10.3390/w16192839.
3. Augas, J.; Foulon, É.; Rousseau, A.N.; Baraër, M. Extension of a Monolayer Energy-Budget Degree-Day Model to a Multilayer One. *Water* **2024**, *16*, 1089. https://doi.org/10.3390/w16081089.
4. Blagojević, B.; Mihailović, V.; Bogojević, A.; Plavšić, J. Detecting Annual and Seasonal Hydrological Change Using Marginal Distributions of Daily Flows. *Water* **2023**, *15*, 2919. https://doi.org/10.3390/w15162919.
5. Wang, W.; Tang, S.; Zou, J.; Li, D.; Ge, X.; Huang, J.; Yin, X. Runoff Prediction in Different Forecast Periods via a Hybrid Machine Learning Model for Ganjiang River Basin, China. *Water* **2024**, *16*, 1589. https://doi.org/10.3390/w16111589.
6. Xu, Y.; Liu, C.; Yu, Q.; Zhao, C.; Quan, L.; Hu, C. Study on a Hybrid Hydrological Forecasting Model SCE-GUH by Coupling SCE-UA Optimization Algorithm and General Unit Hydrograph. *Water* **2023**, *15*, 2783. https://doi.org/10.3390/w15152783.
7. Kim, B.; Lee, G.; Lee, Y.; Kim, S.; Noh, S.J. Assessment of the Impact of Spatial Variability on Streamflow Predictions Using High-Resolution Modeling and Parameter Estimation: Case Study of Geumho River Catchment, South Korea. *Water* **2024**, *16*, 591. https://doi.org/10.3390/w16040591.
8. Simeone, C.; Foks, S.; Towler, E.; Hodson, T.; Over, T. Evaluating Hydrologic Model Performance for Characterizing Streamflow Drought in the Conterminous United States. *Water* **2024**, *16*, 2996. https://doi.org/10.3390/w16202996.
9. Bărbulescu, A.; Mohammed, N. Study of the River Discharge Alteration. *Water* **2024**, *16*, 808. https://doi.org/10.3390/w16060808.
10. Mańko, R. Assessing the Effects of Międzyodrze Area Revitalization on Estuarine Flows in the Odra River. *Water* **2023**, *15*, 2926. https://doi.org/10.3390/w15162926.

References

1. Global Water Partnership (GWP). Policy brief. In *Social Equity: The Need for an Integrated Approach*; Global Water Partnership: Stockholm, Sweden, 2012.
2. Voulvoulis, N.; Arpon, K.D.; Giakoumis, T. The EU Water Framework Directive: From great expectations to problems with implementation. *Sci. Total Environ.* **2020**, *575*, 358–366. [CrossRef] [PubMed]
3. Yifru, B.A.; Lee, S.; Lim, K.J. Calibration and uncertainty analysis of integrated SWAT-MODFLOW model based on iterative ensemble smoother method for watershed scale river-aquifer interactions assessment. *Earth Sci. Inform.* **2023**, *16*, 3545–3561. [CrossRef]
4. Singh, V.P. Hydrologic modeling: Progress and future directions. *Geosci. Lett.* **2018**, *5*, 15. [CrossRef]
5. Mehr, A.D.; Shadkani, S.; Abualigah, L.; Safari, M.J.S.; Migdady, H. A novel stabilized artificial neural network model enhanced by variational mode decomposing. *Heliyon* **2024**, *10*, e34142. [CrossRef]
6. Ehtasham, L.; Sherani, S.H.; Nawaz, F. Acceleration of the hydrological cycle and its impact on water availability over land: An adverse effect of climate change. *Meteorol. Hydrol. Water Manag.* **2024**, *12*, 1–21. [CrossRef]
7. Gemechu, T.M.; Zhao, H.; Bao, S.; Yangzong, C.; Liu, Y.; Li, F.; Li, H. Estimation of Hydrological Components under Current and Future Climate Scenarios in Guder Catchment, Upper Abbay Basin, Ethiopia, Using the SWAT. *Sustainability* **2021**, *13*, 9689. [CrossRef]
8. Nasiri, N.; Asghari, K.; Besalatpour, A.A. Quantitative analysis of the human intervention impacts on hydrological drought in the Zayande-Rud River Basin, Iran. *J. Water Clim. Change* **2022**, *13*, 3473–3495. [CrossRef]

9.   Moges, E.Y.; Demissie, Y.; Li, H. Uncertainty Propagation in Coupled Hydrological Models using Winding Stairs and Null-Space Monte Carlo Methods. *J. Hydrol.* **2020**, *589*, 125341. [CrossRef]
10. Hasan, F.; Medley, P.; Drake, J.; Chen, G. Advancing Hydrology through Machine Learning: Insights, Challenges, and Future Directions Using the CAMELS, Caravan, GRDC, CHIRPS, PERSIANN, NLDAS, GLDAS, and GRACE Datasets. *Water* **2024**, *16*, 1904. [CrossRef]

water

Article

Evaluating Hydrologic Model Performance for Characterizing Streamflow Drought in the Conterminous United States

Caelan Simeone [1,*], Sydney Foks [2], Erin Towler [3], Timothy Hodson [4] and Thomas Over [5]

[1]   Oregon Water Science Center, U.S. Geological Survey, Portland, OR 97204, USA
[2]   Water Resources Mission Area, U.S. Geological Survey, Tacoma, WA 98402, USA; sfoks@usgs.gov
[3]   National Center for Atmospheric Research, Boulder, CO 80307, USA
[4]   Water Resources Mission Area, U.S. Geological Survey, Urbana, IL 61801, USA
[5]   Central Midwest Water Science Center, U.S. Geological Survey, Urbana, IL 61801, USA; tmover@usgs.gov
*   Correspondence: csimeone@usgs.gov

Abstract: Hydrologic models are the primary tools that are used to simulate streamflow drought and assess impacts. However, there is little consensus about how to evaluate the performance of these models, especially as hydrologic modeling moves toward larger spatial domains. This paper presents a comprehensive multi-objective approach to systematically evaluating the critical features in streamflow drought simulations performed by two widely used hydrological models. The evaluation approach captures how well a model classifies observed periods of drought and non-drought, quantifies error components during periods of drought, and assesses the models' simulations of drought severity, duration, and intensity. We apply this approach at 4662 U.S. Geological Survey streamflow gages covering a wide range of hydrologic conditions across the conterminous U.S. from 1985 to 2016 to evaluate streamflow drought using two national-scale hydrologic models: the National Water Model (NWM) and the National Hydrologic Model (NHM); therefore, a benchmark against which to evaluate additional models is provided. Using this approach, we find that generally the NWM better simulates the timing of flows during drought, while the NHM better simulates the magnitude of flows during drought. Both models performed better in wetter eastern regions than in drier western regions. Finally, each model showed increased error when simulating the most severe drought events.

Keywords: surface water; drought; United States; hydrological modeling; evaluation

Citation: Simeone, C.; Foks, S.; Towler, E.; Hodson, T.; Over, T. Evaluating Hydrologic Model Performance for Characterizing Streamflow Drought in the Conterminous United States. *Water* **2024**, *16*, 2996. https://doi.org/10.3390/w16202996

Academic Editor: David Post

Received: 13 August 2024
Revised: 19 September 2024
Accepted: 29 September 2024
Published: 21 October 2024

1. Introduction

Drought is a costly natural disaster, sometimes causing billions of dollars' worth of damage, and has wide-ranging impacts, from agriculture to public health [1–3]. While drought has received much attention in the hydrologic literature [4], drought is a complex phenomenon that is difficult to simulate [5–9]. Droughts are commonly separated into four categories: meteorological, agricultural (or soil moisture), hydrological, and socioeconomic [10–12]. Meteorological, agricultural, and hydrological droughts occur in physical systems, whereas socioeconomic droughts are the social and economic impacts of drought [12,13]. We focus on streamflow drought, which is a subset of hydrological drought, that Van Loon [11] defines as follows: "a lack of water in the hydrologic system manifesting itself in abnormally low streamflow in rivers." Streamflow drought can be costly and negatively impact many sectors that rely on streamflow quantity, quality, and timing, such as ecosystem, agricultural, navigation, and municipal services [1].

Due to the importance of streamflow drought, there have been ongoing efforts to simulate and predict drought occurrence and severity [8,14–19]. Many studies and models have shown skill in simulating certain drought events; however, there are many different types of drought indicators [11,20,21], and the methodologies used to evaluate models are

9

inconsistent [22]. Differences in the methodologies used to evaluate drought simulations can make model intercomparison difficult, as is the case more broadly in hydrology [23,24].

Streamflow drought is different to normal low flows. Although droughts may include periods of low streamflows, a recurring seasonal low-flow event is not necessarily a drought [25]. Numerous studies have aimed to quantify low flows (e.g., 7-day mean low-flow (Q7) [25]) and evaluate how well models simulate low-flow metrics [26–30], but these metrics alone are inadequate for evaluating model simulations of drought as they do not address the differences between low flows and drought, and often target a specific low-flow magnitude like Q7 instead of evaluating the full period of a drought.

Common model evaluation metrics, like the Nash–Sutcliffe efficiency (NSE; [31]), are often more sensitive to high flows than low flows, making them a poor indicator of predictive accuracy for drought [32]. To address this issue, drought studies sometimes use evaluation metrics on standardized drought indices (e.g., standardized streamflow index) or streamflow percentiles [11] instead of directly on the streamflow [6,33,34]. This offers some improvement, but often still focuses a large percentage of the evaluation on non-drought periods. Streamflow droughts are, by definition, abnormal events (often between the 2nd and 30th percentile [35,36]), meaning that, generally, 70–98% of streamflow data represent non-drought conditions. Applying metrics across non-drought periods may inflate or deflate model performance metrics, making them less indicative of a model's performance when simulating drought [32].

Studies also use different methods of identifying drought events, making model intercomparison and benchmarking difficult. For example, drought can be identified with a fixed method (where a river is in a drought when it drops below a single fixed long-term level) or a variable method (where the drought threshold varies seasonally based on how much water is typically available in that season) [7,18,35,37].

Previous efforts to evaluate the simulation of streamflow drought vary substantially (e.g., using different temporal resolutions, periods or seasons of interest, methods for drought characterization, and different methods for evaluation), which makes model evaluation intercomparisons challenging. One reason for this is that the focus of traditional hydrologic modeling has been on local or regional catchment scales [23], where evaluations typically vary study by study, depending on the application or location. Similarly, studies sometimes focus on a particular drought event (e.g., [33,34,38]), which by design is not conducive to model evaluation intercomparisons. However, as hydrologic modeling moves to encompass national and larger spatial scales, and toward longer (multi-year) runs, there is a need to provide systematic, comprehensive, comparable metrics to evaluate the streamflow drought modeled across hydroclimatic regions. Previous model intercomparison studies have provided systematic evaluation approaches for Earth system models [39] and long-term streamflow performance [24], but Towler et al. [24] only looked at one low-flow metric.

Given the shift toward standardized approaches to the construction and evaluation of national-scale hydrologic models and the importance of streamflow drought, we propose a systematic and comprehensive approach to evaluating simulations of streamflow drought. We demonstrate the approach through an assessment of the streamflow drought simulation performance of two large-scale hydrological modeling applications in the conterminous United States (CONUS): the National Water Model version 2.1 application of WRF-Hydro (NWM; [40]; accessed through NOAA's Office of Water Prediction) and the National Hydrologic Model application of the Precipitation-Runoff Modeling System version 1.0 three-step calibration with routing (NHM; [41–43]).

Specifically, we use three categories of metrics to evaluate model simulations of streamflow drought:

1. Classification: How well the models simulate the occurrence of observed drought vs. non-drought periods [44] according to Cohen's kappa [45]. Evaluating model simulation of drought occurrence is one of the simplest but most important measures of model performance in drought simulation.

2. Error Components: How well the models simulate the timing, magnitude, and variability of streamflow during periods of drought according to Spearman's r [46], percent bias, and the ratio of standard deviations, respectively. This approach examines how errors in streamflow are split across these three components [47].

3. Drought Signatures: How well the models simulate drought duration, intensity, and severity according to the normalized mean absolute errors of annualized data. These three drought characteristics are widely used by the research community and play a large role in the impact of droughts.

This suite of metrics captures many facets of streamflow drought simulation and evaluates them across many hydrologic environments. This approach extends existing model evaluations of drought (e.g., [48]), with additional focus on multi-objective evaluation that emphasizes the critical features of streamflow drought which are relevant to understanding error. The evaluation of these models across a wider number and density of gages provides a more robust understanding of models [49] and importantly captures the heterogeneity in drought responses, which, especially in mountainous regions, can have major implications for regional responses and water security [50]. Applying this comprehensive evaluation in a benchmarking framework for the intercomparison of the NWM and NHM modeling application allows for an improved understanding of the differences and potential limitations and benefits of each model [24]. We apply advances from large sample intercomparison studies [24,49,51] to continue to build on studies examining streamflow drought simulations (e.g., [48]).

2. Materials and Methods

Prior to evaluating streamflow drought using the three categories of metrics, we acquired data at stream gages with sufficient historical records, identified stream gages in regions with similar hydroclimatic characteristics, and identified periods of streamflow drought that were suitable for the demonstration of the drought evaluation framework.

2.1. Modeling Applications and Observed Data

2.1.1. Modeling Applications

The NHM and NWM are similar in their temporal and spatial extent (CONUS-wide), in which the hydrologic state and flux variables are simulated, and in their use in hydrologic estimation and research. However, these modeling applications have some key differences, including the spatial and temporal resolution, parameter estimation techniques, parameter datasets, calibrations, and forcings. The NHM is forced with the 1 km Daymet version 3 product [52], whereas the NWM is forced with the NOAA-produced 1 km Analysis of Record for Calibration version 1.0 forcing dataset [53]. The NWM calculates states and fluxes on a 1 km grid, while the NHM uses hydrological response units and a stream network as the primary geospatial structure [24]. The NWM runs on an hourly timestep, while the NHM is daily.

While the model calibration details can be found in [24,43], here we highlight several key aspects of calibration for each model. The NWM calibrates its parameters for hourly streamflow at 1378 observation stations, mostly in natural (unimpaired) basins. The calibration uses a modified Nash–Sutcliffe efficiency (NSE; [31,54]), which includes the standard NSE, as well as a log-transformed NSE. Then, the NWM employs hydrologic similarity to regionalize the parameters for the remaining watersheds. On the other hand, the NHM considers multiple objectives in a stepwise manner in its calibration routine. First, it balances the water budgets in each hydrologic response unit (HRU), considers the streamflow timing based on a statistically generated dataset, and finally calibrates to the observed streamflow at 1417 gage locations.

Both the NWM and the NHM have been evaluated for their performance in simulating streamflow (e.g., [24,27,55–58]), but have not yet been evaluated against other models for simulating streamflow drought; they thus serve as good candidates for our methodology. We used daily mean streamflow simulations direct from the NHM and averaged the hourly streamflow simulations from the NWM into a daily mean streamflow for each gage of interest. A daily timestep was chosen for comparison to preserve as much of the event timing relationship as possible while still accommodating common modeling application timescales (many model applications simulate results at a daily timestep). Further descriptions of these modeling applications and calibration techniques are explained in Towler et al. [24] and Hay et al. [43].

2.1.2. Observed Data

The drought performance of both model applications was evaluated at 4662 U.S. Geological Survey (USGS) stream gages across CONUS. We subset the stream gage dataset for benchmarking the hydrologic modeling applications described in Foks et al. [59] to include gages with at least 16 years of daily observations of flow (a longer record was needed to obtain sufficient data during drought periods). The study period spans the climate years (CYs, April 1–March 31) 1985–2016, which is the overlap between the NWM and the NHM historical simulations. Climate years more consistently contain the entire annual low-flow period than calendar years (January through December) or water years (October through September; [60,61]) in CONUS.

2.1.3. Evaluating Regional Performance

We categorized the stream gages into 12 hydrologic regions (Figure 1), defined by their correlation in monthly flows among the minimally altered gages in the Hydro-Climatic Data Network (HCDN; [62]), following approaches by McCabe and Wolock [63]. We then evaluated the regional differences in model performance when simulating streamflow drought by examining the model performance across all stream gages in each region. These same regions were used in the regional drought analysis by Hammond et al. [35]. We attributed stream gages not within the HCDN to the region of the nearest HCDN gage.

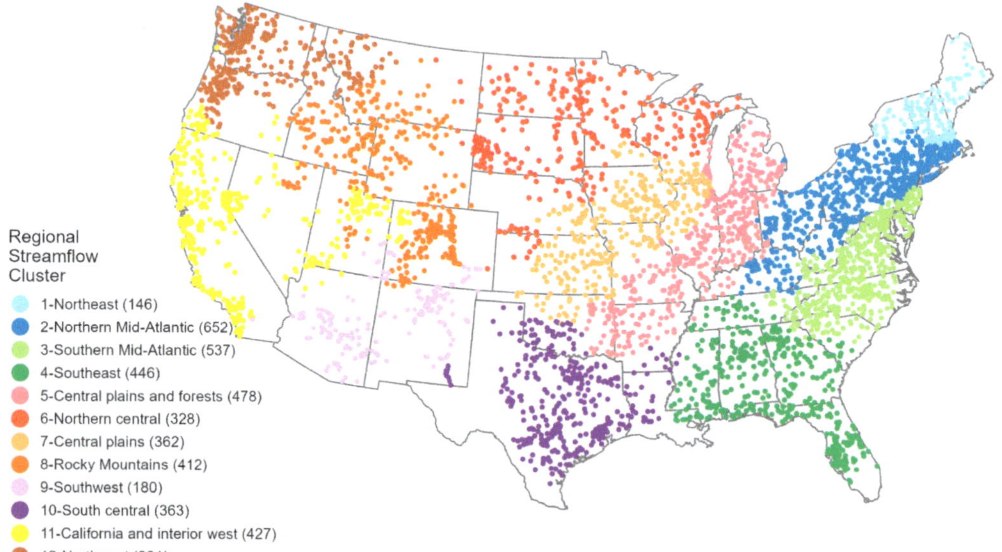

Figure 1. Streamflow gages (n = 4662) colored by regional streamflow cluster. The number of stream gages in each region is in parentheses.

2.2. Identification of Drought

We characterized drought using fixed and variable streamflow percentiles and thresholds at the 5th, 10th, 20th, and 30th percentiles, as implemented by Hammond et al. [35].

We converted the daily streamflow observations and model simulations to percentiles for all gages. Within observations or a single model, percentiles allow for the fair classification of drought across different hydroclimatic regions. Percentiles allow for more direct comparisons of low-flow anomalies (i.e., drought) between observations and models without the influence of model bias, which we evaluate elsewhere. The streamflow percentiles were computed with the Weibull plotting position ($r/(n + 1)$, where r is rank and n is the number of data (e.g., [64])). Two types of percentile-based threshold approaches were used: (1) fixed—all modeled or observed flows in the period on record are used to calculate one fixed threshold, and (2) variable—unique thresholds are calculated for each day of the year using only the values for that day from all years on record (Figure S1a,b). We implemented a modified version [65] of the combined threshold level and continuous dry period methods developed by Van Huijgevoort et al. [66] to handle the zero-flow measurements (<0.00028 cubic meters per second; <0.01 cubic feet per second). This method breaks ties between zero-flow days for percentile rankings based on the number of preceding zero-flow days, with days with more preceding zero-flow days receiving lower percentile rankings. Droughts were classified according to whether the streamflow was below the 5th, 10th, 20th, or 30th percentiles (Figure S1c), which roughly correspond to the extreme, severe, moderate, or abnormal drought classifications used by the U.S. Drought Monitor (https://droughtmonitor.unl.edu/, accessed 1 November 2023). This created a time series of drought presence and absence for each stream gage, threshold, and simulated and observed streamflow dataset. For each climate year, we calculated the following:

1. Drought duration: the total number of days below the threshold (days);
2. Drought severity: the sum of flow departures or deficit below the threshold (cms-days);
3. Drought intensity: the maximum drought intensity (minimum percentiles).

2.3. Evaluation for Drought Performance

We evaluate the performance of model simulations of drought by examining the goodness of the match [44] between modeled and observed events (i.e., event classification) and the quality of the simulation for matched events [44]. To evaluate the quality of the simulations during drought, we examine the error components during drought events (Spearman's r, ratio of standard deviations, and percent bias) and the errors in the important characteristics of drought (drought signatures of duration, severity, and intensity) aggregated to the climate year level. Table 1 shows the streamflow drought statistical metrics included in our systematic evaluation. We evaluated performance in three categories—"Event Classification", "Error Components", and "Drought Signatures"—which are described in the subsections below.

Table 1. Streamflow drought statistical metrics for daily streamflow drought evaluation for different metric categories. Additional calculation details can be found in Text S1.

Category	Statistic	Description	Range (Perfect)	Comments
Event Classification	Cohen's kappa	Cohen's kappa statistic for inter-rater reliability [45]	-1 to 1 (1)	A measure of agreement relative to the probability of achieving results by chance.
Error Components	Spearman's r	Spearman's rank correlation coefficient	-1 to 1 (1)	A nonparametric estimator of correlation for flow timing.
	Ratio of standard deviations	Ratio of simulated to observed standard deviations (for the scorecard this is presented as the absolute deviation from the target of 1)	0 to Inf (1)	Indicates if the flow variability is being over or underestimated.
	Percent bias	Percent bias (simulated minus observed) (for the scorecard this is presented as the absolute percent bias)	-100 to Inf (0)	Indicates if total streamflow volume is being over or underestimated.

Table 1. *Cont.*

Category	Statistic	Description	Range (Perfect)	Comments
Drought Signatures	Drought Duration	Normalized mean absolute error (NMAE) in the annual time series of drought duration, i.e., the sum of days of drought each year for a given threshold.	0 to Inf (0)	Indicates how well the model simulates annual drought durations.
	Drought Intensity	NMAE in the annual time series of the distance the minimum percentile is below the drought threshold, i.e., the overall maximum distance below the threshold for any drought during the year.	0 to Inf (0)	Indicates how well the model simulates annual minimum flow.
	Drought Severity (Flow Deficit Volume)	NMAE in the annual time series of drought deficit volume in cubic meters per second-days (cms-days), i.e., the sum of drought deficits for all droughts during the year.	0 to Inf (0)	Indicates how well the model simulates annual flow deficit. This is a measure of drought severity.

2.3.1. Event Classification Evaluation

Drought and non-drought periods were first classified for modeled and observed time series data. Evaluating if the model can correctly classify drought is one of the simplest but most important measures of model performance in drought simulation. If a model cannot correctly simulate when a drought is occurring, the quality of the rest of its predictions in the context of drought may not be particularly useful. We used Cohen's kappa to evaluate how well the simulated drought periods capture observed drought periods considering each day independently. Cohen's kappa measures the accuracy of the model classification and accounts for class imbalance in cases where categorical results are not evenly balanced (i.e., for the 20th percentile drought threshold, 80% of days will be non-drought). Cohen's kappa compares the relative observed agreement (true positives and true negatives from a contingency table or confusion matrix) to the expected agreement. Landis and Koch [67] provide guidelines for interpreting Cohen's kappa. They describe values < 0 as indicating no agreement and 0–0.20 as slight, 0.21–0.40 as fair, 0.41–0.60 as moderate, 0.61–0.80 as substantial, and 0.81–1 as almost perfect agreement [67]. Cohen's kappa was calculated as follows:

$$k = \frac{p_o - p_e}{1 - p_e} \tag{1}$$

where p_o is the relative observed agreement and p_e is the expected agreement or probability of agreement.

2.3.2. Error Components Evaluation

Commonly used metrics of efficiency like NSE are aggregations of model error components, but more information and insights can often be obtained when these error components are decomposed [47,68,69]. Gupta et al. [47] decomposed NSE and showed that "NSE consists of three distinctive components representing the correlation, the bias, and a measure of relative variability in the simulated and observed value." We evaluated these three individual components of model error rather than a single aggregated metric in order to provide insight into what might be driving model error during periods of drought. To evaluate the model performance during drought periods, we calculated the correlation (which shows errors from timing), bias (which shows errors in the magnitude of the streamflow distribution), and ratio of standard deviations (which evaluates errors in the variability of the streamflow distribution) of the simulated flows corresponding to the observed droughts. We followed a similar approach to that used by Pushpalatha et al. [32] (10th percentile of flow) and Pfannerstill et al. [30] (5th and 20th percentiles of flow) for evaluating the model performance for low flows and calculated our metrics only on values corresponding to drought events.

To evaluate the models' ability to reproduce the sequence of the observed time series, in other words, the timing of streamflow, during drought periods, we calculated Spearman's rank correlation coefficient [70,71]. Being based on the ranks of the flow magnitudes, Spearman's r depends only on the monotonicity of the relation between the observed and simulated flows. As a result, Spearman's r is a better estimator of the timing correlation than the more common Pearson estimator for streamflow data [72], since it is resistant to nonlinearity and skewness. Spearman's r is often used to assess flow timing to determine how well a model reproduces the relative position in time of flow values [24,73]. Our calculations began in the same way as those for standard Spearman's r, where the observed and modeled streamflow are each ranked by magnitude across the entire study period. After calculating ranks, however, we subset the data into periods with observed droughts at the various threshold levels of interest; for example, we took the lowest 20% of flow values for the 20th percentile drought threshold. Spearman's r is calculated on these subsets as follows:

$$\text{Spearman's r} = \frac{cov\left(R(obsQ)_{obs_{drought}}, \; R(simQ)_{obs_{drought}} \right)}{sigma\left(R(obsQ)_{obs_{drought}} \right) * sigma\left(R(simQ)_{obs_{drought}} \right)} \tag{2}$$

where *R(obsQ)* and *R(simQ)* are the ranks of observed and modeled discharges, which are then subset into periods of observed drought. When the streamflow was at zero, we used the continuous dry period method, as described in Section 2.2, to break ties for ranking.

To investigate whether the model over or underestimated the total streamflow volume during drought periods, we calculated the percent bias of the observed flows below the drought threshold versus the modeled flows below the drought threshold ([24]; note that these are thresholds of the equal streamflow percentile, not of the streamflow volume). Our implementation of percent bias for drought flows focused only on flows below a threshold, like the implementation in Yilmaz et al. [74], although they used a 30th percentile threshold while we address a range of thresholds. Percent bias is calculated as follows:

$$PBias = 100 * \frac{mean\left(simQ_{sim_{drought}} \right) - mean\left(obsQ_{obs_{drought}} \right)}{mean\left(obsQ_{obs_{drought}} \right)} \tag{3}$$

To provide a first-order estimate of errors in the statistical distribution of simulated flow magnitudes during periods of drought, we calculated the ratio of standard deviations between the modeled and observed streamflow (rSD) [24],

$$rSD = \frac{\sigma_{sim_{drought}}}{\sigma_{obs_{drought}}} \tag{4}$$

This metric shows the relative variability between simulation and observations [24,47,75] and indicates if the model has over- or under-simulated the variability during periods of drought. The combination of rSD and percent bias evaluates whether the distribution of the simulated drought flows matches the distribution of the observed drought flows. Note that in the scorecard summaries, we present the absolute percent bias and the difference in the rSD from 1 so that both metrics can be presented in a range from better to worse.

2.3.3. Drought Signatures Evaluation

Duration, intensity, and severity are fundamental characteristics of drought events [76–78] and are widely used throughout the literature covering drought. The duration, intensity, and severity of a drought influence the impacts that the event has on socioeconomic and ecological systems [79]. It is important to evaluate how well models simulate these three drought characteristics as they are so heavily used by the research community and play a large role in the impact of droughts. To capture the event characteristics throughout all periods of interest, we aggregated the drought events identified each year to determine the annual signatures of drought in each given climate year. We built on a previous methodology dedicated to hydrologic signatures [80,81] to understand the signatures and characteristics of droughts. The drought signatures we present are duration, severity, and intensity, which capture a range of the important characteristics of drought [35]. Annual resolution signatures provide a continuous and matched time series between simulated and observed data. We then calculated the normalized mean absolute error (NMAE) on this annual time series to compare across stream gages and models:

$$\text{NMAE} = \frac{\sum_{i=1}^{n}|sim_i - obs_i|}{n} \times \frac{1}{mean(obs)} \tag{5}$$

where n is the number of years, and sim_i and obs_i are the simulated and observed drought signature values (duration, severity, or intensity), respectively, for year i. Evaluating several different drought signature attributes is important as different indicators capture different information [35,82].

3. Results

We present our systematic model evaluation results for each metric category in three sections: event classification (Section 3.1), error components (Section 3.2), and drought signatures (Section 3.3). Figure 2 presents a scorecard with an overview of the model performance across all metrics for each model, threshold, and drought characterization method, with more detailed results presented in the following sections. Detailed result data from this systematic evaluation are published for both the NWM [83] and the NHM [84], so that they may be used as benchmarks against which other hydrologic modeling applications can be compared.

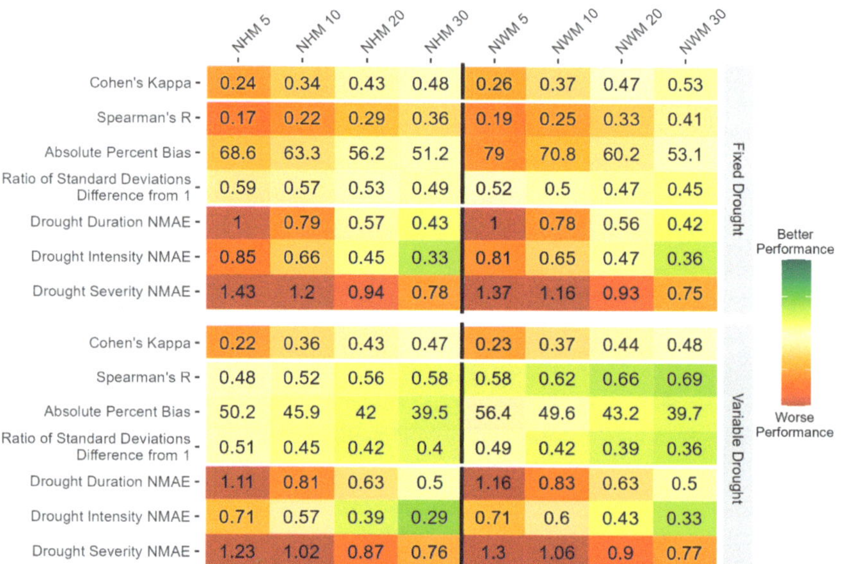

	NHM 5	NHM 10	NHM 20	NHM 30	NWM 5	NWM 10	NWM 20	NWM 30
Cohen's Kappa	0.24	0.34	0.43	0.48	0.26	0.37	0.47	0.53
Spearman's R	0.17	0.22	0.29	0.36	0.19	0.25	0.33	0.41
Absolute Percent Bias	68.6	63.3	56.2	51.2	79	70.8	60.2	53.1
Ratio of Standard Deviations Difference from 1	0.59	0.57	0.53	0.49	0.52	0.5	0.47	0.45
Drought Duration NMAE	1	0.79	0.57	0.43	1	0.78	0.56	0.42
Drought Intensity NMAE	0.85	0.66	0.45	0.33	0.81	0.65	0.47	0.36
Drought Severity NMAE	1.43	1.2	0.94	0.78	1.37	1.16	0.93	0.75
Cohen's Kappa	0.22	0.36	0.43	0.47	0.23	0.37	0.44	0.48
Spearman's R	0.48	0.52	0.56	0.58	0.58	0.62	0.66	0.69
Absolute Percent Bias	50.2	45.9	42	39.5	56.4	49.6	43.2	39.7
Ratio of Standard Deviations Difference from 1	0.51	0.45	0.42	0.4	0.49	0.42	0.39	0.36
Drought Duration NMAE	1.11	0.81	0.63	0.5	1.16	0.83	0.63	0.5
Drought Intensity NMAE	0.71	0.57	0.39	0.29	0.71	0.6	0.43	0.33
Drought Severity NMAE	1.23	1.02	0.87	0.76	1.3	1.06	0.9	0.77

Figure 2. Scorecard comparing median model evaluation results for the National Water Model (NWM) and the National Hydrologic Model (NHM). The scorecard columns are the models at each of the 5th, 10th, 20th, or 30th streamflow percentile thresholds. The rows are the metric categories of drought, namely event classification, error components, and drought signatures (Table 1), for fixed (**top**) and variable (**bottom**) drought methods. The metric values are in the corresponding box. Greener box colors indicate better performance, whereas redder box colors indicate poorer performance.

Percent bias is the absolute value of percent bias. The ratio of standard deviations is the absolute difference in the ratio of standard deviations from 1. Drought signatures are the normalized mean absolute error (NMAE) of drought duration, intensity, and severity.

3.1. Event Classification

The NWM and NHM simulated droughts had moderate agreement with the observed droughts (Cohen's kappa 0.41–0.60) for the fixed and variable methods at the 20th and 30th thresholds, but had only fair agreement (0.21–0.40) when simulating more severe droughts (5th and 10th percentiles; Figure 2). This difference highlights that examining both the fixed and variable methods over a range of thresholds is necessary in relaying the range and patterns of performance. The NWM had slightly but consistently higher Cohen's kappa values than the NHM across both methods (fixed and variable) and all thresholds of characterizing drought (Figure 2). Additionally, the difference in the Cohen's kappa between thresholds (for both the fixed and variable methods) is substantially larger than the difference between the modeling applications, with the NWM only showing the slightly better detection of drought events than the NHM, but there is a two-fold increase in the median between the most extreme threshold (5th percentile) and most moderate threshold (30th percentile) for each model and each method. Both modeling applications classify drought events similarly for fixed and variable drought methods, although the models have slightly more agreement between stream gages for the fixed drought methods (Figure S2).

The examination of model performance over varying hydroclimatic regions is needed to understand if hydrologic models can generalize and classify streamflow drought well across large national scales. We found that the overall streamflow drought classification performance of the NHM and the NWM varies by region, with regions in the wetter, eastern CONUS typically showing the better classification of drought events than regions in the drier, western CONUS (Figure 3). This is consistent across percentile thresholds and drought classification methods. One exception to the generally poorer drought event classification in the west is the wetter, most northwestern region of CONUS (region 12). The northwest region has a Cohen's kappa that is slightly higher than the national median for both the fixed and variable methods for the NHM and generally for the NWM, except for the 5th and 10th percentile fixed thresholds and the 30th percentile variable threshold.

There were some minor differences between the modeling applications in the west and east of CONUS, with the NHM classifying drought occurrence slightly more accurately than the NWM in the Northwest, California, and Interior West for all threshold types. The NWM classifies drought occurrence more accurately at the variable 20th percentile threshold than the NHM for much of central and eastern CONUS, but the NHM performs better in the Northwest and California and Interior West regions.

There was also more regional variability for the fixed threshold than the variable-threshold classification (Figures 3 and S3). The NWM also outperforms the NHM in eastern regions by a wider margin for the fixed method. The median values for each approach were similar across all thresholds, with a Cohen's kappa between 0.41 and 0.47 for the 20th percentile threshold.

To further distinguish the model performance beyond regions, we examine the model performance across various basin characteristics like aridity, the drainage area, and the baseflow index (Figure 4; the baseflow index (BFI) is the long-term fractional contribution of subsurface flow to streamflow). Both modeling applications had a poorer performance with regards to classifying drought in more arid watersheds using the fixed and variable-threshold methods (Figures 4d and S4). We found that both the NHM and NWM showed the worst simulations of drought occurrence for the 20% of stream gages with the highest BFI values (similar to [85,86]) and struggled for the 20% of stream gages with the lowest BFI (Figure 4e). The NHM and NWM also classified drought occurrence more accurately for moderate-sized basins (drainage area greater than 127.9 km^2 and less than 3238 km^2) than for the largest and smallest basins (Figure 4f).

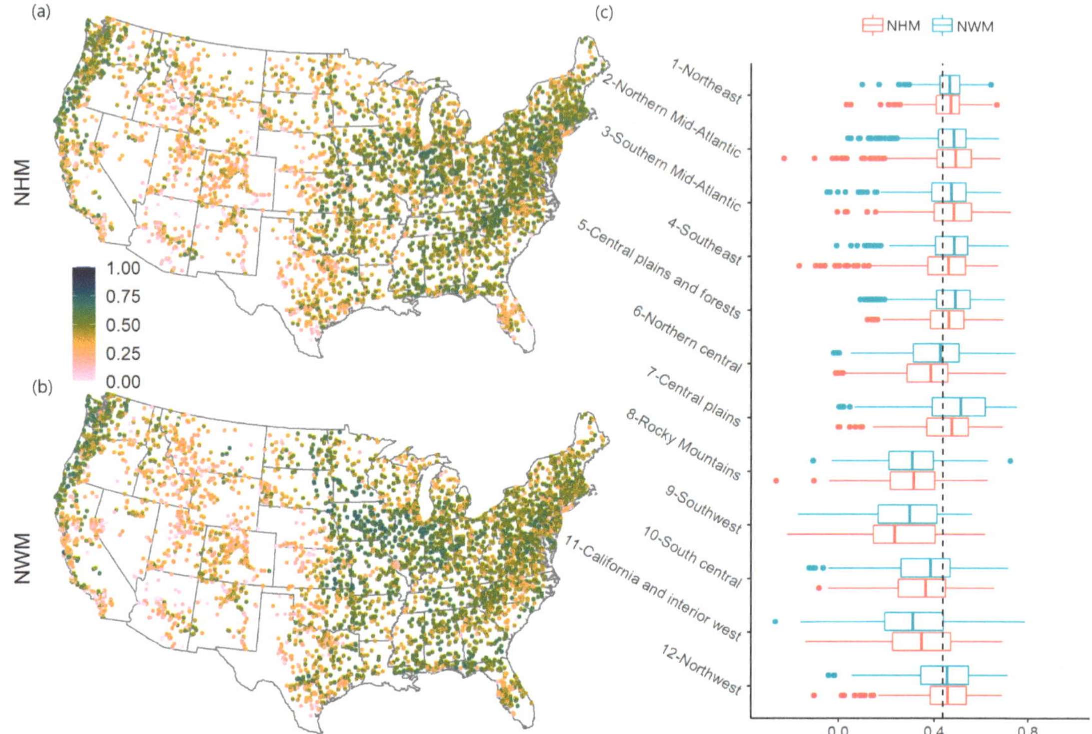

Figure 3. Maps comparing drought event classification results using Cohen's kappa for (**a**) the National Hydrologic Model (NHM) and (**b**) the National Water Model (NWM) for the variable drought method at the 20th percentile threshold. Darker colors indicate more agreement. The (**c**) box plot shows the results by region. The *y*-axes represent the 12 regions described in Figure 1. The *x*-axes represent the respective Cohen's kappa values. The NWM is in blue. The NHM is in red. The vertical black line is the median statistic value across all stream gages for both models. The event classification results using the fixed drought method are presented in the Supplementary Materials.

3.2. Error Components

Across all thresholds, the Spearman's r of the NWM was higher than that for the NHM, indicating that the NWM more accurately simulates the timing of streamflows during drought (Figures 2, S5 and S6). Like drought classification, the differences in Spearman's r between thresholds were larger than the differences between models for the fixed method (Figure 2). The timing of streamflows during drought was also more accurately simulated in wetter than dry regions. The NWM performs better in most regions, especially in the Northeast and Northern Mid-Atlantic regions, although the NHM does better in three out of the four westernmost regions (Northwest, Rocky Mountains, California and Interior West) for fixed-threshold drought (Figure S5). Both the NHM and NWM captured some of the timing of flows during the observed droughts, but not all, with overall moderate correlations for the variable threshold across percentile thresholds (NWM, 0.58–0.69; NHM 0.48–0.58; Figure 2; [87]). Performance for the fixed threshold was worse for both modeling applications, especially the more severe droughts (e.g., 5th percentile) where both models had relatively weak correlations with the observed flows during drought (NWM, 0.26; NHM 0.24).

Overall, the NWM had greater absolute bias than the NHM across all thresholds, indicating that the NHM more accurately simulates the streamflow volume during drought (Figure 2). Both modeling applications overestimate the streamflow during fixed drought

(20th percentile threshold median percent bias across stream gages of 9.4% NHM, and 34% NWM), which is in line with previous studies indicating that models commonly overestimate low flows [88–90]. Both modeling applications have lower median percent biases for streamflow during variable drought, although this bias was much higher for the NWM (20th percentile threshold median percent bias of −3.3% NHM and 22% NWM; Equation (3)). Consistent with the lower median bias, the bias in the NHM was also more balanced across sites: many sites over or underestimate drought flows, whereas the NWM overestimates drought flows for most sites, particularly in the central CONUS (Figure 5). Between modeling applications, the differences in bias were larger for more extreme droughts (5th and 10th percentile thresholds) than for modest droughts (20th and 30th percentile thresholds) (Figure 2). Additionally, the biases for individual stream gages were not strongly correlated between modeling applications (correlation for fixed drought = 0.22, correlation for variable drought = 0.25). The percent bias varies substantially by region, with smaller absolute biases in the wetter Northeastern and Northwest regions, but larger positive biases for dry western regions (No. 6, 9, 10, and 11), especially for the NWM (Figures 5c and S7c). While the median percent biases were well balanced across the CONUS for moderate droughts for the NHM and moderately overestimated flow for the NWM, it is important to note that the median absolute percent biases ranged from 39.5% for the 30th percentile NHM to 79% for the 5th percentile NWM (Figure 2) and that for certain stream gages, particularly in southwestern regions, the percent biases exceeded 400% (Figure 5).

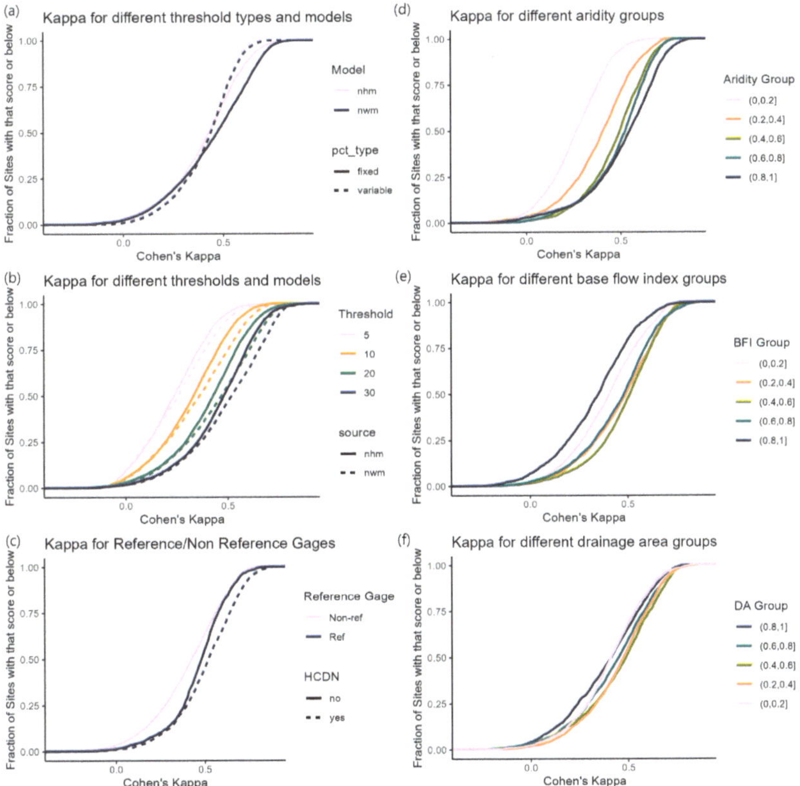

Figure 4. Cumulative distributions of model performance at stream gages for Cohen's kappa exploring model performance under a variety of different conditions and model groupings. Sub figures plot

Cohen's kappa comparing (**a**) the performance between the National Water Model (NWM) and the National Hydrologic Model (NHM) with fixed and variable methods, (**b**) the performance between the NWM and the NHM at 5th, 10th, 20th, and 30th percentile thresholds, (**c**) the performance at reference vs. non-reference [91] and Hydro-Climatic Data Network (HCDN) vs. non-HCDN gages, (**d**) the performance for various quantiles of aridity (Text S2), (**e**) the performance for various quantiles of the baseflow index (BFI; Text S2), and (**f**) the performance for various quantiles of drainage area (DA; Text S2). In cases where it is not explicit, the model performance is shown for the fixed method at the 20th percentile threshold combining both NWM and NHM evaluations.

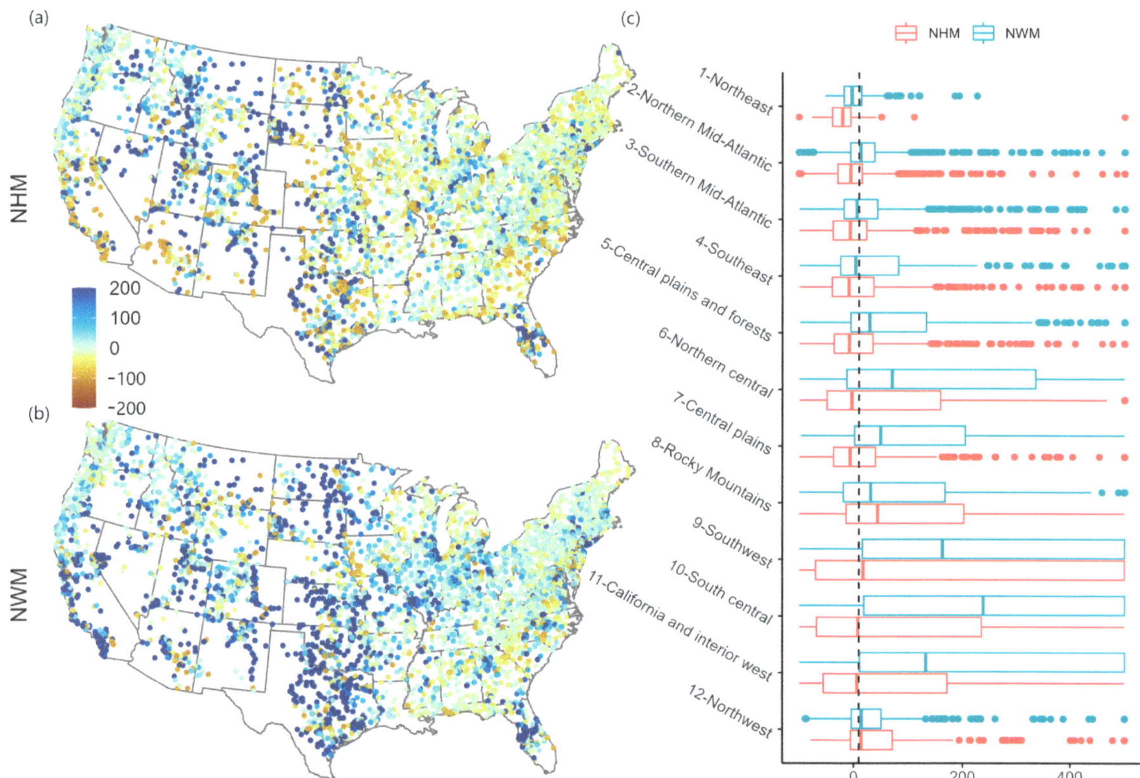

Figure 5. Maps comparing the percent bias of flows during drought for (**a**) the National Hydrologic Model (NHM) and (**b**) the National Water Model (NWM) for the variable drought method at the 20th percentile threshold. Lighter points indicate better results, while blue represents overestimates and red represents underestimates of flow. The (**c**) box plot shows results by region. The *y*-axis represents the 12 regions described in Figure 1. The *x*-axis represents the respective percent bias values. The NWM is in blue. The NHM is in red. The vertical black line is the median statistic value across all stream gages. Note that boxplot values can be above or below the target for the statistic. Additional error component results are shown in the Supplemental Information for Spearman's r using fixed (Figure S5) and variable methods (Figure S6), the percent bias using a fixed method (Figure S7), and the ratio of standard deviations using fixed (Figure S8) and variable methods (Figure S9).

The difference in the ratio of standard deviations from its target value of 1 is higher for the NHM than for the NWM across all drought thresholds (Figure 2). There is relatively little agreement between each model's ratio of standard deviations across stream g ages (fixed drought Spearman's r = 0.27; variable drought Spearman's r = 0.30). The spatial distributions of the ratio of standard deviations across the CONUS are variable

(Figures S8 and S9). We generally see the lowest ratio of standard deviation differences between the modeling applications and the lowest variability among stream gages in the central eastern CONUS. The West, other than the Northwest, has high variability in the ratios of standard deviations across stream gages. The NWM tends to overestimate the variability of drought flows in these regions, while the NHM median ratio of standard deviations is close to 1. Both modeling applications tend to underestimate the variability in the Northeast.

3.3. Drought Signatures

The NHM and the NWM perform similarly when simulating drought signatures, as measured by the NMAE values of the time sequence of the annual drought duration, intensity, and severity values, compared to the variability between different signature types and between different thresholds (Figures 2 and 6). Generally, both modeling applications have lower errors for drought duration and intensity, and higher errors for drought severity (Figure 2).

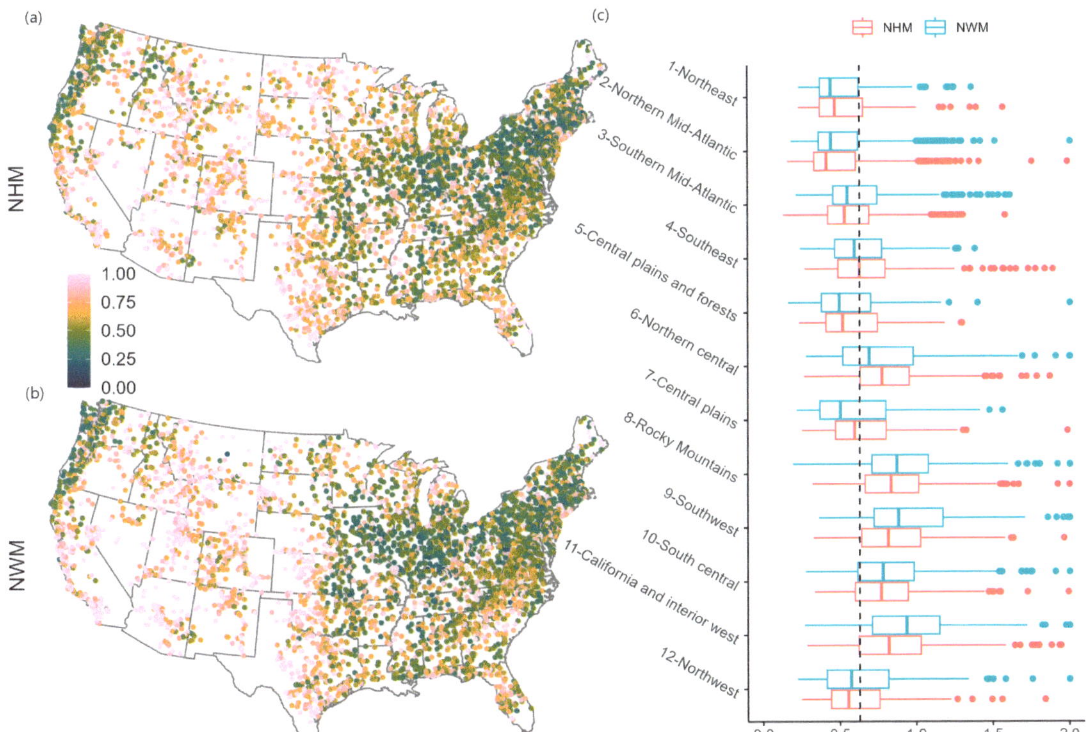

Figure 6. Maps presenting the normalized mean absolute error for (**a**) the National Hydrologic Model (NHM) and (**b**) the National Water Model (NWM) annual drought duration signature calculated with the variable drought method at the 20th percentile threshold. Darker colors represent lower errors between simulated and observed data. The (**c**) box plot shows the results by region. The *y*-axes represent the 12 regions described in Figure 1. The *x*-axes represent the normalized mean absolute error values. The NWM is in blue. The NHM is in red. The vertical black line is the median statistical value across all stream gages. Additional drought signature results are shown in the Supplemental Information for drought severity using fixed (Figure S10) and variable methods (Figure S11), drought intensity using fixed (Figure S12) and variable methods (Figure S13), and drought duration using a fixed method (Figure S14).

Both the NHM and NWM simulate lower drought duration errors using fixed-threshold methods than variable-threshold methods (Figure 2). Despite an overall difference of less than 5% in the NMAE values for drought duration, the correlation between models at individual stream gages is only moderate (fixed drought correlation = 0.81; variable drought correlation = 0.76). Both modeling applications were better at simulating the drought duration in the eastern and northwestern CONUS than in other regions. The better model also varies from region to region. The NWM performs better than the NHM in the Central Plains, but worse in the Southwest and California and Interior West.

Across all percentile thresholds, both the NHM and NWM have higher errors for drought severity than for drought duration and drought intensity, especially in the California and Interior West, Rocky Mountains, and Northern central plains (Figures S10 and S11). The NWM better simulates drought severity using the fixed drought method, while the NHM better simulates drought severity using the variable drought method. Both the NHM and NWM perform better at the 20th percentile threshold than at lower thresholds. Agreement between the NWM and the NHM at individual stream gages is lower for severity than for other drought signatures (fixed drought correlation = 0.56; variable drought correlation = 0.64), likely because drought severity is a measure integrating drought duration and intensity and because our drought duration and intensity signatures are standardized (unitless), unlike our severity signature of flow volume, which is inherently correlated with basin size and flow. Model performance is better in the eastern CONUS than in the western CONUS, with many stream gages in the western CONUS having NMAEs much greater than 1 (Figure 6). When simulating drought severity, the NWM tends to outperform the NHM in wet regions (1, 2, 3, 4, and 12) while the NHM performs better in drier regions (5–11).

Simulations of drought intensity are generally better than those of drought duration. The NHM better simulates drought intensity for both the variable and the fixed 20th percentile thresholds. The NWM, however, better simulates the 5th and 10th percentile fixed-threshold drought intensity. The agreement between the NWM and NHM performance for individual stream gages is greater for variable drought intensity than it is for drought duration (fixed drought correlation = 0.80; variable drought correlation = 0.84). Similar patterns emerge with regard to regional model performance, with a better ability to simulate drought intensity in the Northeast and Northern Mid-Atlantic regions and a poor ability in the western CONUS, apart from the coastal Northwest. The variability in performance is much less for drought intensity than it is for drought severity. With the 20th percentile variable-threshold drought method, the NHM performs better or equal to the NWM in nearly all regions.

Overall, both modeling applications struggle to simulate drought signatures, with a NMAE greater than 50% for all signatures except drought duration at the 30th percentile threshold and drought intensity at the 30th and 20th percentile thresholds. Both modeling applications show high error when simulating drought severity, with NMAE values ranging from 0.76 for the NHM at the 30th percentile variable threshold to 1.43 at the NHM 5th percentile variable threshold.

4. Discussion

4.1. Tradeoffs in Specific Model Performance

Each model has distinct tradeoffs when it comes to simulating aspects of drought. The "better" model depends heavily on the defined drought threshold and the method for determining drought severity (fixed or variable). We generally found that the NWM better reproduces drought occurrence, as well as the timing and variability of flow during droughts (as measured by Cohen's kappa, Spearman's r, and the ratio of the standard deviations, respectively), whereas the NHM better reproduces drought intensity signatures (for moderate and variable drought). Though our findings are consistent with those of Towler et al. [24], the additional streamflow drought metrics used in this study are needed to provide deeper insight into the models' performance during hydroclimatic drought that the low-flow percent bias metric in Yilmaz et al. [74] cannot provide alone.

Tradeoffs in performance for different streamflow error components like magnitude versus timing between hydrologic models are commonly identified in the literature, often with results similar to our findings. Gudmundsson et al. [92] similarly found distinct tradeoffs between models' mean and correlation errors (like metrics of magnitude and timing we use) when simulating European annual runoff cycles. Similar types of tradeoffs extend to simulations of drought. Some of these differences may be due to different initial development priorities for each modeling application (see Towler et al. [24], for modeling application descriptions). For example, the development of the NWM has focused on flood prediction and operates at an hourly timestep [93–95]. This may be the reason why the NWM better simulates the timing of drought events than the NHM. Another consideration is that the NHM was designed for water availability assessments, with a focus on simulating water quantity or magnitude, and operates on a daily timestep [96]. The calibration of the NHM included a step for matching the water balance volumes within each modeling unit, another step with the identified headwater catchments being calibrated using several non-streamflow datasets and a statistically generated streamflow to target the timing of streamflow; finally, the observed streamflow is used within each headwater catchment as a final calibration step [43]. Spatial frameworks are often different between models, where the NWM runs on a 1 km grid, while the NHM is based on hydrologic response units. These differences may have impacts on model performance. The calibration focus on water balance may be the reason why the NHM is better able to simulate streamflow volumes during drought rather than the timing. In the broader perspective, it is important to recognize each model's original purpose when interpreting the comparison results using these streamflow drought metrics, and generally any metrics used in benchmarking.

4.2. Model Performance Exhibits Regional Variation with Better Performance in Wetter Eastern Regions than in Drier Western Regions

The NWM and the NHM both simulate drought more poorly in drier western regions of the CONUS than in wetter eastern regions. Previous studies have also indicated poorer overall and drought-specific model simulations in the western CONUS [24,48,58]. Climate factors [92], including precipitation and aridity [58,97], have been widely found to influence model performance in different regions, so the poorer simulations of drought in drier regions are not surprising. These findings highlight an important area for model improvement since arid regions of the western CONUS are often the regions where drought is a major concern.

We also note that several important factors that influence the water cycle in western CONUS are not well captured by the hydrologic models used in this study. For example, human water use, diversions, and reservoir regulations, which are common throughout western watersheds [98], are not fully represented in either model. Additional processes that are not thoroughly represented, like lake and stream channel evaporation [99] or deep and complex groundwater systems [100], also may impact the model performance in the western CONUS. Other studies have shown that missing groundwater processes and not accounting for human modifications can result in poor model performance [73,101]. Benchmarking both models across our suite of metrics may help guide future development priorities in each model when looking to address issues such as the representation of human influences. If adding reservoir representation to the models, the NHM might benefit from more focus on the improved representation of the impacts of reservoirs on flow regime timing, as it had a poorer performance in simulating flow timing metrics (Spearman's r) during drought relative to the NWM. The NWM, on the other hand, could potentially benefit from more focus on the improved representation of the flow magnitude impacts of reservoirs, as it had a poorer performance when simulating the flow magnitude (percent bias) during drought relative to the NHM.

The relative performance also varies spatially. For example, the NHM simulates drought occurrence (as measured by Cohen's kappa) better in the western coastal regions (11 and 12) despite a worse national performance. In the southcentral CONUS, both modeling applications overestimate the magnitude and variability of drought flows by a large margin (like findings across all flows by Towler et al. [24]). Both modeling applications, especially the NWM, tend to better simulate the timing of flows during drought (Spearman's r) in the southcentral CONUS, with values close to the national median. The Rocky Mountain region is the reverse: both modeling applications simulate the magnitude and variability of flows during drought similarly well to how they do in other regions, but Spearman's r for both models is the lowest of all regions, which indicates difficulties simulating the timing of flows during drought in the Rocky Mountains. The regional differences in which the components of error each model perform well or poorly highlight some of the difficulties and potential tradeoffs between these models. Efforts focusing on improving the simulation of the magnitude of flows during drought may improve drought simulations for the southcentral CONUS, where the model has relatively high biases during periods of drought. In contrast, the Rocky Mountains, where the model simulations have a low Spearman's r with observations, might benefit most from efforts focused on improving the simulations of flow timing during drought.

We see similar tradeoffs between metrics in regions where both modeling applications perform relatively better. The NWM and NHM simulate the timing, magnitude, and variability of flows during drought well in the northeastern CONUS with median statistical values that are either the highest or close to the highest compared to other regions. The NHM performs less well when simulating the occurrence of fixed-threshold drought in the northeast relative to some other regions like the Northern and Southern Mid-Atlantic (although Cohen's kappa values are still at the national median). This indicates that the model's simulations during periods of drought are good, but that the NHM may not always correctly indicate that there is a drought relative to the Mid-Atlantic regions, where the NHM better simulates drought occurrence.

4.3. Larger Differences in Performance Between Thresholds than Between Models Highlights the Importance of Using Consistent Benchmarking Methods

The differences in Cohen's kappa, Spearman's r, the drought duration, and drought severity was greater between thresholds than between modeling applications. Both the NHM and NWM performed worse for more extreme droughts. This may in part reflect similar deficiencies in the climate input data, the applicability of the model equations to arid regions, or the model parameterizations [24]. Once calibrated, models typically have less variance than the original observations, so they tend to overpredict low flows and underpredict high flows [102]. However, we did not see more modeled than observed variability within these drought periods, as measured by the ratio of standard deviations. Both modeling applications may be able to simulate the impact of flow generation processes on moderate drought flows (e.g., 20 and 30%) but have difficulties simulating the processes that drive extreme drought flows (e.g., 5 and 10%). Additionally, these severe flows are hard to calibrate as they occur infrequently.

4.4. Summary

The results show a varied performance between the modeling applications that differs based on the metric being evaluated, the method of identifying drought, and the region or stream gage of interest. These results suggest that these models can provide benefits and useful information in certain contexts and regions, particularly for more moderate droughts in high-performance regions like the eastern CONUS. Users of these modeling applications should exert caution because some evaluation metrics indicate potential issues. For example, some stream gages have percent biases over 400% during periods of drought and some have high NMAE values when simulating drought severity. These factors suggest that the model may be problematic for certain applications, particularly in capturing more severe droughts in certain regions of CONUS such as the drier western regions.

The systematic evaluation of these two modeling applications highlights differences in performance by metric and region. These differences, along with the larger differences between drought thresholds than between modeling applications for many performance statistics, show the importance of the systematic benchmarking of different models and the use of the same methods of drought identification when comparing models. It also shows the importance of evaluating a range of different thresholds, in a range of different regions, using various performance metrics.

The consistent finding that both modeling applications have difficulties in simulating more extreme drought and drought in more arid regions highlights the limitations of both models and suggests a need for continued modeling improvement, as also indicated by others [26,73]. The improved simulation of severe drought flows and drought in arid regions is critical as these flows have large social and ecological impacts. This need for improved simulation may become increasingly important as, in some of these arid regions, extreme drought flows may become more common with a changing climate [103]. It may be especially important for regions such as the Colorado River Basin that experience major droughts with substantial impacts across social and ecological systems in the region [104,105].

4.5. Limitations

This study helps improve the understanding of model performance in simulating drought events, but there are important limitations to note. Our assessment does not control for differences in the model calibration or forcing datasets, which influence the simulated results. For ideal comparisons of the underlying models, both modeling applications would use the same forcings and calibration techniques; however, this can be impracticable for large simulations as they are typically funded by different entities and created for somewhat different purposes. Similarly, the evaluation is performed on simulations at stream gages used for calibration and therefore does not assess out-sample errors, but it is likewise usually impractical in large-domain model implementations to perform the simulations needed to address this issue. Nevertheless, benchmarking the specific implementations of modeling applications or built-for-comparison modeling applications still gives insight into how each can be used for research and improvement. Our study assesses both fixed and variable methods of characterizing drought at four thresholds, but there are many other methods of characterizing streamflow drought that could yield different evaluation results. The autocorrelation of daily values is not considered in our study design, though this is critical to address if planning to perform statistical significance testing between modeling applications. Uncertainty in hydrologic simulations stems from several sources, including uncertainties in the input forcings, as well as from the hydrological model structure, process representation, absence of anthropogenic processes, and parameterization. Although we do not explicitly investigate model uncertainty nor their interactions here, we do note that we use two different hydrological models (NHM and NWM), which could be conducive to developing a multi-model ensemble; studies have shown that the ensemble mean can outperform individual models [106].

5. Conclusions

This study presents a comprehensive approach to evaluating simulations of streamflow drought and applies it to two conterminous U.S.-scale hydrologic modeling applications, the National Water Model (NWM) and the National Hydrologic Model (NHM). Our comparisons between the NWM and the NHM show varied results. The NWM tends to better simulate the timing of streamflow during drought events (measured by Spearman's r) while the NHM tends to better simulate the magnitude of flow during drought events (measured by percent bias). There were also strong spatial trends in the drought simulation performance, with both the NHM and NWM performing better in wetter regions than drier ones, creating a stark east versus west divide in performance. Thus, both modeling applications perform worse in drought simulations for the regions that are most susceptible to drought. Finally, the differences in performance were typically greater between different drought thresholds than between either modeling application, with both the NHM and NWM exhibiting difficulties in simulating the most severe drought events.

Supplementary Materials: The following supporting information can be downloaded at: https://www.mdpi.com/article/10.3390/w16202996/s1, Text S1. Description of Metric Calculations. Text S2. Description of Aridity, Baseflow Index, Drainage Area, and HCDN and Reference Gage Calculations. Figure S1. Figure of our workflow of steps to characterize drought and evaluation model performance. Figure S2. Scatter plot figures comparing National Water Model (NWM) (x-axis) and National Hydrologic Model (NHM) (y-axis) performance for each metric at the 20th percentile threshold. Figure S3. Maps comparing drought event classification results using Cohen's kappa. Figure S4. Cumulative distributions of model performance at stream gages for Cohen's kappa exploring model performance for the National Hydrologic Model (NHM) and National Water Model (NWM) for fixed and variable drought methods, subset into groups based on the aridity quantiles. Figure S5. Maps comparing Spearman's r of flows during drought. Figure S6. Maps comparing Spearman's r of flows during drought. Figure S7. Maps comparing percent bias of flows during drought. Figure S8. Maps comparing the ratio of standard deviations of flows during drought. Figure S9. Maps comparing the ratio of standard deviations of flows during drought. Figure S10. Maps presenting the normalized mean absolute error. Figure S11. Maps presenting the normalized mean absolute error. Figure S12. Maps presenting the normalized mean absolute error. Figure S13. Maps presenting the normalized mean absolute error. Figure S14. Maps presenting the normalized mean absolute error.

Author Contributions: C.S.: Conceptualization, Data Curation, Formal Analysis, Methodology, Investigation, Writing—original draft, Writing—review and editing, Visualization, Project administration. S.F.: Conceptualization, Data Curation, Methodology, Investigation, Writing—original draft, Writing—review and editing, Project administration. E.T.: Conceptualization, Methodology, Writing—original draft, Writing—review and editing. T.H.: Conceptualization, Methodology, Writing—review and editing. T.O.: Conceptualization, Methodology, Writing—review and editing. All authors have read and agreed to the published version of the manuscript.

Funding: This research was supported by the U.S. Geological Survey Water Mission Area Hydro-terrestrial Earth System Testbed (HyTEST) project. Any use of trade, firm, or product names is for descriptive purposes only and does not imply endorsement by the U.S. Government.

Data Availability Statement: All data, software, and additional details of the methodology used in this study are publicly available from Simeone et al. [83,84] and Simeone and Foks [107].

Conflicts of Interest: The authors declare that they have no known competing financial interests or personal relationships that could have appeared to influence the work reported in this paper.

References

1. Wlostowski, A.N.; Jennings, K.S.; Bash, R.E.; Burkhardt, J.; Wobus, C.W.; Aggett, G. Dry landscapes and parched economies: A review of how drought impacts nonagricultural socioeconomic sectors in the US Intermountain West. *Wiley Interdiscip. Rev. Water* **2022**, *9*, e1571. [CrossRef]

2. Smith, A.B.; Matthews, J.L. Quantifying uncertainty and variable sensitivity within the US billion-dollar weather and climate disaster cost estimates. *Nat. Hazards* **2015**, *77*, 1829–1851. [CrossRef]

3. NOAA. National Centers for Environmental Information (NCEI) U.S. Billion-Dollar Weather and Climate Disasters. Retrieved from Billion-Dollar Weather and Climate Disasters | National Centers for Environmental Information (NCEI); 2022. Available online: https://www.ncei.noaa.gov/access/billions/ (accessed on 1 October 2021).

4. Hasan, H.H.; Razali, S.F.M.; Muhammad, N.S.; Ahmad, A. Research trends of hydrological drought: A systematic review. *Water* **2019**, *11*, 2252. [CrossRef]

5. Van Huijgevoort, M.H.J.; Hazenberg, P.; Van Lanen, H.A.J.; Teuling, A.J.; Clark, D.B.; Folwell, S.; Gosling, S.N.; Hanasaki, N.; Heinke, J.; Koirala, S.; et al. Global multimodel analysis of drought in runoff for the second half of the twentieth century. *J. Hydrometeorol.* **2013**, *14*, 1535–1552. [CrossRef]

6. Quintana-Seguí, P.; Barella-Ortiz, A.; Regueiro-Sanfiz, S.; Míguez-Macho, G. The Utility of Land-Surface Model Simulations to Provide Drought Information in a Water Management Context Using Global and Local Forcing Datasets. *Water Resour. Manag.* **2020**, *34*, 2135–2156. [CrossRef]

7. Stahl, K.; Vidal, J.P.; Hannaford, J.; Tijdeman, E.; Laaha, G.; Gauster, T.; Tallaksen, L.M. The challenges of hydrological drought definition, quantification and communication: An interdisciplinary perspective. *Proc. Int. Assoc. Hydrol. Sci.* **2020**, *383*, 291–295. [CrossRef]

8. Brunner, M.I.; Slater, L.; Tallaksen, L.M.; Clark, M. Challenges in modeling and predicting floods and droughts: A review. *WIREs Water* **2021**, *8*, e1520. [CrossRef]

9. Rivera, J.A.; Infanti, J.M.; Kumar, R.; Mutemi, J.N. Challenges of Hydrological Drought Monitoring and Prediction. *Front. Water* **2021**, *3*, 750311. [CrossRef]

10. Tallaksen, L.M.; Van Lanen, H.A. (Eds.) Hydrological Drought: Processes and Estimation Methods for Streamflow and Groundwater. 2004. Available online: https://hdl.handle.net/11311/1256137 (accessed on 1 October 2021).

11. Van Loon, A.F. Hydrological drought explained. *WIREs Water* **2015**, *2*, 359–392. [CrossRef]

12. Wilhite, D.A.; Glantz, M.H. Understanding the drought phenomenon: The role of definitions. *Water Int.* **1985**, *10*, 111–120. [CrossRef]

13. Guo, Y.; Huang, S.; Huang, Q.; Wang, H.; Fang, W.; Yang, Y.; Wang, L. Assessing socioeconomic drought based on an improved multivariate standardized reliability and resilience index. *J. Hydrol.* **2019**, *568*, 904–918. [CrossRef]

14. Mishra, A.K.; Singh, V.P. Drought modeling—A review. *J. Hydrol.* **2011**, *403*, 157–175. [CrossRef]

15. Hao, Z.; Singh, V.P.; Xia, Y. Seasonal drought prediction: Advances, challenges, and future prospects. *Rev. Geophys.* **2018**, *56*, 108–141. [CrossRef]

16. Smith, K.A.; Barker, L.J.; Tanguy, M.; Parry, S.; Harrigan, S.; Legg, T.P.; Prudhomme, C.; Hannaford, J. A multi-objective ensemble approach to hydrological modelling in the UK: An application to historic drought reconstruction. *Hydrol. Earth Syst. Sci.* **2019**, *23*, 3247–3268. [CrossRef]

17. Fung, K.F.; Huang, Y.F.; Koo, C.H.; Soh, Y.W. Drought forecasting: A review of modelling approaches 2007–2017. *J. Water Clim. Change* **2020**, *11*, 771–799. [CrossRef]

18. Sutanto, S.J.; Van Lanen, H.A. Streamflow drought: Implication of drought definitions and its application for drought forecasting. *Hydrol. Earth Syst. Sci.* **2021**, *25*, 3991–4023. [CrossRef]

19. Dyer, J.; Mercer, A.; Raczyński, K. Identifying Spatial Patterns of Hydrologic Drought over the Southeast US Using Retrospective National Water Model Simulations. *Water* **2022**, *14*, 1525. [CrossRef]

20. Yihdego, Y.; Vaheddoost, B.; Al-Weshah, R.A. Drought indices and indicators revisited. *Arab. J. Geosci.* **2019**, *12*, 69. [CrossRef]

21. Faiz, M.A.; Zhang, Y.; Ma, N.; Baig, F.; Naz, F.; Niaz, Y. Drought indices: Aggregation is necessary or is it only the researcher's choice? *Water Supply* **2021**, *21*, 3987–4002. [CrossRef]

22. Alawsi, M.A.; Zubaidi, S.L.; Al-Bdairi, N.S.S.; Al-Ansari, N.; Hashim, K. Drought Forecasting: A Review and Assessment of the Hybrid Techniques and Data Pre-Processing. *Hydrology* **2022**, *9*, 115. [CrossRef]

23. Archfield, S.A.; Clark, M.; Arheimer, B.; Hay, L.E.; McMillan, H.; Kiang, J.E.; Seibert, J.; Hakala, K.; Bock, A.; Wagener, T.; et al. Accelerating advances in continental domain hydrologic modeling. *Water Resour. Res.* **2015**, *51*, 10078–10091. [CrossRef]

24. Towler, E.; Foks, S.S.; Dugger, A.L.; Dickinson, J.E.; Essaid, H.I.; Gochis, D.; Viger, R.J.; Zhang, Y. Benchmarking high-resolution hydrologic model performance of long-term retrospective streamflow simulations in the contiguous United States. *Hydrol. Earth Syst. Sci.* **2023**, *27*, 1809–1825. [CrossRef]

25. Smakhtin, V.U. Low flow hydrology: A review. *J. Hydrol.* **2001**, *240*, 147–186. [CrossRef]

26. Nicolle, P.; Pushpalatha, R.; Perrin, C.; François, D.; Thiéry, D.; Mathevet, T.; Le Lay, M.; Besson, F.; Soubeyroux, J.-M.; Viel, C.; et al. Benchmarking hydrological models for low-flow simulation and forecasting on French catchments. *Hydrol. Earth Syst. Sci.* **2014**, *18*, 2829–2857. [CrossRef]

27. Hodgkins, G.A.; Dudley, R.W.; Russell, A.M.; LaFontaine, J.H. Comparing trends in modeled and observed streamflows at minimally altered basins in the United States. *Water* **2020**, *12*, 1728. [CrossRef]

28. Mubialiwo, A.; Abebe, A.; Onyutha, C. Performance of rainfall–runoff models in reproducing hydrological extremes: A case of the River Malaba sub-catchment. *SN Appl. Sci.* **2021**, *3*, 515. [CrossRef]

29. Worland, S.C.; Farmer, W.H.; Kiang, J.E. Improving predictions of hydrological low-flow indices in ungaged basins using machine learning. *Environ. Model. Softw.* **2018**, *101*, 169–182. [CrossRef]

30. Pfannerstill, M.; Guse, B.; Fohrer, N. Smart low flow signature metrics for an improved overall performance evaluation of hydrological models. *J. Hydrol.* **2014**, *510*, 447–458. [CrossRef]

31. Nash, J.E.; Sutcliffe, J.V. River flow forecasting through conceptual models part I—A discussion of principles. *J. Hydrol.* **1970**, *10*, 282–290. [CrossRef]

32. Pushpalatha, R.; Perrin, C.; Le Moine, N.; Andreassian, V. A review of efficiency criteria suitable for evaluating low-flow simulations. *J. Hydrol.* **2012**, *420*, 171–182. [CrossRef]

33. Dehghani, M.; Saghafian, B.; Rivaz, F.; Khodadadi, A. Evaluation of dynamic regression and artificial neural networks models for real-time hydrological drought forecasting. *Arab. J. Geosci.* **2017**, *10*, 266. [CrossRef]

34. Barella-Ortiz, A.; Quintana-Seguí, P. Evaluation of drought representation and propagation in regional climate model simulations across Spain. *Hydrol. Earth Syst. Sci.* **2019**, *23*, 5111–5131. [CrossRef]

35. Hammond, J.C.; Simeone, C.; Hecht, J.S.; Hodgkins, G.A.; Lombard, M.; McCabe, G.; Wolock, D.; Wieczorek, M.; Olson, C.; Caldwell, T.; et al. Going Beyond Low Flows: Streamflow Drought Deficit and Duration Illuminate Distinct Spatiotemporal Drought Patterns and Trends in the U.S. During the Last Century. *Water Resour. Res.* **2022**, *58*, e2022WR031930. [CrossRef]

36. Heudorfer, B.; Stahl, K. Comparison of different threshold level methods for drought propagation analysis in Germany. *Hydrol. Res.* **2017**, *48*, 1311–1326. [CrossRef]

37. Sarailidis, G.; Vasiliades, L.; Loukas, A. Analysis of streamflow droughts using fixed and variable thresholds. *Hydrol. Process.* **2019**, *33*, 414–431. [CrossRef]

38. Jehanzaib, M.; Bilal Idrees, M.; Kim, D.; Kim, T.W. Comprehensive evaluation of machine learning techniques for hydrological drought forecasting. *J. Irrig. Drain. Eng.* **2021**, *147*, 04021022. [CrossRef]

39. Collier, N.; Hoffman, F.M.; Lawrence, D.M.; Keppel-Aleks, G.; Koven, C.D.; Riley, W.J.; Mu, M.; Randerson, J.T. The International Land Model Benchmarking (ILAMB) System: Design, Theory, and Implementation. *J. Adv. Model. Earth Syst.* **2018**, *10*, 2731–2754. [CrossRef]

40. Gochis, D.J.; Barlage, M.; Cabell, R.; Casali, M.; Dugger, A.; FitzGerald, K.; McAllister, M.; McCreight, J.; RafieeiNasab, A.; Read, L.; et al. The WRF-Hydro® Modeling System Technical Description, (Version 5.1.1). *NCAR Technical Note.* 2020. 107p. Available online: https://ral.ucar.edu/sites/default/files/public/projects/wrf-hydro/technical-description-user-guide/wrf-hydrov5.2technicaldescription.pdf (accessed on 1 October 2021).

41. Regan, R.S.; Markstrom, S.L.; Hay, L.E.; Viger, R.J.; Norton, P.A.; Driscoll, J.M.; LaFontaine, J.H. Description of the national hydrologic model for use with the precipitation-runoff modeling system (prms) (No. 6-B9). In *US Geological Survey Techniques and Methods*; U.S. Geological Survey: Reston, VA, USA, 2018. [CrossRef]

42. Hay, L.E.; LaFontaine, J.H. *Application of the National Hydrologic Model Infrastructure with the Precipitation-Runoff Modeling System (NHM-PRMS), 1980–2016, Daymet Version 3 Calibration [Data Set]*; U.S. Geological Survey: Reston, VA, USA, 2020. [CrossRef]

43. Hay, L.E.; LaFontaine, J.H.; Van Beusekom, A.E.; Norton, P.A.; Farmer, W.H.; Regan, R.S.; Markstrom, S.L.; Dickinson, J.E. Parameter estimation at the conterminous United States scale and streamflow routing enhancements for the National Hydrologic Model infrastructure application of the Precipitation-Runoff Modeling System (NHM-PRMS). In *U.S. Geological Survey Techniques and Methods*; U.S. Geological Survey: Reston, VA, USA, 2023; Chapter B10, 50p. [CrossRef]

44. Zhao, L. Event prediction in the big data era: A systematic survey. *ACM Comput. Surv. (CSUR)* **2021**, *54*, 1–37. [CrossRef]

45. Cohen, J. A coefficient of agreement for nominal scales. *Educ. Psychol. Meas.* **1960**, *20*, 37–46. [CrossRef]

46. Spearman, C. The Proof and Measurement of Association Between Two Things. In *Studies in Individual Differences: The Search for Intelligence*; Jenkins, J.J., Paterson, D.G., Eds.; Appleton-Century-Crofts: New York, NY, USA, 1961; pp. 45–58.

47. Gupta, H.V.; Kling, H.; Yilmaz, K.K.; Martinez, G.F. Decomposition of the mean squared error and NSE performance criteria: Implications for improving hydrological modelling. *J. Hydrol.* **2009**, *377*, 80–91. [CrossRef]

48. Hughes, M.; Jackson, D.L.; Unruh, D.; Wang, H.; Hobbins, M.; Ogden, F.L.; Cifelli, R.; Cosgrove, B.; DeWitt, D.; Dugger, A.; et al. Evaluation of retrospective National Water Model Soil moisture and streamflow for drought-monitoring applications. *J. Geophys. Res. Atmos.* **2024**, *129*, e2023JD038522. [CrossRef]

49. Addor, N.; Do, H.X.; Alvarez-Garreton, C.; Coxon, G.; Fowler, K.; Mendoza, P.A. Large-sample hydrology: Recent progress, guidelines for new datasets and grand challenges. *Hydrol. Sci. J.* **2020**, *65*, 712–725. [CrossRef]

50. Bales, R.C.; Goulden, M.L.; Hunsaker, C.T.; Conklin, M.H.; Hartsough, P.C.; O'Geen, A.T.; Hopmans, J.W.; Safeeq, M. Mechanisms controlling the impact of multi-year drought on mountain hydrology. *Sci. Rep.* **2018**, *8*, 690. [CrossRef] [PubMed]

51. Gupta, H.V.; Perrin, C.; Blöschl, G.; Montanari, A.; Kumar, R.; Clark, M.; Andréassian, V. Large-sample hydrology: A need to balance depth with breadth. *Hydrol. Earth Syst. Sci.* **2014**, *18*, 463–477. [CrossRef]

52. Thornton, P.E.; Thornton, M.M.; Mayer, B.W.; Wei, Y.; Devarakonda, R.; Vose, R.S.; Cook, R.B. *Daymet: Daily Surface Weather Data on a 1-km Grid for North America, Version 3*; ORNL DAAC: Oak Ridge, TN, USA, 2017. [CrossRef]

53. Fall, G.; Kitzmiller, D.; Pavlovic, S.; Zhang, Z.; Patrick, N.; St. Laurent, M.; Trypaluk, C.; Wu, W.; Miller, D. The Office of Water Prediction's Analysis of Record for Calibration, version 1.1: Dataset description and precipitation evaluation. *JAWRA J. Am. Water Resour. Assoc.* **2023**, *59*, 1246–1272. [CrossRef]

54. Krause, P.; Boyle, D.P.; Base, F. Comparison of different efficiency criteria for hydrological model assessment. *Adv. Geosci.* **2005**, *5*, 89–97. [CrossRef]

55. Gochis, D.J.; Cosgrove, B.; Dugger, A.L.; Karsten, L.; Sampson, K.M.; McCreight, J.L.; Flowers, T.; Clark, E.P.; Vukicevic, T.; Salas, F.R.; et al. Multi-variate evaluation of the NOAA National Water Model. In *AGU Fall Meeting*; NSF National Center for Atmospheric Research: Boulder, CO, USA, 2018.

56. Lahmers, T.M.; Hazenberg, P.; Gupta, H.; Castro, C.; Gochis, D.; Dugger, A.; Yates, D.; Read, L.; Karsten, L.; Wang, Y.-H. Evaluation of NOAA National Water Model Parameter Calibration in Semiarid Environments Prone to Channel Infiltration. *J. Hydrometeorol.* **2021**, *22*, 2939–2969.

57. Hodgkins, G.A.; Over, T.M.; Dudley, R.W.; Russell, A.M.; LaFontaine, J.H. The consequences of neglecting reservoir storage in national-scale hydrologic models: An appraisal of key streamflow statistics. *J. Am. Water Resour. Assoc.* **2023**, *60*, 110–131. [CrossRef]

58. Johnson, J.M.; Fang, S.; Sankarasubramanian, A.; Rad, A.M.; da Cunha, L.K.; Jennings, K.S.; Clarke, K.C.; Mazrooei, A.; Yeghiazarian, L. Comprehensive analysis of the NOAA National Water Model: A call for heterogeneous formulations and diagnostic model selection. *J. Geophys. Res. Atmos.* **2023**, *128*, e2023JD038534. [CrossRef]

59. Foks, S.S.; Towler, E.; Hodson, T.O.; Bock, A.R.; Dickinson, J.E.; Dugger, A.L.; Dunne, K.A.; Essaid, H.I.; Miles, K.A.; Over, T.M.; et al. Streamflow benchmark locations for conterminous United States (cobalt gages). In *U.S. Geological Survey Data Release*; U.S. Geological Survey: Reston, VA, USA, 2022. [CrossRef]
60. Carpenter, D.H.; Hayes, D.C. Low-flow characteristics of streams in Maryland and Delaware. *Water-Resour. Investig. Rep.* **1996**, *94*, 4020.
61. Feaster, T.D.; Lee, K.G. Low-flow frequency and flow-duration characteristics of selected streams in Alabama through March 2014. In *U.S. Geological Survey Scientific Investigations Report 2017–5083*; U.S. Geological Survey: Reston, VA, USA, 2017; 371p. [CrossRef]
62. Lins, H.F. USGS hydro-climatic data network 2009 (HCDN-2009). In *US Geological Survey Fact*; U.S. Geological Survey: Reston, VA, USA, 2012; Sheet 2012-3047.
63. McCabe, G.J.; Wolock, D.M. Clusters of monthly streamflow values with similar temporal patterns at 555 HCDN (Hydro-Climatic Data Network) sites for the period 1981 to 2019. In *U.S. Geological Survey Data Release*; U.S. Geological Survey: Reston, VA, USA, 2022. [CrossRef]
64. Laaha, G.; Gauster, T.; Tallaksen, L.M.; Vidal, J.P.; Stahl, K.; Prudhomme, C.; Heudorfer, B.; Vlnas, R.; Ionita, M.; Van Lanen, H.A.J.; et al. The European 2015 drought from a hydrological perspective. *Hydrol. Earth Syst. Sci.* **2017**, *21*, 3001–3024. [CrossRef]
65. Simeone, C.E. Streamflow Drought Metrics for select GAGES-II streamgages for three different time periods from 1921–2020. In *U.S. Geological Survey Data Release*; U.S. Geological Survey: Reston, VA, USA, 2022. [CrossRef]
66. Van Huijgevoort, M.H.J.; Hazenberg, P.; Van Lanen, H.A.J.; Uijlenhoet, R. A generic method for hydrological drought identification across different climate regions. *Hydrol. Earth Syst. Sci.* **2012**, *16*, 2437–2451. [CrossRef]
67. Landis, J.R.; Koch, G.G. The measurement of observer agreement for categorical data. *Biometrics* **1977**, *33*, 159–174. [CrossRef] [PubMed]
68. Clark, M.P.; Vogel, R.M.; Lamontagne, J.R.; Mizukami, N.; Knoben, W.J.; Tang, G.; Gharari, S.; Freer, J.E.; Whitfield, P.H.; Shook, K.R.; et al. The abuse of popular performance metrics in hydrologic modeling. *Water Resour. Res.* **2021**, *57*, e2020WR029001. [CrossRef]
69. Hodson, T.O.; Over, T.M.; Foks, S.S. Mean squared error, deconstructed. *J. Adv. Model. Earth Syst.* **2021**, *13*, e2021MS002681. [CrossRef]
70. Helsel, D.R.; Hirsch, R.M.; Ryberg, K.R.; Archfield, S.A.; Gilroy, E.J. Statistical methods in water resources. In *U.S. Geological Survey Techniques and Methods*; [Supersedes USGS Techniques of Water-Resources Investigations, Book 4, Chapter A3, version 1.1.]; U.S. Geological Survey: Reston, VA, USA, 2020; Book 4, Chapter A3, 458p. [CrossRef]
71. Yue, S.; Pilon, P.; Cavadias, G. Power of the Mann-Kendall and Spearman's rho tests for detecting monotonic trends in hydrological series. *J. Hydrol.* **2002**, *259*, 254–271. [CrossRef]
72. Barber, C.; Lamontagne, J.R.; Vogel, R.M. Improved estimators of correlation and R2 for skewed hydrologic data. *Hydrol. Sci. J.* **2020**, *65*, 87–101. [CrossRef]
73. Tijerina-Kreuzer, D.; Condon, L.; FitzGerald, K.; Dugger, A.; O'neill, M.M.; Sampson, K.; Gochis, D.; Maxwell, R. Continental hydrologic intercomparison project, phase 1: A large-scale hydrologic model comparison over the continental United States. *Water Resour. Res.* **2021**, *57*, e2020WR028931. [CrossRef]
74. Yilmaz, K.K.; Gupta, H.V.; Wagener, T. A process-based diagnostic approach to model evaluation: Application to the NWS distributed hydrologic model. *Water Resour. Res.* **2008**, *44*. [CrossRef]
75. Newman, A.J.; Mizukami, N.; Clark, M.P.; Wood, A.W.; Nijssen, B. Benchmarking of a physically based hydrologic model. *J. Hydrometeorol.* **2017**, *18*, 2215–2225. [CrossRef]
76. Yevjevich, V.M. *An Objective Approach to Definitions and Investigations of Continental Hydrologic Droughts*; Colorado State University: Fort Collins, CO, USA, 1967; Volume 23, p. 25.
77. Dracup, J.A.; Lee, K.S.; Paulson, E.G., Jr. On the statistical characteristics of drought events. *Water Resour. Res.* **1980**, *16*, 289–296. [CrossRef]
78. Mishra, A.K.; Singh, V.P. A review of drought concepts. *J. Hydrol.* **2010**, *391*, 202–216. [CrossRef]
79. Noel, M.; Bathke, D.; Fuchs, B.; Gutzmer, D.; Haigh, T.; Hayes, M.; Poděbradská, M.; Shield, C.; Smith, K.; Svoboda, M. Linking drought impacts to drought severity at the state level. *Bull. Am. Meteorol. Soc.* **2020**, *101*, E1312–E1321. [CrossRef]
80. Addor, N.; Nearing, G.; Prieto, C.; Newman, A.J.; Le Vine, N.; Clark, M.P. A ranking of hydrological signatures based on their predictability in space. *Water Resour. Res.* **2018**, *54*, 8792–8812. [CrossRef]
81. McMillan, H.K. A review of hydrologic signatures and their applications. *WIREs Water* **2021**, *8*, e1499. [CrossRef]
82. Pournasiri Poshtiri, M.; Towler, E.; Pal, I. Characterizing and understanding the variability of streamflow drought indicators within the USA. *Hydrol. Sci. J.* **2018**, *63*, 1791–1803. [CrossRef]
83. Simeone, C.; Leah, S.; Katharine, K. Results of benchmarking National Water Model v2.1 simulations of streamflow drought duration, severity, deficit, and occurrence in the conterminous United States. In *U.S. Geological Survey Data Release*; U.S. Geological Survey: Reston, VA, USA, 2024.
84. Simeone, C.; Leah, S.; Katharine, K. Results of benchmarking National Hydrologic Model application of the Precipitation-Runoff Modeling System (v1.0 byObsMuskingum) simulations of streamflow drought duration, severity, deficit, and occurrence in the conterminous United States. In *U.S. Geological Survey Data Release*; U.S. Geological Survey: Reston, VA, USA, 2024; Chapter B10, 50p.
85. Rudd, A.C.; Bell, V.A.; Kay, A.L. National-scale analysis of simulated hydrological droughts (1891–2015). *J. Hydrol.* **2017**, *550*, 368–385. [CrossRef]

86. Massmann, C. Identification of factors influencing hydrologic model performance using a top-down approach in a large number of US catchments. *Hydrol. Process.* **2020**, *34*, 4–20. [CrossRef]

87. Overholser, B.R.; Sowinski, K.M. Biostatistics primer: Part 2. *Nutr. Clin. Pract.* **2008**, *23*, 76–84. [CrossRef]

88. Farmer, W.H.; Vogel, R.M. On the deterministic and stochastic use of hydrologic models. *Water Resour. Res.* **2016**, *52*, 5619–5633. [CrossRef]

89. Moges, E.; Ruddell, B.L.; Zhang, L.; Driscoll, J.M.; Norton, P.; Perez, F.; Larsen, L.G. HydroBench: Jupyter supported reproducible hydrological model benchmarking and diagnostic tool. *Front. Earth Sci.* **2022**, *10*, 884766. [CrossRef]

90. Wan, T.; Covert, B.H.; Kroll, C.N.; Ferguson, C.R. An Assessment of the National Water Model's Ability to Reproduce Drought Series in the Northeastern United States. *J. Hydrometeorol.* **2022**, *23*, 1929–1943. [CrossRef]

91. Falcone, J.A. *GAGES-II: Geospatial Attributes of Gages for Evaluating Streamflow*; U.S. Geological Survey: Reston, VA, USA, 2011. [CrossRef]

92. Gudmundsson, L.; Wagener, T.; Tallaksen, L.M.; Engeland, K. Evaluation of nine large-scale hydrological models with respect to the seasonal runoff climatology in Europe: Land Surface Models Evaluation. *Water Resour. Res.* **2012**, *48*. [CrossRef]

93. Maidment, D.R. Conceptual Framework for the National Flood Interoperability Experiment. *J. Am. Water Resour. Assoc.* **2016**, *53*, 245–257. [CrossRef]

94. NOAA. National Water Model CONUS Retrospective Dataset. Available online: https://registry.opendata.aws/nwm-archive (accessed on 1 October 2021).

95. UCAR. Supporting the NOAA National Water Model. 2019. Available online: https://ral.ucar.edu/projects/supporting-the-noaa-national-water-model (accessed on 1 October 2021).

96. Regan, R.S.; Juracek, K.E.; Hay, L.E.; Markstrom, S.L.; Viger, R.J.; Driscoll, J.M.; LaFontaine, J.H.; Norton, P.A. The US Geological Survey National Hydrologic Model infrastructure: Rationale, description, and application of a watershed-scale model for the conterminous United States. *Environ. Model. Softw.* **2019**, *111*, 192–203. [CrossRef]

97. Hansen, C.; Shiva, J.S.; McDonald, S.; Nabors, A. Assessing retrospective National Water Model streamflow with respect to droughts and low flows in the Colorado River basin. *JAWRA J. Am. Water Resour. Assoc.* **2019**, *55*, 964–975. [CrossRef]

98. Carlisle, D.; Wolock, D.M.; Konrad, C.P.; McCabe, G.J.; Eng, K.; Grantham, T.E.; Mahler, B. *Flow Modification in the Nation's Streams and Rivers*; US Department of the Interior, US Geological Survey: Reston, VA, USA, 2019.

99. Friedrich, K.; Grossman, R.L.; Huntington, J.; Blanken, P.D.; Lenters, J.; Holman, K.D.; Gochis, D.; Livneh, B.; Prairie, J.; Skeie, E.; et al. Reservoir evaporation in the Western United States: Current science, challenges, and future needs. *Bull. Am. Meteorol. Soc.* **2018**, *99*, 167–187. [CrossRef]

100. Hare, D.K.; Helton, A.M.; Johnson, Z.C.; Lane, J.W.; Briggs, M.A. Continental-scale analysis of shallow and deep groundwater contributions to streams. *Nat. Commun.* **2021**, *12*, 1450. [CrossRef]

101. Lane, R.A.; Coxon, G.; Freer, J.E.; Wagener, T.; Johnes, P.J.; Bloomfield, J.P.; Greene, S.; Macleod, C.J.A.; Reaney, S.M. Benchmarking the predictive capability of hydrological models for river flow and flood peak predictions across over 1000 catchments in Great Britain. *Hydrol. Earth Syst. Sci.* **2019**, *23*, 4011–4032. [CrossRef]

102. Vogel, R.M. Editorial: Stochastic and deterministic world views. *J. Water Resour. Plan. Manag.* **1999**, *125*, 311–313. [CrossRef]

103. Ahmadalipour, A.; Moradkhani, H.; Svoboda, M. Centennial drought outlook over the CONUS using NASA-NEX downscaled climate ensemble. *Int. J. Climatol.* **2017**, *37*, 2477–2491. [CrossRef]

104. Salehabadi, H.; Tarboton, D.G.; Udall, B.; Wheeler, K.G.; Schmidt, J.C. An Assessment of Potential Severe Droughts in the Colorado River Basin. *J. Am. Water Resour. Assoc.* **2022**, *58*, 1053–1075. [CrossRef]

105. Williams, A.P.; Cook, B.I.; Smerdon, J.E. Rapid intensification of the emerging southwestern North American megadrought in 2020–2021. *Nat. Clim. Chang.* **2022**, *12*, 232–234. [CrossRef]

106. Xia, Y.; Mitchell, K.; Ek, M.; Cosgrove, B.; Sheffield, J.; Luo, L.; Alonge, C.; Wei, H.; Meng, J.; Livneh, B.; et al. Continental-scale water and energy flux analysis and validation for North American Land Data Assimilation System project phase 2 (NLDAS-2): 2. Validation of model-simulated streamflow. *J. Geophys. Res. Atmos.* **2012**, *117*, D03110. [CrossRef]

107. Simeone, C.E.; Foks, S.S. HyMED—Hydrologic Model Evaluation for Drought: R package version 1.0.0. In *U.S. Geological Survey Software Release*; U.S. Geological Survey: Reston, VA, USA, 2024.

Article

Evaluation of Groundwater Resources in the Middle and Lower Reaches of Songhua River Based on SWAT Model

Xiao Yang [1,2,3], Changlei Dai [1,2,3,*], Gengwei Liu [1,2,3], Xiang Meng [1,2,3] and Chunyue Li [1,2,3]

[1] School of Hydraulic and Electric Power, Heilongjiang University, Harbin 150080, China; 15045983268@163.com (X.Y.); mengxiang020710@163.com (X.M.); 18845583185@163.com (C.L.)
[2] Institute of Groundwater Cold Region, Heilongjiang University, Harbin 150080, China
[3] International Joint Laboratory of Hydrology and Hydraulic Engineering in Cold Regions of Heilongjiang Province, Harbin 150080, China
* Correspondence: daichanglei@126.com

Citation: Yang, X.; Dai, C.; Liu, G.; Meng, X.; Li, C. Evaluation of Groundwater Resources in the Middle and Lower Reaches of Songhua River Based on SWAT Model. *Water* **2024**, *16*, 2839. https://doi.org/10.3390/w16192839

Academic Editors: Agnieszka Rutkowska and Katarzyna Baran-Gurgul

Received: 6 September 2024
Revised: 28 September 2024
Accepted: 3 October 2024
Published: 6 October 2024

Abstract: The SWAT model primarily investigates sources of water pollution and conducts ecological assessments of surface water in contemporary hydrology and water resources research. To date, there have been limited accomplishments in the study of groundwater resources in China. The MODFLOW model currently primarily simulates groundwater levels and the migration of water quality, depending on the hydrological surface water data in the relevant area. This study aims to investigate the groundwater distribution characteristics of the middle and lower reaches of the Songhua River, a significant agricultural and grain production region in China. The research focuses on the middle and lower reaches of the Songhua River basin in Northeast China and employed the SWAT distributed hydrological model to simulate runoff. The monthly recorded runoff at Tongjiang Station in Jiamusi City was utilized to calibrate the model parameters. Consequently, the MODFLOW model was introduced to compare and assess the simulation outcomes of the SWAT model, ultimately ascertaining the distribution characteristics of shallow groundwater, groundwater recharge, recoverable volume, and groundwater levels in the Songhua River Basin. The findings indicate that: (1) The SWAT model demonstrates efficacy in the study region, achieving R^2 and NS values of 0.81 and 0.76, respectively, thereby fulfilling the fundamental criteria for scientific research. The MODFLOW model exhibits strong performance in the study region, achieving a periodic R^2 of 0.98 and a verification R^2 of 0.97, with the discrepancy between simulated and actual groundwater levels confined to 0.6 m, thereby satisfying the criteria for scientific research. (2) In 2011, 2014, and 2016, the groundwater recharge in the middle and lower sections of the Songhua River was 24.33×10^8 m^3, 30.79×10^8 m^3, and 32.25×10^8 m^3, respectively, aligning closely with the SWAT simulation results, while the average annual groundwater level depth was 8.17 m. (3) In the research area, groundwater recharging occurs primarily by atmospheric precipitation, while drainage predominantly transpires via groundwater as base flow, constituting 81.46%. Secondly, the recharge of shallow groundwater to deep aquifers is around 7.14%, with a minimal share attributed to vadose zone loss, constituting merely 2.1%. (4) From 2010 to 2016, the average groundwater runoff modulus of the middle and lower reaches of the Songhua River basin was 1.005 L/(s·km^2), with a total recharge of 216.58×10^8 m^3 and a total recoverable amount of 105.11×10^8 m^3. The mean yearly supply was 25.11×10^8 m^3. The total groundwater recharge was 26.54×10^8 m^3 in the driest year (2011) and 33.25×10^8 m^3 in the year of most ample water (2016).

Keywords: SWAT model; MODFLOW model; Songhua River basin; Runoff simulation; groundwater distribution characteristics; groundwater resource evaluation

1. Introduction

The Songhua River Basin possesses abundant groundwater resources, supplying fresh water for agricultural crop development in Heilongjiang Province [1]. Specifically, in the middle and lower reaches of the Songhua River basin (Jiamusi Tongyi River basin, the principal stream of the Songhua River), the groundwater resource recharge for the entire agricultural area of Heilongjiang Province constitutes 50.1% [2] of the total natural supply for the Sanjiang Plain. Nonetheless, swift economic advancement over the past three decades has led to over-cultivation of paddy fields and imprudent management of water resources [3], resulting in significant environmental and ecological issues, including the depletion of groundwater levels and land subsidence, which have emerged as critical impediments to the sustainable development of the economy and society. Comprehending the condition and regulatory framework of regional groundwater resources is essential for their judicious development [4]. Consequently, it is imperative to enhance the assessment of groundwater resources in the middle and lower sections of the Songhua River Basin and to diversify the methodologies for water resource evaluation [5–7].

The SWAT model (Soil and Water Assessment Tool) is a distributed hydrological model developed for the United States Department of Agriculture, renowned globally for its robust physical simulation capabilities. This model effectively simulates the hydrological cycle of a basin based on environmental parameters, including basin climate, soil, and land use in the research region [8]. The SWAT model currently emphasizes the assessment of water pollution sources and surface water ecology in hydrological research [9]. Otherwise, there have been limited scientific advancements in the assessment of groundwater resources. Recently, domestic scholars Chen Peiyuan [10] and Zhao Liangjie [11] conducted a preliminary investigation. Chen Peiyuan employed the SWAT model to analyze groundwater distribution features and assess water resources in the Jinghe River Basin. The ultimate runoff rate R^2 and NS attained values of 0.83 and 0.7, respectively, and the computed groundwater resources aligned with the Groundwater Resources Assessment Report published by the local government. The SWAT model was deemed appropriate for simulating shallow groundwater resources and satisfied the fundamental criteria for scientific investigation. Zhao Liangjie initially classified hydrological characteristics using the rainfall guarantee rate, based on runoff simulations conducted using the SWAT model. The groundwater runoff modulus parameter inversion and rainfall infiltration coefficient approach were employed to validate the runoff modulus. In conclusion, utilizing the runoff modulus approach to address groundwater storage simulated by the SWAT model is both possible and effective for calculating groundwater resources. Furthermore, Li Yuanjie et al. [12] developed a MODFLOW model for Linhe District of Bayanzhur City, Inner Mongolia, and assessed water resources utilizing the water balance approach and numerical simulation technique [13]. Zhu Henghua [14] et al. employed MODFLOW to assess the groundwater model of Licheng District in Jinan City, utilizing the PEST module to calibrate the permeability coefficient. The model assessed the potential increase in shallow groundwater, provided that the groundwater level does not decrease further. Zhang Hongwei et al. [14] developed a groundwater flow model for Linqing City, Shandong Province, and examined the variations in groundwater levels under various rainfall scenarios. The findings indicated that the peak groundwater level during rainy years was 3 m more than in dry years. Foreign scientists Nguyen Ngoc [15]and colleagues utilized Visual Modflow to develop a groundwater flow model for Dak Lak. The research findings indicated that the reliability of the MODFLOW model's computed outputs is significantly high, even with limited drilling data. The impacts of recharge and evaporation on groundwater resources and water balance were examined under several climate change scenarios (RCP4.5 and RCP8.5). BUSHIRA K M [16] et al. employed the MODFLOW module in ModelMuse to construct an underground flow model for the Colorado basin in Mexico and calibrated both the steady-state surface and subterranean flow models. Xiaolong Li et al. [17] developed a groundwater flow model for the Manas River basin in China and evaluated the groundwater levels of 43 representative observation wells. The simulation outcome was favorable. During standard operation of the pumping wells, the groundwater level declined at a rate of 0.15 m/d.

Consequently, based on the aforementioned research context and prior experience, this paper focuses on the middle and lower reaches of the Songhua River, a national key agricultural production base, as the subject of study. It analyzes the groundwater recharge outcomes derived from the SWAT and MODFLOW models, validates the feasibility of the results, and subsequently delineates the groundwater levels in the study area. For thorough assessment, this paper employs the groundwater runoff modulus method to calculate various results simulated by the SWAT model, ultimately determining groundwater recharge in the study area and identifying groundwater distribution characteristics in the middle and lower reaches of the Songhua River. This research is highly important for the management of regional water resources and the rehabilitation of regional groundwater levels.

2. Overview of the Study Area

The middle and lower reaches of the Songhua River watershed encompass the segment of the river extending from Jiamusi to Tongjiang, measuring a total length of 267 km. This section traverses the Sanjiang Plain, characterized by alluvial plains on either side, flat topography, abundant vegetation, and relatively unobstructed rivers and banks. The waterways intersect extensively, with riverbanks ranging from 5 to 10 km in width, and numerous shoals present within the river. The location is situated in the northern temperate monsoon climate zone, characterized by prolonged cold winters and warm, wet summers. The annual average temperature ranges from -3 to $5\,^{\circ}$C, with a maximum of $40\,^{\circ}$C and a minimum of $-50\,^{\circ}$C. The basin experiences yearly average precipitation of approximately 500 mm, with an average annual runoff of 762×10^8 m^3, and exhibits a distinct interannual fluctuation characterized by alternating periods of abundance and stagnation [18,19]. The primary river basin is characterized by valley terrace landforms alongside the Songnen Plain, Sanjiang Plain, and additional plain landforms. The soil types primarily consist of basic black soil, calcareous alluvial soil, saturated thin layer soil, soft shallow soil, organic soil, anthropogenic accumulation soil, simple active luvisol, and saturated conical soil [20]. This region encompasses a large area with significant variations in hydrogeological conditions, groundwater depth, distribution, and water quality. The groundwater aquifers mostly consist of loose rock pore water, clastic rock pore fracture water, and bedrock fracture water. Loose rocks constitute the most extensive pore water distribution area and reserves, with the aquifer lithology comprising Quaternary sand, sand gravel, and gravel. The aquifer thickness varies from 10 to 300 m, the groundwater level is typically less than 10 m deep, the water yield is substantial, and the inflow rate of an individual well generally runs from 500 to 3000 cubic meters per day. The predominant chemical composition of the groundwater is either calcium bicarbonate- or sodium calcium-type, with salinity generally below 1 g/L [21]. The second type is pore fissure water found in clastic rock, located beneath the Quaternary aquifer formations in the plains of China (visible in certain regions) and within the meso-Cenozoic depression (fault) basins in mountainous locations such as Mudanjiang, Qitaihe, Shuangyashan, and Jixi. The aquifer consists of Neogene, Paleogene, Cretaceous, and Jurassic sand, sand conglomerate, and coal measures. The lithology, thickness, and burial depth of the aquifers exhibit significant variability, the water abundance is highly inconsistent, and the groundwater possesses a particular pressure. The structural complex contains a greater abundance of water. Bedrock fissure water is predominantly found in extensive bedrock mountainous regions and lava plateau areas. Water-bearing fissures can be categorized into structural fissure water, weathering fissure water, and basalt cavity fissure water, based on their formation and functional properties, with the latter being found in the southeastern region of the middle and lower reaches of the Songhua River. The distribution and degree of water richness in bedrock fissure water are influenced by lithology, topography, hydrology, and meteorological conditions, resulting in considerable variability in their water-rich characteristics, thus rendering them generally unsuitable for large-scale centralized water delivery. The groundwater in the research area is characterized by low salinity and is classified as bicarbonate-type freshwater [22]. The hydrological cycle of the bedrock fissures in the eastern mountainous region is robust, primarily replenished by air precipitation. Following a brief runoff, a portion is replenished by subterranean runoff during transit. The overview of the study area is shown in Figure 1.

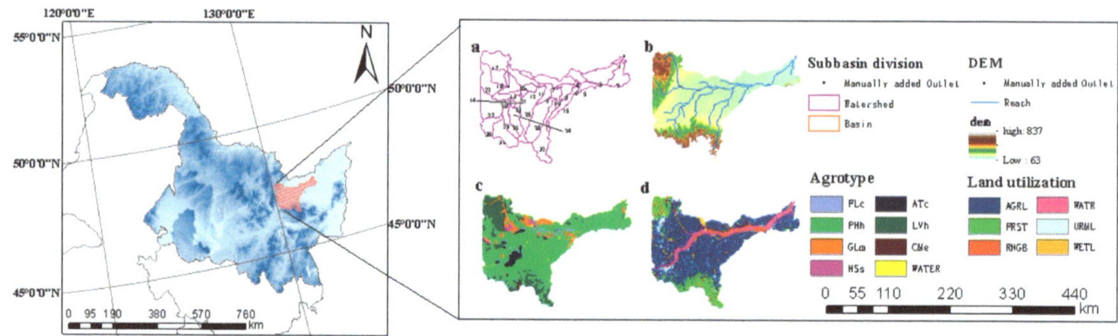

Figure 1. A comprehensive diagram of the study region. (**a**) represents the subwatershed zoning map created using the SWAT model (ArcSWAT2012). (**b**) displays a geographic elevation map of the study region. (**c**,**d**) illustrate the distribution of soil types and land use, respectively, in the study area. The details of (**c**) can be found in Table 1, while the details of (**d**) can be found in Table 5).

Table 1. Comparison table of soil types and abbreviations of land use types in the study area.

Name Abbreviation	Description	Name Abbreviation	Description
ATc	Anthropogenic accumulation	HSs	Organic soil
CMe	Saturated protosol	LVh	Simple high activity luvisols
FLc	Calcareous alluvial soil	PHh	Simple black soil
GLm	Mollic gleysol	WATER	Water body

3. Data and Methods

3.1. Data Sources

3.1.1. Digital Elevation Model (DEM)

Research indicates that when utilizing SWAT for runoff simulation, the elevation map chosen as the operational data source should have a spatial resolution ranging from 20 to 150 m [23]. This study utilized digital elevation model (DEM) data with a spatial resolution of 30 m, obtained from the NASA Earth Science data website, to extract pertinent parameters of the watershed.

3.1.2. Soil Type Data

The soil database included the spatial distribution and physical characteristics of various soils within the research area. This report states that the resolution of the 2010 soil data at 1:100,000 was 1000 km^2. To ensure the model operates efficiently and to streamline the development of the soil database, numerous soil types were reclassified based on their physical parameters, adhering to the principle of maximizing the proportion of soils with identical physical properties. The classification yielded eight distinct soil categories. Table 2 displays the precise proportion of each predominant soil type within the eight categorized soil classifications.

Table 2. Proportion of main soil types %.

FLC	PHh	GLM	HSs	ATc	LVh	CMe
2.28	26.44	6.09	1.02	2.01	13.17	0.38

This study details the derivation of soil data parameters for the SWAT (SWAT2012)model using SPAW [24] (Soil Profile Water Transfer,SPAW software version: 6.02.75) software, where the carbon content in the soil layer must be converted into organic mass before being fed into the SPAW software for analysis. The database includes the quantities of soil gravel, clay loam, and clay. To enable the computation of USLE-K parameters within the model, the substitution formula suggested by Williams was utilized [25]. The precise numbers for soil layer 1 and soil layer 2 from the final calculations are shown in Table 3, while the corresponding explanations of the soil physical coefficients referenced in Table 3 are detailed in Table 4.

Table 3. Soil coefficient and level calculated by SPAW.

Coefficient / Soil Type	SOL_BD1	SOL_AWC1	SOL_K1	SOL_CBN1	SOL_BD2	SOL_AWC2	SOL_K2	SOL_CBN2	Hierarchy
FLc	1.53	0.14	9.32	0.6	1.48	0.14	12.65	0.4	L-L
LPe	1.55	0.1	9.36	1.13	0	0	0	0	L
PHh	1.37	0.14	14.24	1.95	1.52	0.13	8.22	0.67	L-L
GLm	1.41	0.14	13.58	1.65	1.5	0.13	5.2	0.69	L-CL
HSs	1.14	0.13	13.65	39.4	1.18	0.14	22.43	38.46	CL-SaCL
ATc	0.98	0.18	44.52	1.12	1.49	0.14	8.94	0.82	SIL-L
LVh	1.52	0.13	9.33	0.74	1.52	0.13	4.11	0.36	L-CL
CMe	1.49	0.13	10.27	1	1.55	0.12	5.70	0.37	L-L
WATER	1.72	0	260	0	0	0	0	0	-

Table 4. Related descriptions of soil coefficients involved in the calculation of SPAW.

Coefficient	Description	Coefficient	Description
SOL_BD	weight of dried soil, comprising soil particles and intergranular pores, per unit volume. It stands for the moist bulk density of soil (SOILdensity).	CLAY	Clay content, %wt, refers to soil particles < 0.002 mm in diameter.
SOL_AWC	Indicates the effective water content of soil layer, in mm/mm.	SILT	SILT1 refers to the loam content of the soil (%wt), that is, the percentage by weight of soil particles between 0.002 and 0.05 mm in diameter.
SOL_CBN	Organic carbon content (%wt) of the soil layer.	SAND	Sand content, %wt, refers to particles with diameters between 0.05 and 2.0 mm.
SOL_K	Saturated water conductivity/saturated hydraulic conductivity, mm/hr.	ROCK	Gravel content, %wt, refers to particles with a diameter greater than 2 mm.
SOL_ZMS	Represents the maximum root depth of the soil profile, mm.	USLE_K	Erodibility factor

3.1.3. Land Use Type Data

The land use data were derived from the 2022 global land cover dataset with a 30-m resolution, published by the Academy of Aerospace Information Innovation, part of the Chinese Academy of Sciences. The land use types in the research region were divided into six categories: cultivated land, forest land, grassland, water bodies, a combination of urban and rural areas, industrial and mining land and residential land, and fallow land. Table 5 displays the relevant categories of model inputs.

Table 5. SWAT code of land use.

Reclassification Coding	Name	SWAT Coding
1	Cultivated land	AGRL
2	Forest land	FRST
3	Grassland	RNGB
4	Water bodies	WATR
5	Urban and rural, industrial and mining, and residential land	URML
6	Fallow land	WETL

3.1.4. Meteorological Data and Runoff Data

The meteorological database component of the SWAT model consists of two phases: the initial phase involves inputting the recorded meteorological data into the SWAT model's original file; the subsequent phase entails constructing a weather generator based on the research area's parameters and objectives. The primary meteorological data utilized are daily records of precipitation, temperature, relative humidity, sun radiation, and wind velocity. The meteorological data included in this work were CMADSV1.1 data obtained from the National Tibetan Plateau Scientific Data Center [26]. The duration spans from 2008 to 2016, effectively aligning with the temporal parameters of the model's operation. This database is among the most extensively utilized meteorological datasets for the SWAT model. This database satisfies the accuracy criteria for the model's final output outcomes after extensive utilization by numerous scholars [27]. This article utilized DEM data from the middle and lower portions of the Songhua River basin, selecting precipitation, temperature, relative humidity, solar radiation, and wind speed from 60 meteorological stations in the study area as experimental meteorological data. Monthly runoff data from the Tongjiang Hydrological Station, situated at the basin's complete exit, were picked for the period from 2008 to 2016. Table 6 presents the data sources utilized for constructing the SWAT model.

Table 6. Basic geographic data required for the middle and lower reaches of Songhua River basin model.

Data Type	Data Source
Digital Elevation Model (DEM)	NASA Earth Science data website (https://nasadaacs.eos.nasa.gov/) (accessed on 15 July 2024)
Soil type and attribute list	HWSD data downloaded from the National Tibetan Plateau Scientific Data Center (World Soil Database) (accessed on 15 July 2024)
Land type use data	Institute of Aerospace Information Innovation, Chinese Academy of Sciences
Meteorological data	CMADS (V1.1) downloaded from the National Tibetan Plateau Scientific Data Center (accessed on 18 July 2024)
Runoff data	Tongjiang city hydrology station

3.2. Research Methods

3.2.1. SWAT Model

The hydrological process of the SWAT model comprises two components: the surface simulation stage and the subsurface simulation stage [28]. The surface simulation phase comprises two stages: runoff production and slope confluence, which regulate the influx of water and solute from each sub-basin to the principal river. The water surface simulation phase involves the confluence of rivers and reservoirs, modeling the transport dynamics of water and solutes to the basin's total output. The water balance equation utilized to simulate the hydrological cycle is presented in Equation (1).

$$SW_t = SW_0 + \sum_{i=1}^{t} \left(R_{day,i} - Q_{swf,i} - E_{a,j} - W_{seep,i} - Q_{gw,i} \right) \tag{1}$$

where SW_t is the soil water content at the end of the period, mm; SW_0 is the soil water content at the beginning of the period, mm; t is the calculation period; $R_{day,i}$ is the rainfall on day i, mm; $S_{surf,i}$ is the surface runoff on day i, mm; $E_{a,i}$ is the evaporation amount on day i, mm; $W_{seep,i}$ is the permeability on day i, mm; Q_{gw} is the underground runoff on day i, mm.

SWAT models are capable of simulating surface water, soil water, and groundwater dynamics. The basin can be further divided into several natural sub-basins according to its actual topography, thereby mitigating the influence of spatio-temporal variations in natural factors on simulation outcomes, and additionally delineating relevant hydrological units within each sub-basin for collaborative simulation of changing features [29]. Figure 2 illustrates the schematic diagram of its principle.

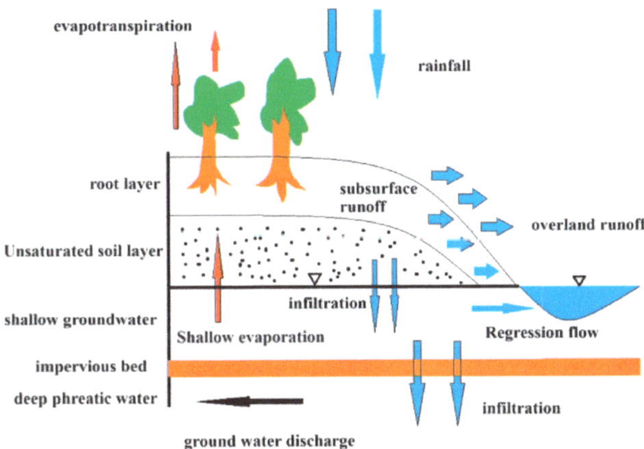

Figure 2. Schematic diagram of SWAT model.

3.2.2. Shallow Aquifer Reservoir Variable Calculation Method

Based on the calculation principle of groundwater balance [30], the calculation formula for shallow aquifer reservoir variables is as follows:

$$\Delta S_{gw} = PERC - GWQ - REVAP - DARC \tag{2}$$

where ΔS_{gw} is the shallow aquifer reservoir variable (mm); $PERC$ is the leakage water in the vadose zone (mm); $REVAP$ is the water quantity retained in the aquifer vadose zone (mm); $DARC$ is the seepage volume of the deep aquifer (mm); GWQ is the contribution of underground runoff to the main river course (mm).

The groundwater runoff modulus method [31] is used to calculate the natural supply of the basin, and the specific formula is as follows:

$$Q = M \times F \times t \times 10^{-7} \tag{3}$$

where Q is the natural supply amount (10^4 m^3); M is the groundwater runoff modulus (L/(s·km^2)); F is the catchment area (km^2); t is time (s). The groundwater runoff modulus is M calculated as follows:

$$M_{year} = {W_{year}} \Big/ {F} \times t \times 1000 \tag{4}$$

where W_{year} is the average underground runoff in the time step (m^3); F is the catchment area (km^2); t is time (s).

3.2.3. Construction of MODFLOW Model

The fundamental principle of groundwater numerical simulation dates back to 1856, introduced by the French engineer Darcy through Darcy's law. As groundwater numerical simulation theory and computer software advance, groundwater numerical simulation software continues to evolve and mature. Currently, its primary simulation techniques encompass the finite difference method, finite element approach, and others. These methods were extensively employed in groundwater numerical simulation during the 1960s. The predominant simulation software encompasses Visual MODFLOW, FEFLOW, and GMS, among others, although certain simulation programs are incorporated into open-source compilation platforms as toolkits, exemplified by the FloPy toolkit in Python (Python 3.12.3). The ongoing advancement and enhancement of this program render the numerical simulation of groundwater more precise and dependable. This work utilized the MOD-FLOW module inside GMS software (GMS10.8), characterized by its user-friendly interface and effective 3D visualization, to develop a groundwater model for the middle and lower portions of the Songhua River basin. This paper used the MODFLOW NWT(Version number of the GMSMODFLOW model used: V2.1.1) program within the module to realize the groundwater numerical model.

MODFLOW NWT is a Modflow-2005 adaptation of Newton's formula created by the United States Geological Survey to more effectively manage unconfined aquifers.

The basic governing equation of MODLFOW is:

$$\frac{\partial}{\partial x}\left(K_{xx}\frac{\partial h}{\partial y}\right) + \frac{\partial}{\partial y}\left(K_{yy}\frac{\partial h}{\partial y}\right) + \frac{\partial}{\partial z}\left(K_{zz}\frac{\partial h}{\partial z}\right) - W = S_s\frac{\partial h}{\partial t} \tag{5}$$

where K_x, K_y, K_z, are the permeability coefficients (m/d) along the x, y and z axes; h is the water head (m); W is the groundwater source sink item (m/d), including precipitation infiltration recharge, irrigation return water, diving evaporation, mechanical well exploitation, water exchange between aquifer and river, and water exchange between diving and confined water, up to the unit volume flow through medium and isotropic soil in a non-equilibrium state; S_s is the specific water storage coefficient of the porous medium; t is time (d).

(1)   Aquifer generalization

The groundwater simulation range option aligns with the SWAT model. The shallow aquifers in the studied area consist predominantly of Quaternary Holocene sand and gravel, with a thickness ranging from 100 to 200 m in most regions. The riverbed and floodplain of the Songhua River's main stream and its tributaries are predominantly constituted of Holocene (Q4) deposits, specifically a thin layer of yellow clay and sub-clay, together with sand and gravel in the lower section of the basin. The terrace of the interriver zone in the basin primarily consists of Upper Pleistocene (Q3) yellow-brown sand and gravel, with discontinuous sub-clay overlaying it. The flat region corresponds to the Middle Pleistocene (Q2), characterized by gray-brown, gray-black silty sand, sand, and sand gravel, with the lower section interspersed with sub-clay and silty sub-clay. The basin's base consists of Lower Pleistocene (Q1) deposits, characterized by yellow-green and gray-green medium sand, fine sand, silty sand, and sand gravel. Consequently, the characteristics of the research region were delineated based on the aforementioned fundamental lithology, as illustrated in Figure 3, with the specific values presented in Table 7. The model's roof elevation was derived from the interpolation of 30 m precision DEM elevation data, while the upper floor was determined using the aquifer thickness indicated by borehole and pumping well data, supplemented by the approximate aquifer thickness documented in the hydrogeological data for the study area. Figure 4 illustrates a schematic representation of the elevation points at the top and bottom inside the study region.

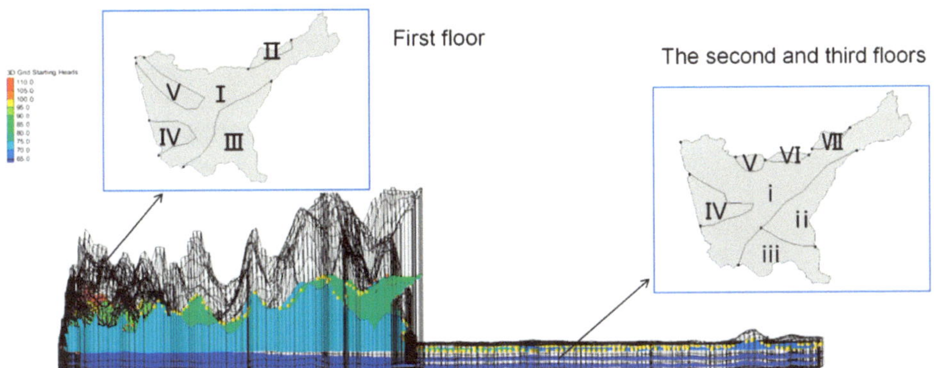

Figure 3. Study area parameter partition map. (The Roman numerals in the figure are the partition number of the permeability coefficient).

Table 7. Schematic diagram of top and bottom elevation points in the study area.

Partition Number	Initial Range of Permeability Coefficient (m/d)	Initial Value Range of Water Supply Degree
I	20~25	0.1~0.2
II	15~20	0.15~0.20
III	15~20	0.10~0.15
IV	1~5	0~0.1
V	15~20	0.1~0.2
i	20.0~25.0	0.001~0.002
ii	10.0~15.0	0.01~0.02
iii	15.0~20.0	0.01~0.02
iv	10.0~15.0	0.001~0.002
v	20.0~25.0	0.01~0.02
vi	18.0~20.0	0.001~0.002
vii	15.0~20.0	0.001~0.002

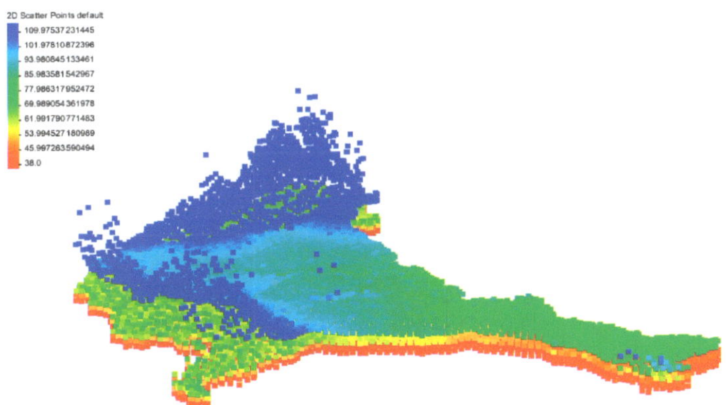

Figure 4. Schematic diagram of top and bottom elevation points in the study area.

(2) Generalization of boundary conditions

  The northwestern and southern edges of the research area have a significant hydraulic connection with the basin, and the mountains in these regions receive lateral recharge, thereby categorizing them as lateral inflow boundaries. The northern section of the study area represents the convergence of the Heilongjiang basin and the Songhua River basin, which aligns approximately parallel to the isowater line, and thus was categorized as the lateral inflow boundary. Conversely, the eastern boundary was classified as the zero flow boundary due to the minimal vertical flow observed. The western boundary of the study area features numerous outflowing tributaries, including the Wutong River and Anbang River, hence it was classified as a continuous head boundary.

(3)   The model's space-time dispersion

Grid division: This study utilized the watershed area derived from the SWAT model as the operational boundary for the MODFLOW model, encompassing an effective calculation area of 10,788.1 km^2. The study area was segmented into a 1000 m $\times$ 1000 m square grid, comprising 168 rows, 198 columns, and 3 layers, encompassing a total of 99,792 effective grids. Figure 5 shows the grid differentiation of the study area in the model.

Figure 5. Discrete graphic of model space.

(4)   Determination of initial conditions

This simulation utilized the iso-water level in the middle and lower sections of the Songhua River basin on 31 January 2008 as the model's initial water level (refer to the picture below). The simulation period was designated from January 2008 to December 2018, with each month serving as the stress period. Figure 6 below shows the initial flow field input for model operation.

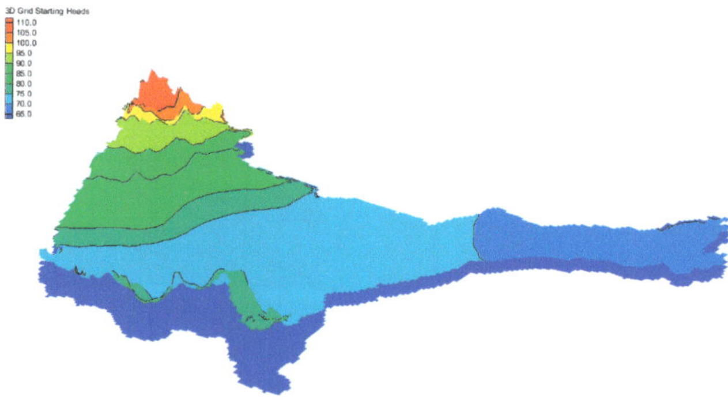

Figure 6. Preliminary water level chart.

4. Results and Analysis

4.1. Subwatershed Division and HRU Unit Based on SWAT Model

The partitioning of subwatersheds is a crucial component in surface runoff simulation in the SWAT model. Based on the imported DEM topographic map and the actual river system vector map, the optimal threshold value for accumulated water area (21,588.43 Ha) was utilized in this context, and the total outflow at the Tongjiang Hydrological Station in Tongjiang City, automatically generated by the SWAT model, was selected. A total of 32 sub-basins were delineated. Each Hydrologic Response Unit (HRU) possesses a distinct land use, soil type, and slope classification, constituting the smallest fundamental surface unit. The quantity of HRUs is dictated by the number of subbasins, land use, soil type, topographic slope, and reclassification threshold. This research established a minimum area ratio of land use, soil type, and slope categorization at 10%, resulting in a division into 112 Hydrologic Response Units (HRU).

4.2. Calibration and Verification of SWAT Model Parameters

The SWAT model has numerous parameters, and since the middle and lower portions of the Songhua River basin are situated in a cold temperate zone, parameters exhibiting a high correlation sensitivity coefficient were chosen for model adjustment. Twenty-two parameters were selected, and their sensitivity was assessed using SWAT-CUP(Version of SWAT-CUP software used: 5.2.1.1) software. This study employed a global sensitivity analysis. The t-test and p-value significance test were employed to assess the sensitivity of the parameters. Following the establishment of the parameters and the selection of their initial range, 500 iterative computations were executed utilizing the SU-F2 sampling technique integrated into the model [32]. The determination coefficient (R^2) and Nash efficiency coefficient (NS) were employed to assess the model's adequacy. The ideal parameters are presented in Table 8.

Table 8. Sensitivity analysis table of parameters.

Encoding	Parameter Name	Parameter Meaning	Optimal Parameter (Basin No. 1)
1	r__CN2.mgt	SCS runoff curve value	0.80
2	v__GW_DELAY.gw	Groundwater delay time (h)	793.90
3	v__GWQMN.gw	Level threshold of shallow aquifers when groundwater enters the main channel (mm)	2.17
4	v__REVAPMN.gw	Shallow groundwater evaporation depth threshold (mm)	954.40
5	v__SOL_AWC().sol	Surface water availability (mm)	−0.52
6	v__CH_K2.rte	Effective permeability coefficient (mm/h)	795.29
7	v__RCHRG_DP.gw	Permeability coefficient of deep aquifer	0.67
8	r__SOL_K().sol	Soil saturated water conductivity (mm/h)	1.104
9	r__SOL_ALB().sol	Moist soil albedo	0.29
10	v__ALPHA_BNK.rte	Base flow regression constant	0.31
11	v__SLSUBBSN.hru	Average slope length (m)	1.91
12	r__HRU_SLP.hru	Average slope (m/m)	2.25
13	v__CANMX.hru	Maximum canopy water storage (mm)	227.5
14	v__SFTMP.bsn	Average air temperature on snowfall days (°C)	10.9
15	v__SMTMP.bsn	Average temperature on snowfall days (°C)	13.7
16	v__SMFMX.bsn	Snowmelt factor	34.7
17	v__TIMP.bsn	Temperature lag coefficient of snow cover	2.93
18	v__SNOCOVMX.bsn	Snow depth threshold/cm	992.29
19	v__TLAPS.sub	Temperature lapse rate (°C/km)	4.51
20	v__ESCO.hru	Soil evaporation compensation coefficient	1.41
21	v__EPCO.hru	Plant absorption compensation coefficient	0.89
22	v__ALPHA_BF.gw	Base flow alpha factor (1/day)	1.29

Upon reinserting the amended parameters into the model, the worksheet was revised and the validation executed once more. The findings of runoff rate determination and verification are presented in Figure 7.

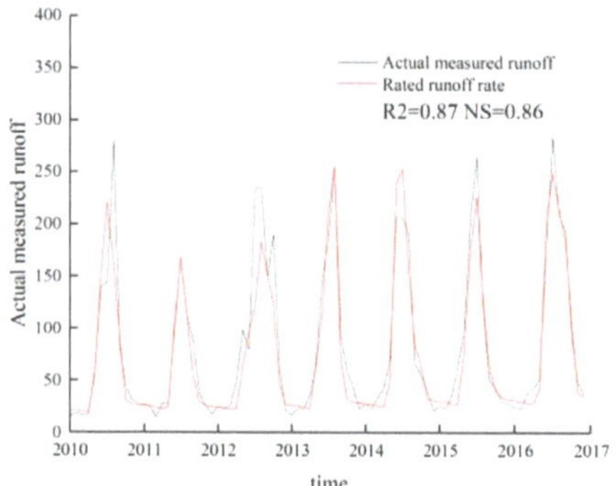
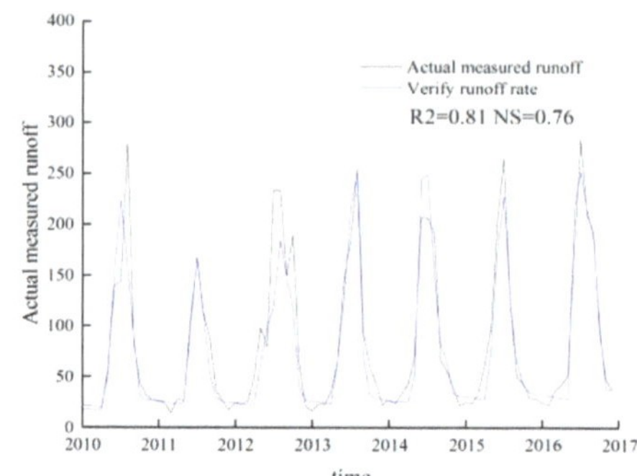

Figure 7. Determination and verification of the runoff rate of the model (the longitudinal coordinate indicates the runoff unit: m^3).

The evaluation criteria employed in this experiment were the determination coefficient (R^2) and Nash efficiency coefficient (NSE), which are widely utilized in research. A higher secondary coefficient indicates a stronger correlation between the simulated value and the measured value, resulting in a more favorable outcome. The dependability distribution is often presented as in the Table 9 below [33] under normal conditions.

Table 9. R^2 and NSE confidence comparison table.

Model Reliability	R^2	NSE
Equivalent to gold	$0.80 < R^2 \leq 1.00$	$0.75 < NSE \leq 1.00$
Excellent	$0.70 < R^2 \leq 0.80$	$0.65 < NSE \leq 0.75$
Typical	$0.50 < R^2 \leq 0.70$	$0.50 < NSE \leq 0.65$
Not satisfactory	$R^2 \leq 0.50$	$NSE \leq 0.50$

The experimental results show that the runoff simulation for Tongjiang hydrological station is ideal with $R^2 > 0.8$, NSE > 0.75.

4.3. Calibration and Validation of Parameters Utilizing the MODFLOW Model

This paper utilized measured well data from January 2008 to December 2012 to evaluate the parameters, whereas data from January 2012 to December 2016 were employed to validate the parameters. The hydrogeological parameters after calibration are shown in Table 10 below.

Table 10. Conclusive values of hydrogeological parameters.

Partition Number	Value of the Permeability Coefficient (m/d)	Initial Value of Water Supply
I	22	0.18
II	16	0.13
III	16	0.12
IV	2	0~0.1
V	17	0.15
i	23.0	0.0014
ii	13.0	0.009
iii	15	0.009
iv	14	0.008
v	23	0.0014
vi	17.0	0.0011
vii	17.0	0.0011

Figure 8 illustrates a schematic representation of all observed well simulations during the training and verification intervals. The graphic illustrates that the model's outcomes during the training and verification periods align closely with the measured findings. This publication selected a total of 11 observation wells in the study region. Figure 9 illustrates a comparative diagram of simulated and observed water levels for a single well during the training and verification phases. The disparity between the actual water level and the simulated water level is within 0.6 m, satisfying the fundamental criteria for scientific research and providing a more accurate representation of the conditions in the study area.

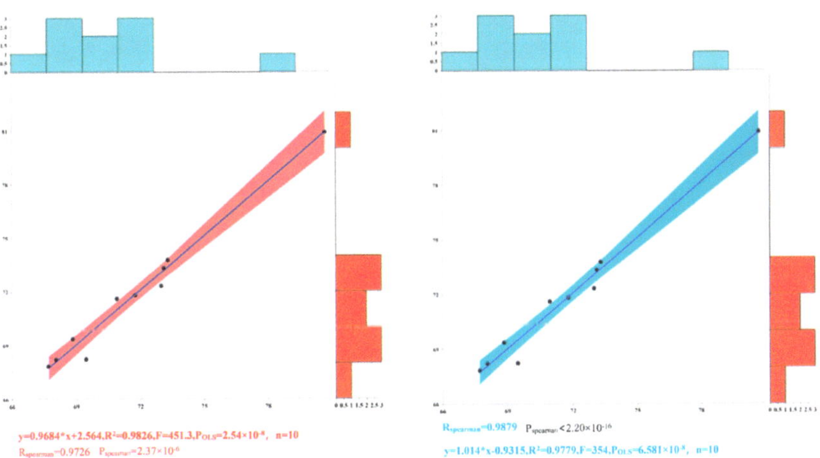

Figure 8. Evaluation of all recorded well simulation outcomes during the calibration and validation phases.

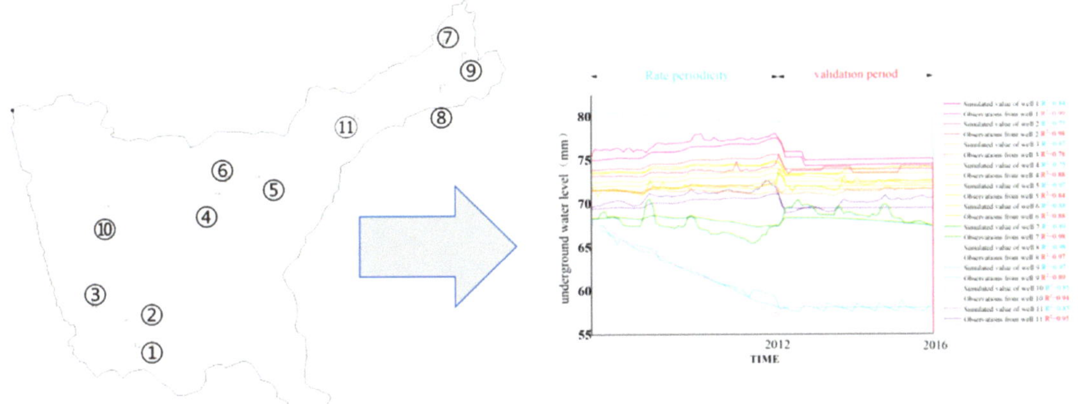

Figure 9. Comparison of simulated and actual water levels in a single well throughout calibration and validation intervals.(The serial number in the picture represents the observation well number assigned to each observation well in order to facilitate the experiment.)

4.4. Evaluation of Groundwater Resources Based on SWAT Model

The water storage statistics for shallow aquifers at the end of each year as simulated by the model were correlated with the 32 sub-basins in the aforementioned division (refer to Figures 10 and 11). The long-term changes in shallow aquifer water storage align closely with the findings presented in the 2021 Songhua River Basin Health Assessment

Report by the Heilongjiang Institute of Water Resources Science, indicating strong model applicability. The substantial rise in shallow aquifer water storage in sub-basins 1, 2, 3, 4, and 5 corresponds with the regional precipitation distribution trend from 2008 to 2016, as illustrated in Figure 12 (the sub-basin serial numbers are indicated on the survey map of the study area). Overall, the allocation of water storage in shallow aquifers within the basin exhibits significant variability. The primary trend is centered on the Jiamusi area, with a gradual decline towards the northeast and southeast. The net discharge trend of groundwater in the simulated results aligns closely with actual conditions, generally accumulating in the southeast and northwest directions while decreasing from the center to the periphery. Sub-basins 1, 2, 3, 5, and 7 constitute the shallow aquifers within the primary flow region of the Songhua River, with water resources predominantly derived from surface water.

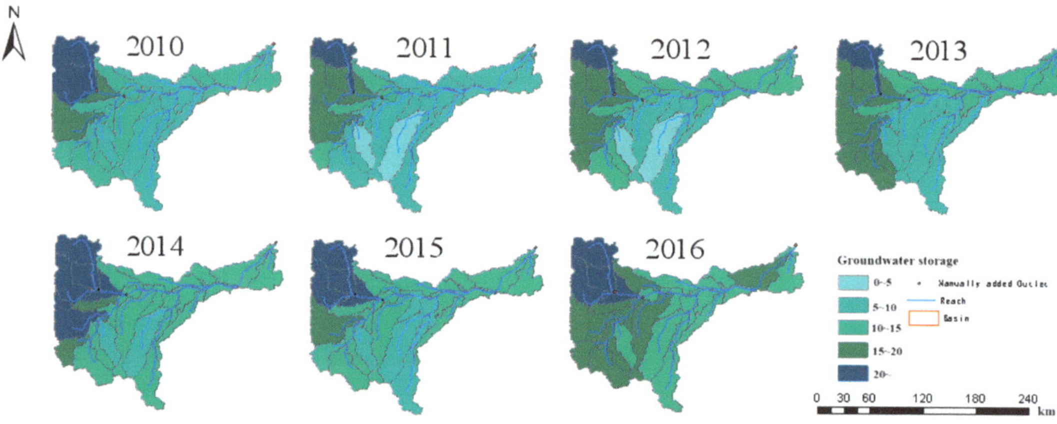

Figure 10. Annual mean water storage of phreatic beds in the sub-basin from 2010 to 2016 (unit: $\times 10^8$ m^3).

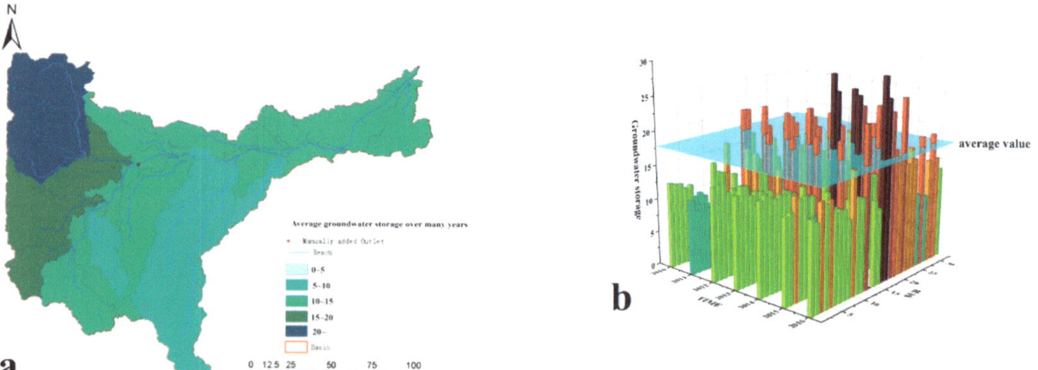

Figure 11. Variation map of water storage in a shallow aquifer in a sub-watershed. (**a**) shows the geographical distribution characteristics of the average annual water storage in the shallow aquifer, while (**b**) illustrates a schematic diagram of the changes in average annual water storage in the shallow aquifer. Unit: 10^8 m^3).

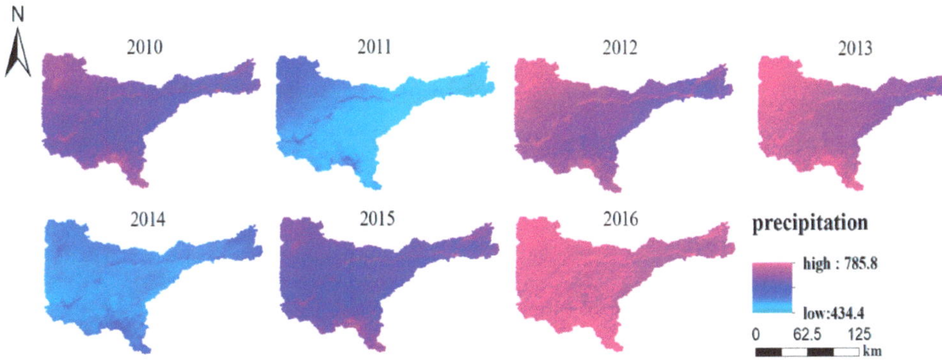

Figure 12. Distribution trend of precipitation in the middle and lower reaches of Songhua River from 2008 to 2016 (Unit: mm).

The groundwater balance was employed to ascertain groundwater reserves in the middle and lower portions of the Songhua River basin from 2010 to 2016 (refer to Table 11). The average storage variable is 838 million m^3/year, indicating that the studied region is predominantly in a healthy extraction condition. In 2016, the simulated runoff reached its peak, with a storage variable of 872 million m^3/year. It is in a state of positive equilibrium. The primary groundwater discharge occurs as base flow to replenish the river, constituting 81.46%. The second, around 7.14%, primarily represents the infiltration recharge from the shallow aquifer to the deep aquifer. The smallest percentage of discharge is the loss of water flow via the vadose zone, constituting 2.1%.

Table 11. Groundwater storage variable units in the middle and lower reaches of Songhua River basin, in 10^8 m^3.

Year	Supply Term		Excretion Term		Subtotal	ΔSgw
	PERC	REVAP	GWQ	DARCHG		
2010	25.9	2.0	16.10	8.63	6.93	+6.93
2011	13.1	2.51	17.3	9.05	15.76	−15.76
2012	28.1	1.48	17.44	0.93	8.25	8.25
2013	23.77	0.1	20.6	1.08	1.99	1.99
2014	26.65	0.1122	22	1.16	3.38	3.38
2015	18.1	0	22.08	1.15	5.13	−5.13
2016	32.01	0.06	22.06	1.17	8.72	8.72
Mean value	23.94	0.60	17.18	3.31	50.16	8.38
Discharge percentage/%		2.8	81.46	15.69		

Owing to the intricate geology of the middle and lower sections of the Songhua River and the varying hydrogeological conditions across the molecular basins, a characteristic hydrological year was chosen for the assessment of water resources based on the delineation of these basins. The monthly recorded runoff data from the Tongjiang Hydrology Station, which represents the total discharge of the middle and lower reaches of the Songhua River, spanning from 2008 to 2016 were utilized. Empirical frequency analysis [34] was employed to determine the driest year, 2011 ($p = 75\%$), within the simulation period, yielding an annual runoff of 662 million m^3. In 2014, with a probability of 50%, the annual runoff amounted to 981 million cubic meters. In 2016, with a precipitation rate of 25%, the annual runoff amounted to 1.216 billion cubic meters. The recoverable quantity of groundwater is assessed in average years.

This research used the groundwater runoff modulus approach to calculate the natural recharge of the groundwater system within the watershed. Initially, the model computed the output of each sub-basin for a typical year, then converted the average runoff of each sub-basin into the groundwater runoff modulus, followed by calculating the natural recharge for each sub-basin. Furthermore, to determine the exploitable groundwater volume inside the basin at the study area's size, the exploitable coefficient approach was employed to estimate the groundwater availability in representative years for each sub-basin.

To enhance the accuracy of the estimation results, the average extraction coefficient ($\rho = 0.45$) for several hydrogeological zones in Jiamusi City was utilized for computation, with the findings presented in Table 12.

Table 12. Subsurface runoff modulus and recharge in sub-basins of the study area.

Subcatchment	Dry Year (2011)			Normal Water Year (2014)			Wet Year (2016)		
	Runoff Modulus ($l \cdot s^{-1} \cdot km^2$)	Supply ($10^4\ m^3 \cdot a^{-1}$)	Recoverable Amount ($10^4\ m^3 \cdot a^{-1}$)	Runoff Modulus ($l \cdot s^{-1} \cdot km^2$)	Supply ($10^4\ m^3 \cdot a^{-1}$)	Recoverable Amount ($10^4\ m^3 \cdot a^{-1}$)	Runoff Modulus ($l \cdot s^{-1} \cdot km^2$)	Supply ($10^4\ m^3 \cdot a^{-1}$)	Recoverable Amount ($10^4\ m^3 \cdot a^{-1}$)
1	0.44	838.89	377.50	0.54	4091.70	1841.27	0.51	3968.50	1785.83
2	0.43	5441.23	2448.55	0.68	7283.90	3277.76	0.62	6427.05	2892.17
3	0.41	7825.90	3521.66	0.60	9888.50	4449.83	0.64	10,191.80	4586.31
4	1.09	5532.00	2489.40	0.98	6996.00	3148.20	1.02	7864.50	3539.03
5	0.88	5938.80	2672.46	0.88	7543.97	3394.79	0.94	8346.50	3755.93
6	0.24	784.61	353.07	0.47	1164.85	524.18	0.40	1070.60	481.77
7	1.75	27,620.44	12,429.20	2.13	37,619.41	16,928.73	1.89	34,748.30	15,636.74
8	1.44	9113.07	4100.88	1.65	11,380.22	5121.10	1.77	12,513.80	5631.21
9	1.72	10,855.50	4884.98	1.93	12,809.49	5764.27	1.90	12,440.40	5598.18
10	1.88	26,439.69	11,897.86	1.68	19,364.28	8713.93	1.71	20,369.70	9166.37
11	0.15	1653.00	743.85	0.15	1118.82	503.47	0.19	1129.20	508.14
12	0.38	4341.00	1953.45	0.26	1650.06	742.53	0.26	1697.80	764.01
13	1.61	25,585.30	11,513.39	1.60	21,921.37	9864.62	1.63	23,753.40	10,689.03
14	0.00	0.00	0.00	0.00	0.00	0.00	0.00	0.00	0.00
15	0.55	5388.50	2424.83	0.65	9579.00	4310.55	0.66	9931.70	4469.27
16	0.12	1682.10	756.95	0.31	3143.90	1414.76	0.28	2222.77	1000.25
17	1.88	114,631.00	51,583.95	1.98	144,797.70	65,158.97	2.01	152,842.08	68,778.94
18	1.45	20,271.00	9121.95	1.58	26,564.70	11,954.12	1.49	25,445.30	11,450.39
19	0.19	125.00	56.25	0.23	153.16	68.92	0.25	166.14	74.76
20	0.21	600.80	270.36	0.48	4344.30	1954.94	0.50	4390.52	1975.73
21	0.55	1417.20	637.74	0.63	1811.23	815.05	0.66	1942.50	874.13
22	1.44	77,046.78	34,671.05	1.71	92,524.61	41,636.07	1.81	94,238.00	42,407.10
23	0.34	3621.00	1629.45	0.41	4637.70	2086.97	1.55	4976.37	2239.37
24	0.55	1725.00	776.25	0.74	2150.50	967.73	0.74	2127.50	957.38
25	1.27	33,780.00	15,201.00	1.49	42,752.40	19,238.58	1.51	45,773.50	20,598.08
26	0.37	7780.00	3501.00	0.61	11,707.48	5268.37	0.68	12,226.17	5501.78
27	1.71	32,163.60	14,473.62	1.86	36,586.00	16,463.70	1.77	32,806.80	14,763.06
28	0.68	18,201.00	8190.45	1.21	27,736.00	12,481.20	1.14	26,522.55	11,935.15
29	0.73	15,840.00	7128.00	1.15	21,519.36	9683.71	1.10	18,530.56	8338.75
30	0.58	9028.00	4062.60	0.71	13,296.80	5983.56	0.68	12,000.30	5400.14
31	1.69	38,314.00	17,241.30	1.96	51,930.90	23,368.91	1.85	44,141.26	19,863.57
32	1.88	17,256.00	7765.20	2.13	24,190.60	10,885.77	2.32	30,336.84	13,651.58
total	—	530,840.41	238,878.18	—	662,258.91	298,016.51	—	665,142.41	299,314.08

4.5. Prediction of Groundwater Recharge Based on MODFLOW Model

This work employed the MODFLOW model to simulate groundwater recharge, and the validity of the research findings was substantiated through comparison with the simulation results of the SWAT model. This work delineated the research area into a water equilibrium zone characterized by an annual equilibrium period. Figure 13 below shows trends in groundwater recharge from 2008 to 2016. The annual groundwater recharge is as follows: $24.06 \times 10^8\ m^3$, $24.33 \times 10^8\ m^3$, $21.80 \times 10^8\ m^3$, $24.33 \times 10^8\ m^3$, $22.04 \times 10^8\ m^3$, $30.11 \times 10^8\ m^3$, $30.79 \times 10^8\ m^3$, $25.15 \times 10^8\ m^3$, and $32.25 \times 10^8\ m^3$, respectively. The simulation outcomes resemble those of the SWAT model.

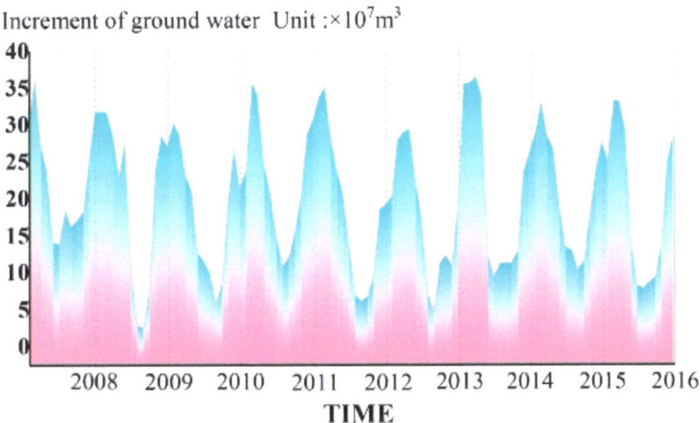

Figure 13. Trend chart depicting groundwater recharge from 2008 to 2016.

Figure 14 illustrates the distinctive trend of yearly mean groundwater level changes predicted by the MODFLOW model. The figure illustrates that the groundwater level distribution trend diminishes progressively from the northwest to the east and from the south to the north, with the water level fluctuations aligning closely with the groundwater storage variations simulated by the SWAT model. The mean depth of the groundwater level over several years is 8.15 m.

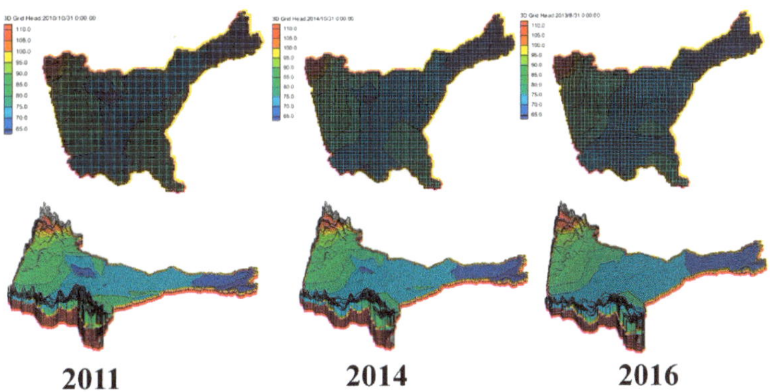

Figure 14. Diagram of the characteristic annual mean water table.

5. Discussion

5.1. Relevance and Constraints of the Model in Groundwater Resource Assessment Research and Potential for Future Developments

This paper compares the groundwater storage and recharge results simulated by the SWAT model with those published in the "Songhua River Basin Health Report 2021" and the results simulated by the MODFLOW model. The SWAT model was deemed appropriate for simulating groundwater resources. The primary factors influencing the formation and evolution of groundwater resources include precipitation, evaporation, infiltration, runoff, freeze–thaw cycles, soil water movement, and groundwater recharge, among others [35]. The SWAT model operates based on the aforementioned fundamental physical processes to simulate functionality. While a singular SWAT model can only mitigate errors in the freeze–thaw cycle through parameter adjustments, its influence is comparatively minor within the broader hydrologic cycle physical process group, and these errors can be rectified using appropriate parameters. Furthermore, the SWAT model necessitates extensive input of meteorological, terrain, soil, land use, and other data, which may have previously been challenging to acquire. However, advancements in remote sensing technology and geographic information systems have facilitated data collection, rendering the application of the SWAT model in groundwater resource evaluation increasingly viable [36].

The SWAT model [37] has been extensively developed over more than 20 years since 1990 and is relatively mature. It is a semi-distributed model, making it easier to comprehend and utilize compared to fully distributed models. Additionally, it offers faster computation than conceptual models, effectively representing the physical mechanisms of the water cycle with improved accuracy. Secondly, the model thoroughly accounts for the material cycle, simulates various types of material migration, and effectively utilizes land use and other remote sensing data to compile extensive foundational databases, including information on crops, pesticides, and fertilizers [38]. Nonetheless, in comparison to other hydrological models, its capacity to elucidate the water cycle mechanism is inadequate. The research area discussed in this paper is situated in northeast China, characterized by pronounced seasonal climate variations and distinct climatic, topographical, and soil attributes compared to other regions [39]. Cold regions exhibit prolonged winters, low temperatures, varied precipitation types, and intricate soil freezing and melting processes. The development and application of the SWAT model may not fully account for the characteristics of cold regions. This study is situated in the cold temperate zone of the North Temperate Zone (Songhua River basin), where freeze–thaw conditions are moderate, resulting in relatively consistent simulation outcomes; however, it is also located in a cold permafrost region. In regions such as Russia and Alaska, ensuring model accuracy is challenging.

This paper integrated the SWAT model with the MODFLOW model to perform dual-model auxiliary verification in order to address the aforementioned issues [40,41]. Despite the enhanced reliability of the results, the watershed simulated by the SWAT model was inappropriately employed as the boundary condition for MODFLOW, resulting in considerable errors and numerous instabilities. The watershed boundary was utilized as a no-flow water boundary, resulting in significant discrepancies with the actual conditions. A LU-SWAT-MODFLOW model should be developed in the future to calibrate and validate the model using multi-source data to mitigate the impact of temperature.

Land use change significantly affects hydrology and non-point source pollution simulation. In the processes of flow, sediment, and non-point source pollution simulation, it is essential to dynamically update land use data to recalibrate the threshold, thereby enhancing the model's simulation accuracy in the context of land use change. Consequently, dynamic land use input must be incorporated into the model [42].

The Songnen Plain and Sanjiang Plain in northeast China represent the regions with the highest agricultural grain yield in the country [43]. The topography in this area is intricate, featuring diverse land use and soil classifications. The existing SWAT model exhibits inadequate processing capabilities for high-precision and multi-transformation terrain data, leading to significant inaccuracies in simulated groundwater recharge, discharge, and shallow groundwater storage [44]. It should be enhanced according to the planting configuration. Accurate planting structure data, derived from the integration of high-resolution drone imagery, remote sensing images, and ground-measured data, serve as the input land use data for the SWAT model. This approach aims to enhance the simulation of the migration and transformation processes of agricultural non-point source pollution, thereby improving the accuracy of watershed runoff simulations.

5.2. Analysis of Groundwater Recharge and Distribution Characteristics

Analysis of the simulation results indicates that precipitation is the primary source of groundwater recharge in the middle and lower reaches of the Songhua River. The predominant component of drainage is groundwater recharge as base flow, while the least significant is loss in the regression vadose zone. The primary mechanism of water loss is absorption by plant roots from the superficial water layer. The Songhua River basin experiences a temperate continental climate characterized by significant annual temperature variation. The disparity between evapotranspiration and precipitation is excessive.

Forecasts indicate that the potential evapotranspiration of the Songhua River basin will rise in the 21st century. Although the region is primarily replenished by atmospheric precipitation, the supply is inadequate. The primary reason is the extended duration of the cold season, during which the river freezes, leading to significant groundwater recharge into the river [45].

The groundwater storage in the study area exhibited a gradual increase over 10 to 16 years, based on its distribution characteristics. The primary distribution trend centers on the Jiamusi area, with a gradual decline towards the northeast and southeast directions. Seventy-five percent of the land in the northeast and southeast quadrants of the study area is allocated for agricultural cultivation, resulting in substantial water extraction for irrigation purposes. Furthermore, the southeast direction is distanced from the primary river trunk, resulting in a comparatively limited water supply. The simulation results indicate a significant alteration in the southeastern region of the basin. In recent years, the optimization of paddy field planting systems has effectively allocated water demand and alleviated the strain on groundwater extraction, leading to a substantial increase in the water reserves of the phreatic aquifers in this region [46]. Besides the annual rise in precipitation, the gradual augmentation of water reserves is attributable to Fujin City's optimization of its flood control and diversion irrigation management system for the three adjacent reservoirs since 2014, thereby diminishing the groundwater demand for agricultural irrigation [47].

6. Conclusions

1. The application of the SWAT and MODFLOW models for assessing groundwater resources yielded favorable simulation results in this region. The runoff simulation at the Tongjiang Hydrological Station, located at the basin's total water outlet, was exemplary. R^2 exceeded 0.8, NSE surpassed 0.75, and the R^2 values for simulation and verification of groundwater levels were 0.98 and 0.97, respectively. The discrepancy between the simulated value and the actual value was less than 0.6 m.

2. The study area is predominantly characterized by a robust extraction sector. In 2016, the simulated runoff reached its peak, with a storage variable of 872 million m^3/a. It is in a state of positive equilibrium. The primary source of groundwater in the discharge item, represented as base flow recharge from the river, constituted 81.46%. The second factor accounts for approximately 7.14%, primarily attributed to the replenishment of deep aquifers, while the least significant factor, the loss to the vadose zone, constitutes merely 2.1%.

3. From 2010 to 2016, the average groundwater runoff modulus in the middle and lower reaches of the Songhua River basin was 1.005 $L/(s \cdot km^2)$, with a total recharge of $216.58 \times 10^8 \ m^3$ and a total recoverable amount of $105.11 \times 10^8 \ m^3$. The mean annual recharge was $25.11 \times 10^8 \ m^3$, while the total groundwater recharge was $26.54 \times 10^8 \ m^3$, $33.11 \times 10^8 \ m^3$, and $33.25 \times 10^8 \ m^3$ in the super dry year (2011), normal year (2014), and high water year (2016), respectively, with the groundwater recharge in the high water year being 1.25 times greater.

4. The MODFLOW model was employed to simulate groundwater recharge in the middle and lower reaches of the Songhua River for the years 2011, 2014, and 2016. The discrepancies in results compared to the SWAT model were $2.22 \times 10^8 \ m^3$, $2.32 \times 10^8 \ m^3$, and $1.0 \times 10^8 \ m^3$, respectively, with a minimal relative error base. The SWAT model effectively simulates groundwater resource assessment in cold regions.

Author Contributions: Conceptualization, X.Y.; Methodology, X.Y.; Software, X.Y.; Validation, X.Y.; Formal analysis, X.Y.; Data curation, C.L. and X.M.; Writing—original draft, X.Y.; Writing—review & editing, C.D. and G.L. All authors have read and agreed to the published version of the manuscript.

Funding: This research was funded by: (1) [Research and analysis of Sino-Russian glacial flow measurement technology in Heilongjiang (Amur River) and suggestions on survey schemes]. (2) [Yunnan Key Laboratory of International Rivers and Transboundary Eco-Security] grant number [2022KF03] and the APC was funded by [2022KF03].

References

1. Guo, L.; Zheng, G.G.; Li, R.Y.; Fu, C.Y.; Chen, J.; Meng, Y.C.; Pan, Y.; Hu, P. An integrated approach for quantifying trace metal sources in surface soils of a typical farmland in the three rivers plain, China. *Environ. Pollut.* **2023**, *337*, 136077.
2. You, L.; Wang, S.; Tao, Y.; Liu, Y. Appropriate Method to Estimate Farmland Drainage Coefficient in the Wanyan River Surface Waterlogged Area in Suibin County of the Sanjiang Plain, China. *Appl. Sci.* **2023**, *13*, 2769. [CrossRef]
3. Liu, H.; Zhao, H.; Feng, X.; Han, Y. Tradeoff and synergy of farmland ecosystem services in Heilongjiang Province. *J. China Agric.* **2023**, *28*, 160–171. (In Chinese)
4. Chen, Y.; Yang, Q.; Luo, Y.; Shen, Y.; Pan, X.; Li, L.; Li, Z. Water resources in arid region of Northwest China. *Arid. Land Geogr.* **2012**, *35*, 1–9. (In Chinese)
5. Hu, C.; Lu, C.; Wu, B. Practice and thinking on the Development of water-saving agriculture in Songhua River Basin in the new era. In Chinese Hydraulic Society. In Proceedings of the 2023 Chinese Water Conservancy Academic Conference (Volume 3), Zhengzhou, Henan, China, 4 November 2023; Yellow River Water Conservancy Press: Zhengzhou, China, 2023; Volume 5.
6. Ye, X.; Cui, R.; Wang, L.; Du, X. The influence of riverbank filtration on regional water resources: A case study in the Second Songhua River catchment, China. *Water Supply* **2020**, *20*, 1425–1438. [CrossRef]
7. Liu, S.; Zhou, Z.; Liu, J.; Li, J.; Xie, M.; Jiang, Y.; Wang, H. Based on the songhua river basin water resources evolution regularity of permafrost hydrology simulation. *J. South-North Water Divers. Water Conserv. Sci. Technol.* **2023**, *21*, 127–136, (Both in English and Chinese).
8. Li, J. Optimization Design of SWAT Model and Its Application in Chayu River Basin. Master's Thesis, Lanzhou University, Lanzhou, China, 2024.
9. Liu, H.Y. Research and application progress of SWAT model. *Subtrop. Soil Water Conserv.* **2019**, *31*, 34–37. (In Chinese)
10. Chen, P.; Li, J.; Yu, Q.; Guo, J.; Mang, J. Groundwater distribution characteristics and water resources evaluation in Jinghe River Basin based on SWAT model. *J. Irrig. Drain.* **2021**, *40*, 102–109+126.
11. Zhao, L.; Wang, Y.; Zhou, Y. Groundwater resources evaluation in the Pearl River Basin based on SWAT model. *Earth Sci.* **2024**, *49*, 1876–1890.
12. Li, Y.-J.; Jiang, X.-H.; Chen, J.; Yang, Z.; Wang, G.; Ding, H. Prediction of groundwater flow field in riverside area of Bayannur City based on GMS. *South--North Water Divers. Hydraul. Sci. Technol.* **2016**, *14*, 36–41+83.
13. Zhu, H.; Zhang, X.; Wang, W.; Jiang, C.; Liu, Z.; Liu, B. Numerical simulation and prediction of groundwater in Licheng District of Jinan City. *Groundwater* **2021**, *43*, 52–54+70. (In Chinese)
14. Zhang, H.; Zhang, S.; Zhang, Y.; Liu, K.; Pan, W.; Xu, J.; Bian, Z.; Yang, S.; Pan, Y.; Xu, Z. Nitrogen control of shallow groundwater based on Visual MODFLOW software. *J. Jinan Univ. (Nat. Sci. Ed.)* **2022**, *36*, 56–64.
15. An, N.N.; Nhut, H.S.; Phuong, T.A.; Huy, V.Q.; Hanh, N.C.; Thao, G.T.P.; Trinh, P.T.; Hoa, P.V.; Bình, N.A. Groundwater simulation in Dak Lak province based on MODFLOW model and climate change scenarios. *Front. Eng. Built Environ.* **2022**, *2*, 55–67. [CrossRef]
16. Bushira, K.M.; Hernandez, J.R.; Sheng, Z. Surface and groundwater flow modeling for calibrating steady state using MODFLOW in Colorado River Delta, Baja California, Mexico. *Model. Earth Syst. Environ.* **2017**, *3*, 815–824. [CrossRef]
17. Li, X.; He, X.; Yang, G.; Zhao, L.; Chen, S.; Wang, C.; Chen, J.; Yang, M. Study of groundwater using visual MODFLOW in the Manas River Basin, China. *Water Policy* **2016**, *18*, 1139–1154. [CrossRef]
18. Jia, X. Songhua river basin in recent 40 years of meteorological drought evolution characteristics. *J. Northeast. Water Conserv. Water Electr.* **2024**, *42*, 42–45+71.
19. Liu, Z.-T.; Hao, Z.-C.; Xu, H.-Q.; Ju, Q.; Wu, G.-Y.; Xing, R.-F. Climate modeling and prediction in Songhua River Basin based on model integration. *J. Hydroelectr. Power* **2019**, *46*, 14–18.
20. Ye, X.-Y.; Wu, Y.-M.; Du, X.-Q.; Li, W.-L. Comprehensive identification method of dynamic genetic types of subsurface water: A case study of Songhua River Basin, Sanjiang Plain. *Prog. Water Sci.* **2022**, *33*, 68–78.
21. Zhao, S. Based on the Suitable Water Level of Sanjiang Plain Groundwater Exploitation Groundwater Regulation Scheme Research. Master's Thesis, Jilin University, Jilin, China, 2023. [CrossRef]
22. Wu, Z.; Wu, Y.; Yu, Y.; Zhang, G.-X.; Sun, Y.-N.; Fu, Q.-N. Groundwater environment change and its controlling factors in Songhua River Basin, Sanjiang Plain. *J. Hydropower Energy* **2024**, *42*, 38–42.
23. Ridwansyah, I.; Yulianti, M.; Apip; Onodera, S.I.; Shimizu, Y.; Wibowo, H.; Fakhrudin, M. The impact of land use and climate change on surface runoff and groundwater in Cimanuk watershed. *Indones. Limnol.* **2020**, *21*, 487–498. [CrossRef]
24. Xu, A.; Mu, Z. G four river basin upstream of the SWAT model database construction. *J. Yangtze River Technol. Econ.* **2021**, *5*, 114–116. [CrossRef]
25. Zhu, Y. *Study on Water Resource Vulnerability in Songhua River Basin under Changing Environment*; Jilin University: Jilin, China, 2024.
26. Chen, K.; Wang, L.; Liu, Y.; Liu, J.-X. Applicability evaluation of CMADS dataset in Hulan River Basin. *J. Irrig. Drain.* **2019**, *43*, 60–68.

27. Gu, X.; Yang, G.; He, X.; Zhao, L.; Li, X.; Li, P.; Liu, B.; Gao, Y.; Xue, L.; Long, A. Simulation of hydrological processes in Manas River Basin based on CMADS and SWAT model. *J. Water Resour. Water Eng.* **2021**, *32*, 116–123.

28. Dai, J.; Cui, Y. Swat-based distributed hydrological model of irrigation district—I. Principle and method of model construction. *J. Hydraul. Eng.* **2009**, *40*, 145–152.

29. Zhao, X.; Yan, X. Study on the Change Trend of Soil Water Storage in Heilongjiang Province in the Past 20 Years. *Geogr. Sci.* **2006**, *5*, 5569–5573.

30. Wang, Z.; Liu, C.; Huang, Y. Research on the principle, structure and application of SWAT model. *Prog. Geogr.* **2003**, *1*, 79–86.

31. Cui, J. Application of water balance method in groundwater resource assessment. *Groundwater* **2022**, *44*, 51–52+91. (In Chinese)

32. Saurette, D.D.; Biswas, A.; Heck, J.R.; Gillespie, A.W.; Berg, A.A. Determining minimum sample size for the conditioned Latin hypercube sampling algorithm. *Pedosphere* **2024**, *34*, 530–539. [CrossRef]

33. Bao, Y.; Chen, J.; Shao, S. Comparison on thermal and hydraulic performances of transverse mini-channel based on a novel comprehensive evaluation factor. *Appl. Therm. Eng.* **2024**, *253*, 123795. [CrossRef]

34. Hou, M.; Dong, A.Z.; Ying, H.Y.; Wang, Z.-H. Influence of parameter change of empirical frequency formula on hydrological frequency analysis. *J. Zhejiang Univ. Water Resour. Hydropower* **2023**, *35*, 24–30.

35. Wu, L.; Lu, T.; Guo, J.; Lu, Y.; Li, H. High fluoride groundwater distribution and formation mechanism of jilantai desert basin. *J. Irrig. Drain. Lancet* **2024**, *9*, 113–120.

36. Liu, Y.; Wu, T.; Wang, S.; Ju, M.-D. Study on the efficiency of coastal vegetation belt in reducing non-point source pollution by coupled remote sensing and SWAT model. *Nat. Resour. Remote Sens.* **2024**, 1–10. Available online: http://kns.cnki.net/kcms/detail/10.1759.P.20230531.1032.019.html (accessed on 5 August 2024).

37. Zhao, H. Simulation and Prediction of Snowmelt Runoff in Northeast Cold Region Based on Improved SWAT. Master's Thesis, Jilin University, Jilin, China, 2023.

38. Lai, G.; Wu, D.; Zhong, Y.; Zeng, F.-H.; Chen, J.; Zhang, W.-L. Development and Application of SWAT Model. *J. Hohai Univ. (Nat. Sci. Ed.)* **2012**, *40*, 243–251.

39. Xin, Y.; Zhang, W.; Zou, L.; Zhou, T.-J.; Zhao, Y. High-resolution regional model prediction of precipitation erosivity affected by climate change in Northeast China. *Chin. J. Atmos. Sci.* **2019**, *48*, 480–496.

40. Yifru, B.A.; Lee, S.; Lim, K.J. Calibration and uncertainty analysis of integrated SWAT-MODFLOW model based on iterative ensemble smoother method for watershed scale river-aquifer interactions assessment. *Earth Sci. Inform.* **2023**, *16*, 3545–3561. [CrossRef]

41. Deng, Z.; Ma, Q.; Zhang, J.; Feng, Q.; Niu, Z.; Zhu, G.; Jin, X.; Chen, M.; Chen, H. A New Socio-Hydrology System Based on System Dynamics and a SWAT-MODFLOW Coupling Model for Solving Water Resource Management in Nanchang City, China. *Sustainability* **2023**, *15*, 16079. [CrossRef]

42. Li, L.F.; Chen, J.Y.; Chen, L.; Chen, S.-B.; Xu, Y.-Z.; Zhi, X.-S.; Shen, Z.-Y.; Yang, D.-W. Improvement of SWAT model reservoir module and simulation of nitrogen and phosphorus flux in watershed. *Chin. J. Environ. Sci.* **2023**, *43*, 427–439. (In Chinese)

43. Cui, H.; Wu, T.; Liu, J.; Liu, W.-P.; Li, Z.-H.; Cheng, X.-X.; Liu, W.-P. Effect of succession process on groundwater in marsh wetland and farmland in Sanjiang Plain. *Hydrogeol. Eng. Geol.* **2023**, *50*, 51–58.

44. Zhang, D.; Zhang, W.; Zhu, L.; Zhu, Q.-A. Research on improvement and application of the hydrophysical model of distributed SWAT basin. *Sci. Geogr. Sin.* **2005**, *4*, 52–58.

45. Jia, C.C.; Guo, L.; Cui, S.; Fu, Q.; Liu, D. Temporal and spatial evolution of NPP and its response mechanism to extreme climate in Songhua River Basin. *South--North Water Divers. Water Sci. Technol.* **2019**, *22*, 131–147. (In Chinese)

46. Chu, G.; Chen, S.; Xu, C.M.; Liu, Y.-H.; Wang, D.-Y.; Zhang, X.-F. Evolution and prospect of paddy planting system in China. *Chin. Rice* **2021**, *27*, 63–65+70. (In Chinese)

47. Yu, Y. Study on the impact of engineering water resources utilization on aquatic ecology in Jinxi Irrigation area, Fujin City, Heilongjiang Province. *Sci. Technol. Innov.* **2020**, *13*, 118–119.

Article

Runoff Prediction in Different Forecast Periods via a Hybrid Machine Learning Model for Ganjiang River Basin, China

Wei Wang [1], Shinan Tang [2], Jiacheng Zou [3,*], Dong Li [3], Xiaobin Ge [3], Jianchu Huang [3] and Xin Yin [4]

[1] National Institute of Natural Hazards, Ministry of Emergency Management of China, Beijing 100085, China; weiwang@ninhm.ac.cn

[2] General Institute of Water Resources and Hydropower Planning and Design, Ministry of Water Resources, Beijing 100120, China; tangshinan@giwp.org.cn

[3] Hydrology and Water Resources Monitoring Center of Lower Ganjiang River, Yichun 336000, China; swjld@foxmail.com (D.L.); ycswgxb@foxmail.com (X.G.); hjcsw@foxmail.com (J.H.)

[4] Nanjing Hydraulic Research Institute, Nanjing 210029, China; xyin@nhri.cn

* Correspondence: zoujiacheng26@foxmail.com; Tel.: +86-13766445857

Abstract: Accurate forecasting of monthly runoff is essential for efficient management, allocation, and utilization of water resources. To improve the prediction accuracy of monthly runoff, the long and short memory neural networks (LSTM) coupled with variational mode decomposition (VMD) and principal component analysis (PCA), namely VMD-PCA-LSTM, was developed and applied at the Waizhou station in the Ganjiang River Basin. The process begins with identifying the main forecasting factors from 130 atmospheric circulation indexes using the PCA method and extracting the stationary components from the original monthly runoff series using the VMD method. Then, the correlation coefficient method is used to determine the lag of the above factors. Lastly, the monthly runoff is simulated by combining the stationary components and key forecasting factors via the LSTM model. Results show that the VMD-PCA-LSTM model effectively addresses the issue of low prediction accuracy at high flows caused by a limited number of samples. Compared to the single LSTM and VMD-LSTM models, this comprehensive approach significantly enhances the model's predictive accuracy, particularly during the flood season.

Keywords: monthly runoff forecasting; factor selection; variable modal decomposition; principal component analysis; long short-term memory neural network

Citation: Wang, W.; Tang, S.; Zou, J.; Li, D.; Ge, X.; Huang, J.; Yin, X. Runoff Prediction in Different Forecast Periods via a Hybrid Machine Learning Model for Ganjiang River Basin, China. *Water* **2024**, *16*, 1589. https://doi.org/10.3390/w16111589

Academic Editors: Juraj Parajka and Alexander Shiklomanov

Received: 7 April 2024
Revised: 25 May 2024
Accepted: 30 May 2024
Published: 1 June 2024

1. Introduction

Runoff prediction is essential for water resource management, allocation, and effective utilization [1,2]. Accurate runoff prediction, especially in medium- and long-term timescales, can provide effective scientific support for agricultural irrigation, industrial and domestic water use, reservoir optimization, water conservancy project planning and design, etc. [3,4]. Another significant aspect of runoff prediction is providing early warnings for floods and drought, which are strongly correlated with certain meteorological factors, such as precipitation and runoff, as indicated by relevant studies [5,6]. However, due to its vulnerability to climate change and human activities, runoff has the characteristics of being highly nonlinear, unstable and complicated [7]. Therefore, it is still a challenge to obtain high-precision runoff-prediction results.

Currently, runoff prediction models can be divided into two types: process-driven models [8–10] and data-driven models [11–13]. Process-driven models are modeled to simulate complex non-linear physical hydrological process through a series of mathematical equations based on an understanding and simplification of the principles of the natural water system [14,15]. For example, the Xin'anjiang model [16], the Soil and Water Assessment Tool [17], and Sacramento Soil Moisture Accounting [18] are the most widely used physically driven models. Although they can reveal the physical mechanism of runoff generation [19,20], there is still another important factor that can affect the hydrological processes, e.g., human activities such as hydropower, which are processes hardly modeled using traditional physically based models. Additionally, the modeling demands a great deal of accurate and reliable information on hydrological processes (e.g., precipitation and evapotranspiration), which leads to common shortcomings such as difficulty in determining the parameters and poor versatility of the model. With improvements in computing power, data-driven models have shown great potential in capturing the rainfall–runoff relationship and predicting the runoff in a given basin. Without considering the hydrological physical processes, data-driven models are widely employed to address a variety of classification and regression problems by establishing the statistical relationships between inputs and outputs. For the prediction of medium- and long-term runoff, researchers have used data-driven models such as artificial neural networks (ANNs), support vector machines (SVMs), and long short-term memory (LSTM) networks to capture the nonlinear and unsteady characteristics of runoff time series [21–23], and some studies have achieved better performance than those using traditional process-driven models [24,25]. As a branch of data-driven models, deep learning models can better address the insufficient ability of classical data-driven models to deal with nonlinear relationships in difficult situations. Recently, with the rapid development of deep learning models, it has been increasingly studied in the simulation and prediction of hydrological elements such as runoff, evapotranspiration, and soil moisture [26–28]. For instance, Castangia et al. [29] explored the applicability of the transformer model to flood forecasting and found that the model has higher prediction accuracy than recurrent neural networks. Although many new deep learning models have been applied to the hydrologic forecasting field, long short-term memory (LSTM) still keeps a wide application in runoff prediction [30], especially for monthly runoff prediction.

In addition to the selection of appropriate hydrological models, identifying the key forecasting factors that drive runoff variability is another aspect of building a reliable forecast model [31]. Since runoff is a non-stationary component with periodicity, stochasticity, and trend, the accuracy of direct prediction using the above models is limited [32]. The signal decomposition technique can decompose the runoff series into several relatively stable components to reduce the non-stationarity of the time series with high complexity and strong nonlinear, which can help the model better capture the change patterns of the runoff series and improve the prediction accuracy [33,34]. For example, Wang et al. [35] found that the runoff prediction results of the auto-regressive integrated moving average (ARIMA) model combined with the ensemble empirical mode decomposition (EEMD) are more accurate and stable than that of a single ARIMA model. Zuo et al. [36] developed a single-model forecasting (SF) scheme based on variational mode decomposition (VMD) and LSTM to predict daily runoff with a lead time of 1–7 days, and found that the SF-VMD-LSTM can effectively capture the unsteady and nonlinear nature of the runoff. Additionally, previous studies have shown that the rainfall–runoff process is also closely connected with climatic conditions and human activities except for traditional meteorological factors such as precipitation and potential evapotranspiration [37,38]. Champagne et al. [39] quantified the contribution of atmospheric circulation on runoff response for four basins in southern Ontario and found that the temporal increase in high pressure contributed more than 40% to the increase in runoff in winter. To improve the model performance of runoff simulation, factors such as EI Nino, LaNina, and atmospheric circulation affecting the regional hydrological cycle were selected as model inputs. For example, Yan et al. [40] found that considering atmospheric circulation anomaly factors can effectively reduce the influence of extreme weather and climate anomalies on the prediction accuracy of medium- and long-term runoff. Mostaghimzadeh et al. [41] studied the impact of climate–atmospheric indices on runoff predictions and found that runoff is highly correlated with the Pacific STT in the Great Karon system. However, many studies based on the LSTM model only consider a single forecast period, and whether decomposition technology can improve the performance of the LSTM model in multiple forecast periods is not clear. Moreover, the effect of atmospheric circulation indexes on the model runoff prediction based on decomposition technology remains to be investigated.

Therefore, a hybrid machine learning model coupled with LSTM, VMD, and PCA was created in this study to predict monthly runoff in the Ganjiang River Basin, aiming to explore the effect of decomposition technology and atmospheric circulation indexes on the performance of the hybrid machine learning model in multiple forecast periods. Key forecasting factors were extracted from 130 atmospheric circulation indexes using the PCA method; meanwhile, stationary components were derived from the original monthly runoff series using the VMD method. Subsequently, the lag time of the above factors was determined by the correlation coefficient method. Lastly, the impact of VMD decomposition and the incorporation of atmospheric circulation on the runoff prediction of the LSTM model were investigated. The paper is organized as follows. Section 2 describes the model and evaluation indicators. Section 3 delineates the study area and data preprocessing. Section 4 provides an analysis and discussion of the results. The main findings and conclusions are given in Section 5.

2. Methodology

2.1. Variational Mode Decomposition

The VMD algorithm is an adaptive, completely non-recursive mode variational and signal processing method with the core of constructing and solving variational problems [42]. It overcomes the endpoint effect and the problem of modal component overlapping in the empirical mode decomposition (EMD) method by determining the number and the best center frequency of modal decompositions of the sequence according to the actual situation, and effectively obtains multiple smooth subsequences with different frequencies. Assuming that the original signal f is decomposed into k modes with finite bandwidth and center frequency, to ensure that the sum of the estimated bandwidths of each mode is minimum and all modes' sum is kept constant, the constraint variational problem can be shown as follows:

$$\min_{\{u_k\},\{u_k\}} \left\{ \sum_k \| \partial_t[(\delta(t) + j/\pi t) * u_k(t)]e^{-jw_k t} \|_2^2 \right\} \tag{1}$$

$$s.t. \sum_{k=1}^{K} u_k = f \tag{2}$$

where k is the number of decomposed modes, u_k and ω_k correspond to the kth modal component and the center frequency after decomposition, $\delta(t)$ is the Dirac function, $*$ is the convolution operator, and f is the original time series. See reference [43] for a detailed solving process.

2.2. Principal Component Analysis

PCA method is a data dimensionality reduction algorithm that transforms multiple variables into a few composite variables through orthogonal transformations with minimal loss of data information [44]. It screens principal forecasting factors by standardizing variables and calculating the covariance matrix and its eigenvectors and eigenvalues. A smaller variance contribution means less information for the selected factors. The equation for the extraction of the forecasting factors can be described as follows:

$$\begin{cases} y_1 = a_{11}x_1 + a_{12}x_2 + \cdots + a_{1n}x_n \\ y_2 = a_{21}x_1 + a_{22}x_2 + \cdots + a_{2n}x_n \\ \cdots \\ y_m = a_{m1}x_1 + a_{m2}x_2 + \cdots + a_{mn}x_n \end{cases} \tag{3}$$

where A is the feature vector matrix composed of coefficient a. y_1 is the linear combination of $x_1, x_2, \cdots, x_n$ with the largest variance among all linear combinations. Similarly,

y_m is the linear combination of $x_1, x_2, \cdots, x_n$ with the mth largest variance among all linear combinations.

2.3. Long Short-Term Memory Network

The LSTM model (see Figure 1) is a special form of RNN with cell state and gate structure as the core [45]. The cell state plays a role in the transmission of information, while the gate structure determines the retention and forgetting of information, and the interaction of the two ensures the efficient transfer of information through the sequence. It overcomes the problem of long-term dependencies and is more suitable for dealing with time series forecasting problems. The computation process of the LSTM unit is described in Equations (4)–(8):

$$Input\ gate:\ I_t = \sigma(w_i \cdot G[h_{t-1}, x_t] + b_f) \tag{4}$$

$$Forget\ gate:\ F_t = \sigma\left(w_f \cdot G[h_{t-1}, x_t] + b_f\right) \tag{5}$$

$$Output\ gate:\ O_t = \sigma(w_O \cdot G[h_{t-1}, x_t] + b_O) \tag{6}$$

$$Cell\ state:\ \begin{cases} \widetilde{G_t} = \tanh\left(w_g \cdot [h_{t-1}, x_t] + b_g\right) \\ G_t = F_t \cdot G_{t-1} + I_t \cdot \widetilde{G_t} \end{cases} \tag{7}$$

$$Output\ vector:\ h_t = O_t \cdot \tanh(C_t) \tag{8}$$

where h represents time output; w is the weights of gates; b is the bias of gates; C is the cell state; x is the input; σ denotes the sigmoid function; $\widetilde{G_t}$ is the information status through the input gate; t represents the time step.

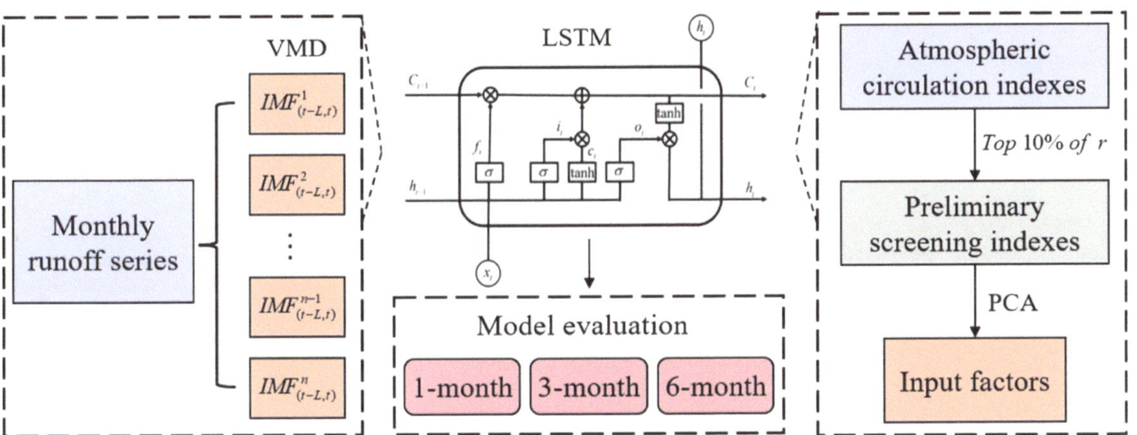

Figure 1. The process of monthly runoff prediction via the VMD-PCA-LSTM model.

The LSTM model's hyperparameters including hidden layer nodes, learning rate, dropout rate, and batch size are determined by the Bayesian optimization (BO) algorithm [46] in the training period. The rest refer to the previous research [47]. The initial point and iteration times of the BO algorithm are set to 20 and 30 times, respectively.

2.4. VMD-PCA-LSTM

The hybrid VMD-PCA-LSTM model mainly includes the following three steps (see Figure 1):

(1) Multiple stationary intrinsic modal components (IMF) and a residual component (residual) were obtained by decomposing the runoff series according to the VMD method;

(2) The PCA method was used to reduce the dimension of the atmospheric circulation indexes, and then principal components with a cumulative contribution rate greater than 90% were selected as forecasting factors;

(3) Normalized processing and determinations of the inputs and outputs of the LSTM model were carried out.

Each inherent modal component $(IMF^1_{(t-L,t)}, IMF^2_{(t-L,t)}, \cdots IMF^n_{(t-L,t)})$ and trend component $(Residual_{(t-L,t)})$ are used as predictors to predict the different forecasting periods of runoff (R_{t+1}, R_{t+3}, R_{t+6}) in Waizhou station. In the LSTM model, L represents the lag time. According to the periodic variation law of monthly runoff, the L of a single LSTM model can be directly set to 12. However, the interannual variation law of runoff decomposed by VMD has changed, which means that the L value of the hybrid model cannot be set to 12 directly, and the optimal value of L needs to be determined by repeated debugging. Likewise, considering that the effect of atmospheric circulation on runoff has a lag time [48,49], the optimal value of L needs to be determined by repeated debugging.

2.5. Evaluation Metrics

The evaluation metrics used in this study consist of Nash–Sutcliffe efficiency (NSE), root mean square error ($RMSE$), correlation coefficient (r) and volume error (VE). The closer the NSE and r values are to 1, the smaller the $RMSE$, and the closer the VE value is to 0, the more accurate the runoff predictions. These metrics can be represented mathematically:

$$NSE = 1 - \frac{\sum_{t=1}^{n}(Q_{sim,t} - Q_{obs,t})^2}{\sum_{t=1}^{n}(Q_{obs,t} - \overline{Q}_{obs})^2} \tag{9}$$

$$RMSE = \sqrt{\frac{\sum_{t=1}^{n}(Q_{sim,t} - Q_{obs,t})^2}{n}} \tag{10}$$

$$r = \frac{\sum_{t=1}^{n}(Q_{obs,t} - \overline{Q}_{obs})(Q_{sim,t} - \overline{Q}_{sim})}{\sqrt{\sum_{i=1}^{n}(Q_{obs,t} - \overline{Q}_{obs})^2 \times \sum_{i=1}^{n}(Q_{sim,t} - \overline{Q}_{sim})^2}} \tag{11}$$

$$VE = 1 - \frac{\sum_{t=1}^{n} Q_{sim,t}}{\sum_{t=1}^{n} Q_{obs,t}} \tag{12}$$

where Q_{sim} and Q_{obs} are the simulated and observed monthly runoff, respectively; $\overline{Q}_{sim}$ and $\overline{Q}_{obs}$ are the mean value of the time series; t denotes the tth month; n is the length of the series.

3. Study Area and Data Preprocessing

3.1. Gangjiang River Basin

The Ganjiang River, which originates from Huangzhuling in Wuyi Mountain, is the main river in the Poyang Lake basin, accounting for 51% of its area (see Figure 2). The basin is located between the longitudes 113°45′–114°45′ E and latitudes 25°55′–26°35′ N and has a total drainage area of 80,948 km^2, all within Jiangxi Province. The landscape is mainly mountainous and hilly, with a terraced distribution from the south to the north and an altitude of 23–2103 m above sea level. The basin belongs to the subtropical humid monsoon climate, characterized by abundant rainfall, while the spatiotemporal distribution of precipitation is unevenly affected by the terrain and monsoon, with 50% of the precipitation concentrated from April to June. The average annual rainfall, potential evapotranspiration, and runoff are about 1550 mm, 1070 mm, and 870 mm, respectively [50]. According to local government documents, April to September is defined as the flood season of the basin.

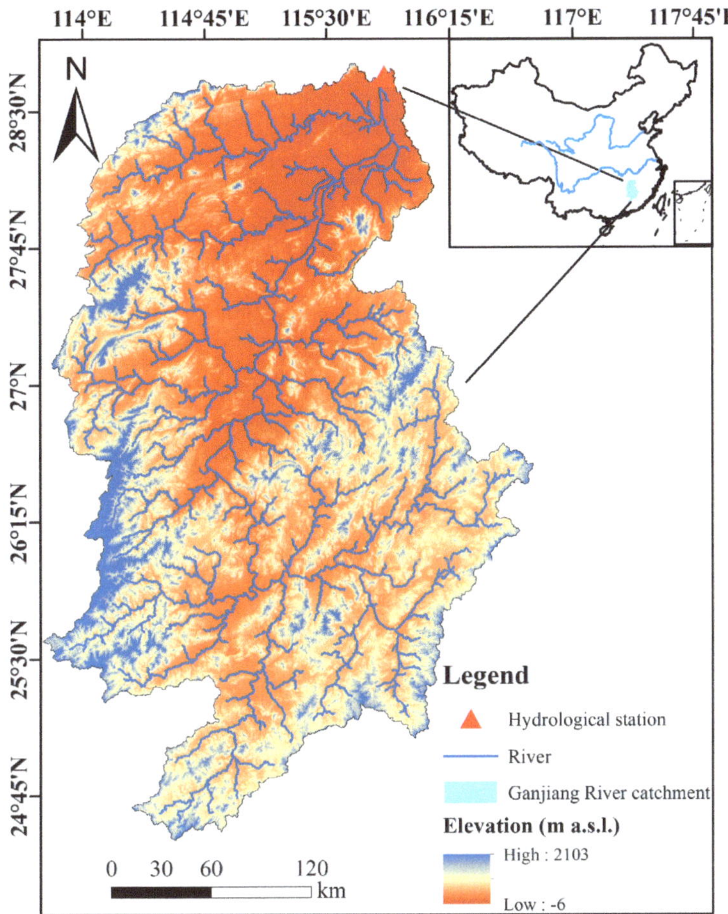

Figure 2. Location of the Ganjiang River Basin.

The monthly runoff data of Waizhou Station are collected from the local hydrological departments with time ranges from 1957 to 2016. Likewise, Atmospheric circulation indexes consist of 88 atmospheric circulation indexes, 26 sea temperature indexes, and 16 other indexes provided by the National Climate Center of China Meteorological Administration (http://cmdp.ncc-cma.net/Monitoring/cn_indexes_130.php accessed on 4 December 2023). Note that the first 80% of the above sequence data are used for training, while the last 20% are used for verifying.

3.2. Monthly Runoff from the VMD Decomposition

The decomposition effect of VMD is largely influenced by the number of mode decompositions (K), and the subsequent prediction effect will be affected if the selecting value of K is unreasonable. In general, the appropriate K value can be preliminarily selected according to the distribution of center frequency under different modes. To further determine K, the correlation of the decomposed adjacent mode components is analyzed, as shown in Table 1. In the table, r_{n-m} represents the correlation coefficient of the decomposed nth mode and the mth mode. It can be seen that when the K is less than 6 or larger than 8, the correlation coefficients of adjacent modal components fluctuate greatly, which indicates that the mode component is stacked, leading to the over-decomposition of the runoff signal. When K is between 6 and 8, the correlation coefficient of the adjacent mode components is stable and less than 0.2, and each mode shows the runoff signal characteristics of the corresponding central frequency. Therefore, the K value of Waizhou station is selected as 8.

Table 1. Correlation coefficients of adjacent modes.

K	r_{1-2}	r_{2-3}	r_{3-4}	r_{4-5}	r_{5-6}	r_{6-7}	r_{7-8}	r_{8-9}
2	0.128	-	-	-	-	-	-	-
3	0.011	0.113	-	-	-	-	-	-
4	0.009	0.071	0.203	-	-	-	-	-
5	0.031	0.035	0.050	0.184	-	-	-	-
6	0.044	0.029	0.050	0.174	0.170	-	-	-
7	0.076	0.082	0.024	0.045	0.170	0.169	-	-
8	0.092	0.075	0.081	0.018	0.042	0.168	0.169	-
9	0.085	0.089	0.051	0.153	0.026	0.035	0.166	0.169

The results of monthly runoff from Waizhou station after VMD decomposition are shown in Figure 3. The original monthly runoff series is decomposed into eight stationary components (IMF) and a residual term representing the trend, which not only reduces the noise but also helps the model identify the internal transformation law of the runoff series.

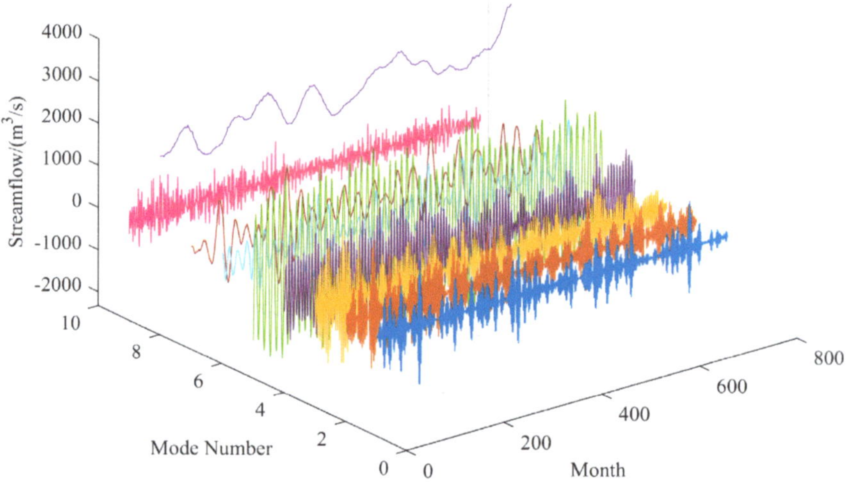

Figure 3. The VMD decomposition of monthly runoff at Waizhou station.

4. Results and Discussion

4.1. Determining Forecasting Factors and Model Parameter

At present, there is no specific principle on how to screen the input elements of machine learning models, and the correlation coefficient method is commonly used in previous studies [30,51]. However, considering VMD decomposition causing the change in runoff interannual variation law, the optimal value of lag time (L) is selected according to the value of correlation coefficients (r) between the IMF and the original runoff series. Taking L equal to 1 month as an example, Figure 4 shows the value of r between each IMF and the original runoff series under different L; red and yellow shading indicate a good related degree, and blue shading represents poor. It can be seen that when the mode number is larger than 5, the value of r decreases as L increases, and when the mode number is less than 5, the value of r slightly fluctuates; meanwhile, the r of IMF5 presents a decrease and then an increase trend with increasing L. Additionally, to further determine the optimal value of L, the sum of r is obtained by each IMF in the same L, and the value of L is finally determined as 1 by selecting the L corresponding to the maximum sum of r.

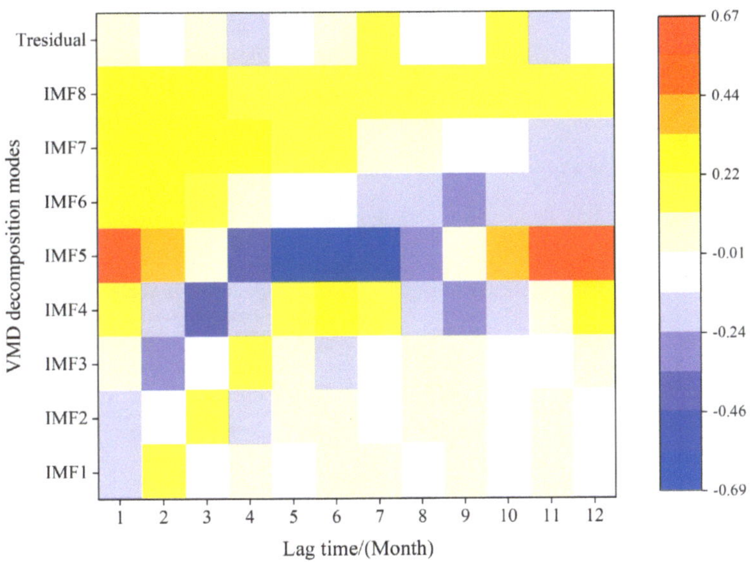

Figure 4. Heatmap of the correlation coefficient (r) between each IMF and the original runoff series under different lag times.

Similarly, the optimal value of lag time (L) is determined according to the value of r between atmospheric circulation indexes and the original runoff series. In order to select circulation indexes that are as strongly correlated with monthly runoff as much as possible, the top 1% of r among circulation indexes that pass the 0.01 significance test is selected as model additional input factors. It can be found from Table 2 that L is mainly equal to 7 and 8, and the r of North African Subtropical High Ridge Position Indexes, Indian Subtropical High Ridge Position Indexes, and Western Pacific Subtropical High Ridge Position Indexes separately are the top three, which confirms previous studies about the effect of atmospheric circulation indexes on moisture transport in eastern China [52]. Additionally, considering that excessive input factors in the machine learning model will cause an overfitting phenomenon, the PCA method is used to reduce the input dimensions of the above circulation indexes. The ranking of variance contribution for the partial components is shown in Table 3. In general, the greater the variance contribution of the principal component, the more information on the selected factor. It should be noted that the variance contribution of the first principal component reaches 85.42%, indicating that it contains most of the information of the selected circulation indexes. According to the cumulative contribution rate threshold set in Section 2.4, the first two principal components are finally determined as the additional forecasting factors.

Table 2. The rank of correlation coefficient (r) between the atmospheric circulation indexes and the original runoff series in the optimal lag time.

Rank of r	Factor Type	Lag Time /(Month)
6	Northern Hemisphere Subtropical High Ridge Position Indexes	
3	Western Pacific Subtropical High Ridge Position Indexes	
10	South China Sea Subtropical High Ridge Position Indexes	7
5	Pacific Subtropical High Ridge Position Indexes	
11	North African-North Atlantic-North American Subtropical High Area Indexes	
13	North American Subtropical High Area Indexes	
9	Atlantic Subtropical High Area Indexes	
8	North American-Atlantic Subtropical High Area Indexes	
1	North African Subtropical High Ridge Position Indexes	8
12	North African-North Atlantic-North American Subtropical High Ridge Position Indexes	
2	Indian Subtropical High Ridge Position Indexes	
7	Northern Hemisphere Polar Vortex Central Intensity Indexes	
4	East Asian Trough Intensity Indexes	

Table 3. The ranking of variance contribution for the partial components.

Component	Total	Variance/(%)	Cumulative Variance/(%)
1	11.10	85.42	85.42
2	0.90	6.94	92.36
3	0.35	2.68	95.04
4	0.17	1.34	96.38
5	0.11	0.83	97.21
6	0.098	0.75	97.96

4.2. Effect of VMD Decomposition on Runoff Prediction of LSTM Model

To investigate the influence of the VMD decomposition method on the runoff prediction results of the LSTM model, the single LSTM model and the VMD-LSTM model are used to predict the monthly runoff of Waizhou hydrographic stations. The runoff prediction results of different forecast periods are shown in Table 4.

Table 4. The performances of LSTM and VMD-LSTM model during the validation period.

Forecast Period	Model	*NSE*	*RMSE*/(m^3/s)	*VE*/(%)
1 month	LSTM	0.518	1185	−0.77
	VMD-LSTM	0.954	366	−1.97
3 months	LSTM	0.430	1292	0.76
	VMD-LSTM	0.931	450	7.81
6 months	LSTM	0.424	1299	1.62
	VMD-LSTM	0.828	710	4.21

The VMD-LSTM model demonstrates significant improvements over the single LSTM model. Specifically, it increases the *NSE* by 0.404–0.501 and decreases the *RMSE* by 589–842 m^3/s. This suggests that the VMD method significantly enhances the predictive performance of the single LSTM model. Figure 5a further shows the performance of the single LSTM model and the VMD-LSTM model in predicting the monthly runoff for different forecasting periods. A great improvement occurs in predicting the monthly runoff for 3 months, with the *NSE* increasing by 116.7% and *RMSE* decreasing by 922.1%. However, it is worth noting that the *VE* obtained by the VMD-LSTM model increases by 45.3–69.1% compared with the single LSTM model. The reason can be inferred from Figure 6, that the single LSTM model's overestimation of medium and low flow compensates for its underestimation of high flow.

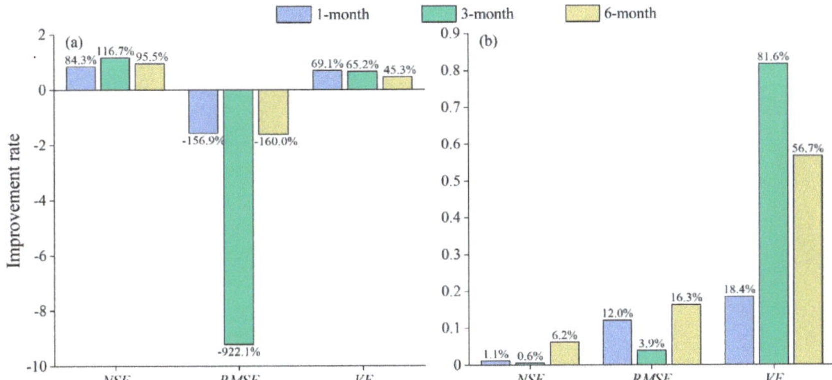

Figure 5. Comparison of runoff prediction in different forecast periods: (**a**) LSTM model and VMD-LSTM model; (**b**) VMD-LSTM model and VMD-PCA-LSTM model.

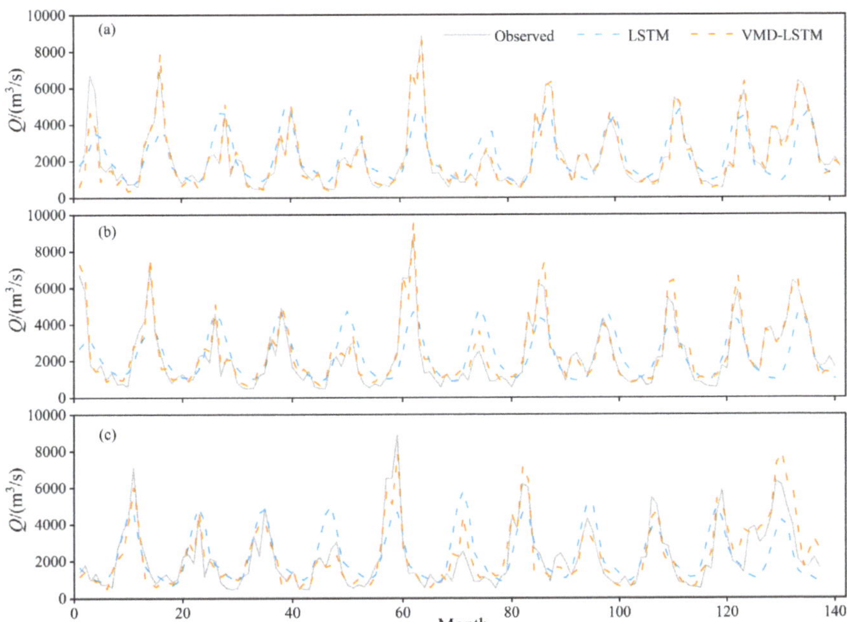

Figure 6. The monthly runoff prediction results during different forecast periods for LSTM and VMD-LSTM models: (**a**) 1 month; (**b**) 3 months; (**c**) 6 months.

Additionally, the prediction accuracy of all models gradually degenerates as the forecast period prolongs, while the VMD-LSTM model degenerates less; for example, the *NSE* of the LSTM and VMD-LSTM model decreases by 18.1% and 13.2% in the foresight period from 1 month to 6 months, respectively. In comparison, when the forecast period increases from 1 month to 3 months, the *NSE* of the VMD-LSTM model shows no significant change, indicating that the VMD decomposition method not only improves the prediction performance of the LSTM model but also extends the forecast period in a certain extent. As shown in Figure 6, the prediction performance of the VMD-LSTM model for high flow is significantly better than that of the single LSTM model in the same forecast period. It is worth noting that the *NSE* of the VMD-LSTM is greater than 0.8, as the forecast period is prolonged, which further explains that the prediction ability improvement of the LSTM model by the VMD decomposition method is mainly reflected in the improvement of the prediction of the high flow.

4.3. Effect of Considering Atmospheric Circulation on Runoff Prediction of LSTM Model

The above research shows that the VMD-LSTM model can simulate the runoff with a lead time of 1, 3 and 6 months more accurately than the single LSTM model. To further enhance the prediction performance, the first two principal components of the atmospheric circulation indexes screened by the PCA method are used as the additional input of the VMD-LSTM model for the simulation of the monthly runoff with a lead time of 1, 3 and 6 months. The runoff prediction results for different forecast periods before and after integrating atmospheric circulation indexes are shown in Table 5. Different from only considering the VMD method, all metrics obtained by the VMD-PCA-LSTM model are better than the VMD-LSTM model, which means that considering the atmospheric circulation indexes as the forecasting factors can comprehensively enhance the prediction performance of the VMD-LSTM model.

Table 5. Predicted results before and after integrating atmospheric circulation indexes. Note: the data on the left and right of the arrow represent the predicted results before and after integrating the atmospheric circulation indexes, respectively.

Forecast Period	*NSE*	*RMSE*/(m^3/s)	*VE*/(%)
1 month	0.954→0.964	366→322	−1.97→−1.61
3 months	0.931→0.936	450→432	7.81→1.43
6 months	0.828→0.879	710→595	4.21→−1.82

Figure 5b further shows the prediction performance of the VMD-LSTM and VMD-PCA-LSTM models for different forecasting periods. It can be found that the improvement degree of the VMD-LSTM model after integrating atmospheric circulation indexes becomes more significant as the forecast period prolongs, particularly when the forecast period is 6 months, the *NSE* and *RMSE* have the most significant improvement with increasing by 6.2% and 16.3%, respectively. The reason may be that the long-term continuous influence of atmospheric circulation on regional climate leads to the fact that the atmospheric circulation indexes of the previous period still affect the climate for a long time in the future, and then affect the runoff through the water cycle process. Therefore, with the increase in the forecast period, the effect of historical runoff gradually weakens, and the prediction performance of the model gradually decreases, while the atmospheric circulation factors still play a certain role in the runoff prediction in the following months, which makes the accuracy of the model improve more significantly with the increase in the forecast period after the integration of atmospheric circulation indexes.

Figure 7 intuitively displays the forecast performance of the VMD-LSTM and VMD-PCA-LSTM models in different forecast periods. When the forecast period is one month, all models can predict well the future change trend of runoff. With the increase in the forecast period, the predicted runoff of the VMD-PCA-LSTM model still maintains a high degree of correspondence with the observed, while the predicted runoff of the VMD-LSTM model has a significant deviation, mainly manifesting as an overestimation of high flow prediction. Due to the small number of samples at the high flow, the accurate prediction of high flow becomes a difficult problem in runoff prediction. The results indicate that adding atmospheric circulation indexes to the model input can effectively solve the problem of low prediction accuracy at high flow caused by a small number of samples.

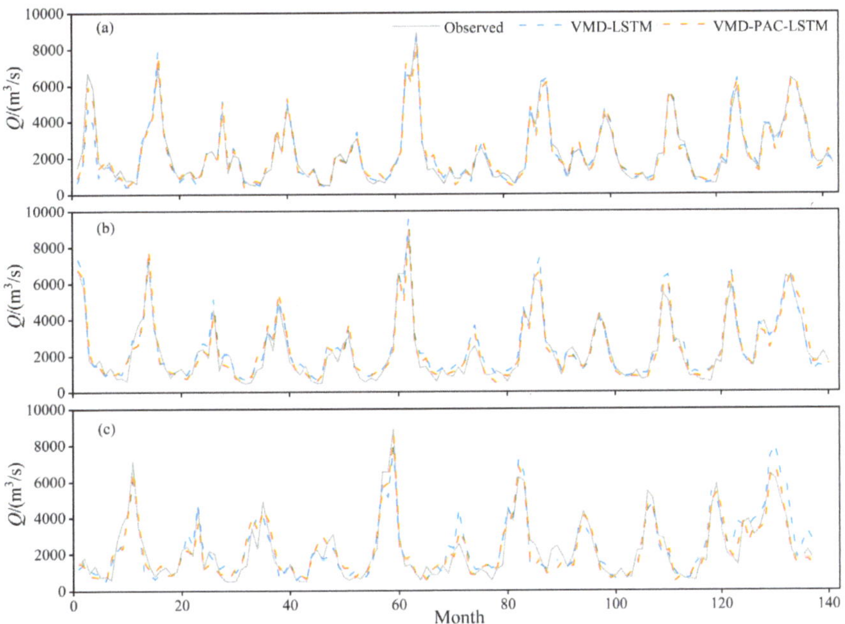

Figure 7. The monthly runoff prediction results during different forecast periods for VMD-LSTM and VMD-PAC-LSTM models: (**a**) 1-month; (**b**) 3-month; (**c**) 6-month.

4.4. Performance of Runoff Prediction in Flood and Non-Flood Season

As a result of the uneven spatiotemporal distribution of precipitation in the basin, issues such as flood disasters, drought, and mismatch between water supply and demand have become progressively prominent. Therefore, the accurate prediction of runoff in flood and non-flood seasons can provide a scientific basis for effectively reducing the risk of flood damage and mitigating the mismatch between water supply and demand.

The prediction results for three models during different forecast periods are displayed in Figure 8. It is quite obvious that the LSTM model slightly overestimates the low flow and strongly underestimates the high flow at all times, which is resolved by the introduction of the VMD method, and the forecast period equaling 1 month is especially significantly improved, while the difference between the VMD-LSTM and VMD-PCA-LSTM models is not obvious. Table 6 demonstrates the runoff prediction results of the VMD-LSTM and VMD-PCA-LSTM models. Compared with the VMD-LSTM model, the performance of the VMD-PCA-LSTM model is better in flood season and degrades in non-flood season, with the r and *RMSE* decreased by 1.7–5.8% and increased 0.7–6.5% in non-flood season and increased by 1.7–5.8% and decreased 0.7–25.1% in flood season, respectively. The results indicate that consideration of atmospheric circulation indexes is not a comprehensive improvement of the model's runoff prediction ability, but rather focuses only on the flood season, particularly for high flows.

Figure 9 presents the results of predicted monthly mean runoff in different forecast periods. There is no significant difference between all models from November to February, and only considering VMD decomposition can improve the LSTM model accuracy of other monthly runoff predictions, while the VMD-PCA-LSTM model slightly improves the VMD-LSTM model overestimation of runoff in all forecast period. In addition, the VMD-LSTM and VMD-PCA-LSTM models have good robustness with increasing forecast periods compared to the LSTM model.

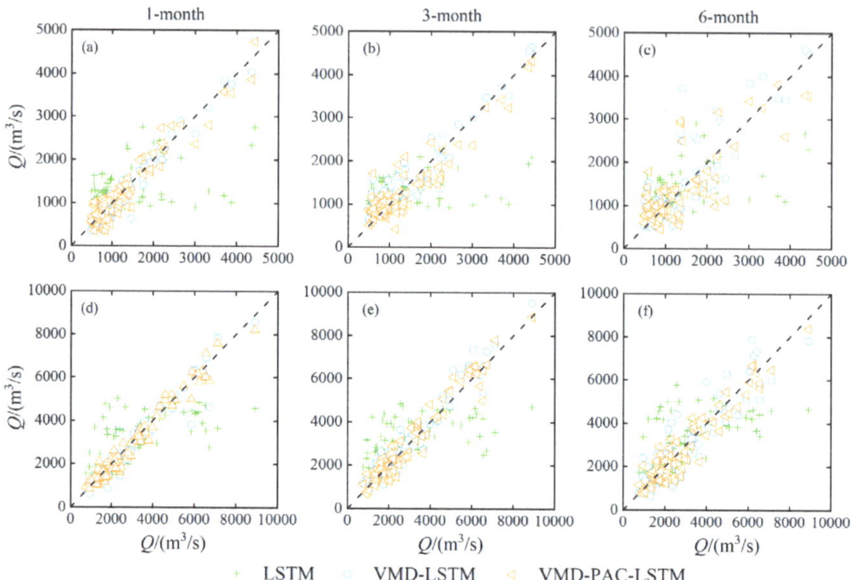

Figure 8. Comparison of predicted results for three models during different forecast periods: (**a–c**) non-flood season; (**d–f**) flood season.

Table 6. Predicted results of VMD-LSTM and VMD-PCA-LSTM model for different forecast periods during the flood and non-flood seasons. Note: the data on the left and right of the arrow represent the predicted results of the VMD-LSTM and VMD-PCA-LSTM models, respectively.

Season	Forecast Period	r	$RMSE/(\mathrm{m^3/s})$
non-flood season	1 month	0.974→0.957	215→269
	3 months	0.940→0.923	343→359
	6 months	0.846→0.797	564→568
flood season	1 month	0.978→0.982	469→366
	3 months	0.966→0.966	532→491
	6 months	0.907→0.941	833→589

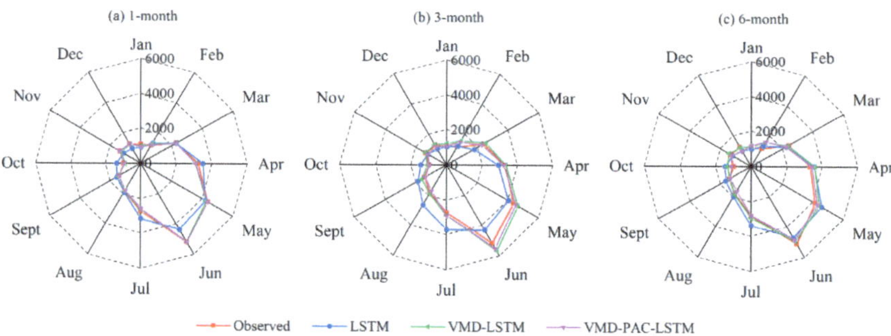

Figure 9. Radar chart of the predicted and observed monthly mean runoff in different forecast periods.

4.5. Discussion

This study uses an LSTM model coupled with VMD and PCA methods to predict the monthly runoff and finds that the hybrid model can enhance the model's predictive accuracy, particularly during the flood season. This is consistent with previous studies that reported an improvement based on the deep learning and decomposition technique for the runoff prediction [53,54]. It is worth noting that considering the structural differences of the model, different results can be obtained under different deep learning models. For example, Li et al. [55] found that the prediction results of back propagation neural network (BPNN), SVM and LSTM had significant differences, and the performance of the LSTM model was the best, especially for the peak flow forecasting. Meanwhile, many researchers often use the LSTM model for hydrological simulation and prediction. Therefore, our study only tests the effect of the LSTM model under VMD and PCA methods and does not further explore the effects of different models.

On the other hand, with the rapid development of decomposition methods, new methods such as the wavelet packet decomposition (WPD), complete ensemble empirical mode decomposition with adaptive noise (CEEMDAN) and singular spectrum analysis (SSA) demonstrate some advantages in medium- and long-term term runoff forecasting [56,57]. However, it cannot simply be concluded that new methods will always produce the best simulations under all conditions. Wang et al. [58] pointed out that selecting an appropriate data pre-processing method based on the specific characteristics of the study area can lead to more accurate prediction results. Therefore, further research is needed on the application of various decomposition techniques in the Ganjiang River Basin.

In addition, the goodness of the result for screening the input factors also influences the model performance. Global sensitivity analysis is a feasible method for selecting medium- and long-term runoff prediction factors based on physical causes, while it is not computationally efficient for complex models and large amounts of data [59]. The correlation coefficient method used in this study is simple and effective based on statistical relationships, but it still requires a subjective setting of the threshold [30]. Recently, attention mechanisms have shown great application potential in identifying key input factors of runoff prediction by automatically assigning weights to all factors [13]. Consequently, it is necessary to further study the LSTM model of coupled attention mechanism and advanced optimization algorithm in the future.

5. Conclusions

This study introduces a hybrid machine learning model built upon the LSTM model coupled with the VMD and PCA for monthly runoff prediction. The VMD decomposition was employed to reduce the noise in the runoff series, while correlation analysis determined the lag time for each IMF and the atmospheric circulation indexes. The PCA method was then utilized to select the forecasting factors from the atmospheric circulation indexes. Finally, the Bayesian optimization algorithm was used to optimize the LSTM network parameters. The constructed hybrid LSTM model was applied to the Waizhou station, considering lead times of 1, 3 and 6 months, aiming ultimately to investigate the impact of VMD decomposition and the inclusion of atmospheric circulation indexes on the runoff prediction accuracy of the LSTM model. The main conclusions are presented below:

(1) For Waizhou station, the number of mode decomposition K is 8, with lag time (L) equaling 1 month. The L of atmospheric circulation indexes is mainly equal to 7 and 8, and the r of North African Subtropical High Ridge Position Indexes, Indian Subtropical High Ridge Position Indexes, and Western Pacific Subtropical High Ridge Position Indexes separately are the top three. The first two principal components are selected as the forecasting factors from the above atmospheric circulation indexes by the PCA method.

(2) The VMD decomposition method can significantly improve the prediction accuracy of the single LSTM model, especially concentrating on the prediction of high flow during the flood and non-flood seasons, and the improvement rate of *NSE* and *RMSE* are 84.3–116.7% and 156.9–922.1% except the *VE*. Additionally, as the forecast period increases, the prediction accuracy of the VMD-LSTM model degenerates less, indicating that the VMD-LSTM model has good robustness. Only considering VMD decomposition can improve the LSTM model accuracy of other monthly runoff pre-

dictions except from November to February, which is not significantly different from the VMD-PCA-LSTM model.

(3) Considering the atmospheric circulation indexes as the forecasting factors, compared to the VMD-LSTM model, significantly enhances prediction accuracy in high flow caused by a small number of samples, especially the decrease in *VE* of up to 81.6%. With the increase in the forecast period, the improvement after integrating atmospheric circulation indexes becomes more significant, especially when the forecast period is 6 months. The *NSE* and *RMSE* have the most significant improvement increasing by 6.2% and 16.3%. However, it is worth noting that the VMD-PCA-LSTM model does not offer a comprehensive enhancement over the VMD-LSTM model in all periods, but rather focuses only on the flood season, particularly for high flows.

Author Contributions: Conceptualization, W.W., S.T. and J.Z.; Data curation, W.W and S.T.; Formal analysis, S.T. and J.Z.; Funding acquisition, X.Y.; Investigation, D.L.; Methodology, W.W., S.T. and J.Z.; Supervision, X.G. and J.H.; Visualization, S.T. and J.Z.; Writing—original draft, W.W., S.T. and J.Z.; Writing—review and editing, W.W., S.T., J.Z and X.Y. All authors have read and agreed to the published version of the manuscript.

Funding: This research was funded by the National Key Research and Development Program of China (2023YFC3206804), the National Natural Science Foundation of China (52394234), and the Jiangxi Province "Science and Technology + Water Conservancy" Joint Plan Project (2022KSG01006).

Data Availability Statement: Data are available from the corresponding author upon reasonable request.

Acknowledgments: The anonymous reviewers and the editor are thanked for providing insightful and detailed reviews that greatly improved the manuscript.

Conflicts of Interest: The authors declare no conflicts of interest.

References

1. Niu, W.-J.; Feng, Z.-K. Evaluating the performances of several artificial intelligence methods in forecasting daily streamflow time series for sustainable water resources management. *Sustain. Cities Soc.* **2021**, *64*, 102562. [CrossRef]
2. Feng, Z.-K.; Niu, W.-J.; Tang, Z.-Y.; Jiang, Z.-Q.; Xu, Y.; Liu, Y.; Zhang, H.-R. Monthly runoff time series prediction by variational mode decomposition and support vector machine based on quantum-behaved particle swarm optimization. *J. Hydrol.* **2020**, *583*, 124627. [CrossRef]
3. Tikhamarine, Y.; Souag-Gamane, D.; Ahmed, A.N.; Sammen, S.S.; Kisi, O.; Huang, Y.F.; El-Shafie, A. Rainfall-runoff modelling using improved machine learning methods: Harris hawks optimizer vs. particle swarm optimization. *J. Hydrol.* **2020**, *589*, 125133. [CrossRef]
4. Akbarian, M.; Saghafian, B.; Golian, S. Monthly streamflow forecasting by machine learning methods using dynamic weather prediction model outputs over Iran. *J. Hydrol.* **2023**, *620*, 129480. [CrossRef]
5. Dung, N.B.; Long, N.Q.; Goyal, R.; An, D.T.; Minh, D.T. The Role of Factors Affecting Flood Hazard Zoning Using Analytical Hierarchy Process: A Review. *Earth Syst. Environ.* **2022**, *6*, 697–713. [CrossRef]
6. Stergiadi, M.; Di Marco, N.; Avesani, D.; Righetti, M.; Borga, M. Impact of Geology on Seasonal Hydrological Predictability in Alpine Regions by a Sensitivity Analysis Framework. *Water* **2020**, *12*, 2255. [CrossRef]
7. Yang, Q.; Zhang, H.; Wang, G.; Luo, S.; Chen, D.; Peng, W.; Shao, J. Dynamic runoff simulation in a changing environment: A data stream approach. *Environ. Model. Softw.* **2019**, *112*, 157–165. [CrossRef]
8. Deng, C.; Wang, W. Runoff Predicting and Variation Analysis in Upper Ganjiang Basin under Projected Climate Changes. *Sustainability* **2019**, *11*, 5885. [CrossRef]
9. Chen, X.; Zhang, K.; Luo, Y.; Zhang, Q.; Zhou, J.; Fan, Y.; Huang, P.; Yao, C.; Chao, L.; Bao, H. A distributed hydrological model for semi-humid watersheds with a thick unsaturated zone under strong anthropogenic impacts: A case study in Haihe River Basin. *J. Hydrol.* **2023**, *623*, 129765. [CrossRef]
10. Kirsta, Y.B.; Troshkova, I.A. High-Performance Forecasting of Spring Flood in Mountain River Basins with Complex Landscape Structure. *Water* **2023**, *15*, 1080. [CrossRef]
11. Xu, Y.; Hu, C.; Wu, Q.; Jian, S.; Li, Z.; Chen, Y.; Zhang, G.; Zhang, Z.; Wang, S. Research on particle swarm optimization in LSTM neural networks for rainfall-runoff simulation. *J. Hydrol.* **2022**, *608*, 127553. [CrossRef]
12. Meng, J.; Dong, Z.; Shao, Y.; Zhu, S.; Wu, S. Monthly Runoff Forecasting Based on Interval Sliding Window and Ensemble Learning. *Sustainability* **2023**, *15*, 100. [CrossRef]
13. Han, D.Y.; Liu, P.; Xie, K.; Li, H.; Xia, Q.; Cheng, Q.; Wang, Y.B.; Yang, Z.K.; Zhang, Y.J.; Xia, J. An attention-based LSTM model for long-term runoff forecasting and factor recognition. *Environ. Res. Lett.* **2023**, *18*, 13. [CrossRef]

14. Kim, T.; Yang, T.; Gao, S.; Zhang, L.; Ding, Z.; Wen, X.; Gourley, J.J.; Hong, Y. Can artificial intelligence and data-driven machine learning models match or even replace process-driven hydrologic models for streamflow simulation?: A case study of four watersheds with different hydro-climatic regions across the CONUS. *J. Hydrol.* **2021**, *598*, 126423. [CrossRef]

15. Zhang, S.; Gan, T.Y.; Bush, A.B.G.; Zhang, G. Evaluation of the impact of climate change on the streamflow of major pan-Arctic river basins through machine learning models. *J. Hydrol.* **2023**, *619*, 129295. [CrossRef]

16. Zang, S.; Li, Z.; Zhang, K.; Yao, C.; Liu, Z.; Wang, J.; Huang, Y.; Wang, S. Improving the flood prediction capability of the Xin'anjiang model by formulating a new physics-based routing framework and a key routing parameter estimation method. *J. Hydrol.* **2021**, *603*, 126867. [CrossRef]

17. Abbaspour, K.C.; Rouholahnejad, E.; Vaghefi, S.; Srinivasan, R.; Yang, H.; Kløve, B. A continental-scale hydrology and water quality model for Europe: Calibration and uncertainty of a high-resolution large-scale SWAT model. *J. Hydrol.* **2015**, *524*, 733–752. [CrossRef]

18. Behrangi, A.; Khakbaz, B.; Jaw, T.C.; AghaKouchak, A.; Hsu, K.; Sorooshian, S. Hydrologic evaluation of satellite precipitation products over a mid-size basin. *J. Hydrol.* **2011**, *397*, 225–237. [CrossRef]

19. Avesani, D.; Galletti, A.; Piccolroaz, S.; Bellin, A.; Majone, B. A dual-layer MPI continuous large-scale hydrological model including Human Systems. *Environ. Model. Softw.* **2021**, *139*, 105003. [CrossRef]

20. Nazemi, A.; Wheater, H.S. On inclusion of water resource management in Earth system models—Part 1: Problem definition and representation of water demand. *Hydrol. Earth Syst. Sci.* **2015**, *19*, 33–61. [CrossRef]

21. Bozorg-Haddad, O.; Zarezadeh-Mehrizi, M.; Abdi-Dehkordi, M.; Loáiciga, H.A.; Mariño, M.A. A self-tuning ANN model for simulation and forecasting of surface flows. *Water Resour. Manag.* **2016**, *30*, 2907–2929. [CrossRef]

22. Hagen, J.S.; Leblois, E.; Lawrence, D.; Solomatine, D.; Sorteberg, A. Identifying major drivers of daily streamflow from large-scale atmospheric circulation with machine learning. *J. Hydrol.* **2021**, *596*, 126086. [CrossRef]

23. Liu, P.; Wang, J.; Sangaiah, A.K.; Xie, Y.; Yin, X. Analysis and Prediction of Water Quality Using LSTM Deep Neural Networks in IoT Environment. *Sustainability* **2019**, *11*, 2058. [CrossRef]

24. Yu, Q.; Jiang, L.; Wang, Y.; Liu, J. Enhancing streamflow simulation using hybridized machine learning models in a semi-arid basin of the Chinese loess Plateau. *J. Hydrol.* **2023**, *617*, 129115. [CrossRef]

25. Wang, X.; Sun, W.; Lu, F.; Zuo, R. Combining Satellite Optical and Radar Image Data for Streamflow Estimation Using a Machine Learning Method. *Remote Sens.* **2023**, *15*, 5184. [CrossRef]

26. Kim, T.; Shin, J.Y.; Kim, H.; Heo, J.H. Ensemble-Based Neural Network Modeling for Hydrologic Forecasts: Addressing Uncertainty in the Model Structure and Input Variable Selection. *Water Resour. Res.* **2020**, *56*, 19. [CrossRef]

27. Bedi, J. Transfer learning augmented enhanced memory network models for reference evapotranspiration estimation. *Knowl.-Based Syst.* **2022**, *237*, 107717. [CrossRef]

28. Xu, L.; Yu, H.; Chen, Z.; Du, W.; Chen, N.; Huang, M. Hybrid Deep Learning and S2S Model for Improved Sub-Seasonal Surface and Root-Zone Soil Moisture Forecasting. *Remote Sens.* **2023**, *15*, 3410. [CrossRef]

29. Castangia, M.; Grajales, L.M.M.; Aliberti, A.; Rossi, C.; Macii, A.; Macii, E.; Patti, E. Transformer neural networks for interpretable flood forecasting. *Environ. Model. Softw.* **2023**, *160*, 105581. [CrossRef]

30. Yao, Z.; Wang, Z.; Wang, D.; Wu, J.; Chen, L. An ensemble CNN-LSTM and GRU adaptive weighting model based improved sparrow search algorithm for predicting runoff using historical meteorological and runoff data as input. *J. Hydrol.* **2023**, *625*, 129977. [CrossRef]

31. Song, P.; Liu, W.; Sun, J.; Wang, C.; Kong, L.; Nong, Z.; Lei, X.; Wang, H. Annual Runoff Forecasting Based on Multi-Model Information Fusion and Residual Error Correction in the Ganjiang River Basin. *Water* **2020**, *12*, 2086. [CrossRef]

32. Zhao, X.; Lv, H.; Lv, S.; Sang, Y.; Wei, Y.; Zhu, X. Enhancing robustness of monthly streamflow forecasting model using gated recurrent unit based on improved grey wolf optimizer. *J. Hydrol.* **2021**, *601*, 126607. [CrossRef]

33. Fang, W.; Huang, S.; Ren, K.; Huang, Q.; Huang, G.; Cheng, G.; Li, K. Examining the applicability of different sampling techniques in the development of decomposition-based streamflow forecasting models. *J. Hydrol.* **2019**, *568*, 534–550. [CrossRef]

34. Zuo, G.G.; Luo, J.G.; Wang, N.; Lian, Y.N.; He, X.X. Two-stage variational mode decomposition and support vector regression for streamflow forecasting. *Hydrol. Earth Syst. Sci.* **2020**, *24*, 5491–5518. [CrossRef]

35. Wang, W.-C.; Chau, K.-W.; Xu, D.-M.; Chen, X.-Y. Improving Forecasting Accuracy of Annual Runoff Time Series Using ARIMA Based on EEMD Decomposition. *Water Resour. Manag.* **2015**, *29*, 2655–2675. [CrossRef]

36. Zuo, G.; Luo, J.; Wang, N.; Lian, Y.; He, X. Decomposition ensemble model based on variational mode decomposition and long short-term memory for streamflow forecasting. *J. Hydrol.* **2020**, *585*, 124776. [CrossRef]

37. Kirono, D.G.C.; Chiew, F.H.S.; Kent, D.M. Identification of best predictors for forecasting seasonal rainfall and runoff in Australia. *Hydrol. Process.* **2010**, *24*, 1237–1247. [CrossRef]

38. Chavasse, D.I.; Seoane, R.S. Assessing and predicting the impact of El Nino southern oscillation (ENSO) events on runoff from the Chopim River basin, Brazil. *Hydrol. Process.* **2009**, *23*, 3261–3266. [CrossRef]

39. Champagne, O.; Arain, M.A.; Coulibaly, P. Atmospheric circulation amplifies shift of winter streamflow in southern Ontario. *J. Hydrol.* **2019**, *578*, 124051. [CrossRef]

40. Yan, X.; Chang, Y.; Yang, Y.; Liu, X. Monthly runoff prediction using modified CEEMD-based weighted integrated model. *J. Water Clim. Change* **2020**, *12*, 1744–1760. [CrossRef]

41. Mostaghimzadeh, E.; Ashrafi, S.M.; Adib, A.; Geem, Z.W. A long lead time forecast model applying an ensemble approach for managing the great Karun multi-reservoir system. *Appl. Water Sci.* **2023**, *13*, 124. [CrossRef]

42. Dragomiretskiy, K.; Zosso, D. Variational Mode Decomposition. *IEEE Trans. Signal Process.* **2014**, *62*, 531–544. [CrossRef]

43. Zhang, X.; Liu, F.; Yin, Q.; Qi, Y.; Sun, S. A runoff prediction method based on hyperparameter optimisation of a kernel extreme learning machine with multi-step decomposition. *Sci. Rep.* **2023**, *13*, 19341. [CrossRef]

44. Jolliffe, I.T.; Cadima, J. Principal component analysis: A review and recent developments. *Philos. Trans. R. Soc. A-Math. Phys. Eng. Sci.* **2016**, *374*, 16. [CrossRef]

45. Hochreiter, S.; Schmidhuber, J. Long Short-Term Memory. *Neural Comput.* **1997**, *9*, 1735–1780. [CrossRef]

46. Thiemann, M.; Trosset, M.; Gupta, H.; Sorooshian, S. Bayesian recursive parameter estimation for hydrologic models. *Water Resour. Res.* **2001**, *37*, 2521–2535. [CrossRef]

47. Kratzert, F.; Klotz, D.; Brenner, C.; Schulz, K.; Herrnegger, M. Rainfall-runoff modelling using Long Short-Term Memory (LSTM) networks. *Hydrol. Earth Syst. Sci.* **2018**, *22*, 6005–6022. [CrossRef]

48. Ling, H.; Xu, H.; Shi, W.; Zhang, Q. Regional climate change and its effects on the runoff of Manas River, Xinjiang, China. *Environ. Earth Sci.* **2011**, *64*, 2203–2213. [CrossRef]

49. Mo, R.; Xu, B.; Zhong, P.-A.; Dong, Y.; Wang, H.; Yue, H.; Zhu, J.; Wang, H.; Wang, G.; Zhang, J. Long-term probabilistic streamflow forecast model with "inputs–structure–parameters" hierarchical optimization framework. *J. Hydrol.* **2023**, *622*, 129736. [CrossRef]

50. Deng, C.; Yin, X.; Zou, J.; Wang, M.; Hou, Y. Assessment of the impact of climate change on streamflow of Ganjiang River catchment via LSTM-based models. *J. Hydrol. Reg. Stud.* **2024**, *52*, 101716. [CrossRef]

51. Wang, N.; Zhang, D.; Chang, H.; Li, H. Deep learning of subsurface flow via theory-guided neural network. *J. Hydrol.* **2020**, *584*, 124700. [CrossRef]

52. Tong, X.; Yan, Z.; Xia, J.; Lou, X. Decisive Atmospheric Circulation Indices for July–August Precipitation in North China Based on Tree Models. *J. Hydrometeorol.* **2019**, *20*, 1707–1720. [CrossRef]

53. Amini, A.; Dolatshahi, M.; Kerachian, R. Real-time rainfall and runoff prediction by integrating BC-MODWT and automatically-tuned DNNs: Comparing different deep learning models. *J. Hydrol.* **2024**, *631*, 130804. [CrossRef]

54. Ma, K.; He, D.; Liu, S.; Ji, X.; Li, Y.; Jiang, H. Novel time-lag informed deep learning framework for enhanced streamflow prediction and flood early warning in large-scale catchments. *J. Hydrol.* **2024**, *631*, 130841. [CrossRef]

55. Li, B.-J.; Sun, G.-L.; Liu, Y.; Wang, W.-C.; Huang, X.-D. Monthly Runoff Forecasting Using Variational Mode Decomposition Coupled with Gray Wolf Optimizer-Based Long Short-term Memory Neural Networks. *Water Resour. Manag.* **2022**, *36*, 2095–2115. [CrossRef]

56. Seo, Y.; Kim, S.; Kisi, O.; Singh, V.P.; Parasuraman, K. River Stage Forecasting Using Wavelet Packet Decomposition and Machine Learning Models. *Water Resour. Manag.* **2016**, *30*, 4011–4035. [CrossRef]

57. Yang, C.; Jiang, Y.; Liu, Y.; Liu, S.; Liu, F. A novel model for runoff prediction based on the ICEEMDAN-NGO-LSTM coupling. *Environ. Sci. Pollut. Res.* **2023**, *30*, 82179–82188. [CrossRef]

58. Wang, W.-C.; Du, Y.-J.; Chau, K.-W.; Cheng, C.-T.; Xu, D.-M.; Zhuang, W.-T. Evaluating the Performance of Several Data Preprocessing Methods Based on GRU in Forecasting Monthly Runoff Time Series. *Water Resour. Manag.* **2024**. [CrossRef]

59. Li, H.Y.; Xie, M.; Jiang, S. Recognition method for mid- to long-term runoff forecasting factors based on global sensitivity analysis in the Nenjiang River Basin. *Hydrol. Process.* **2012**, *26*, 2827–2837. [CrossRef]

water

MDPI

Article

Extension of a Monolayer Energy-Budget Degree-Day Model to a Multilayer One

Julien Augas [1,*], Etienne Foulon [1], Alain N. Rousseau [1] and Michel Baraër [2]

1   Institut National de la Recherche Scientifique, Centre Eau Terre Environnement, 490, rue de la Couronne, Québec, QC G1K 9A9, Canada; etienne.foulon@inrs.ca (E.F.); alain.rousseau@inrs.ca (A.N.R.)
2   Département de Génie de la Construction, École de Technologie Supérieure, Montréal, QC H3C 1K3, Canada
*   Correspondence: julien.augas@inrs.ca

Abstract: This paper presents the extension of the monolayer snow model of a semi-distributed hydrological model (HYDROTEL) to a multilayer model that considers snow to be a combination of ice and air, while accounting for freezing rain. For two stations in Yukon and one station in northern Quebec, Canada, the multilayer model achieves high performances during calibration periods yet similar to the those of the monolayer model, with KGEs of up to 0.9. However, it increases the KGE values by up to 0.2 during the validation periods. The multilayer model provides more accurate estimations of maximum *SWE* and total spring snowmelt dates. This is due to its increased sensitivity to thermal atmospheric conditions. Although the multilayer model improves the estimation of snow heights overall, it exhibits excessive snow densities during spring snowmelt. Future research should aim to refine the representation of snow densities to enhance the accuracy of the multilayer model. Nevertheless, this model has the potential to improve the simulation of spring snowmelt, addressing a common limitation of the monolayer model.

Keywords: multilayer structure; snow water equivalent; ice/air mixture; snow modeling; snowmelt; sensitivity analysis; snow height; winter snow peak

check for updates

Citation: Augas, J.; Foulon, E.; Rousseau, A.N.; Baraër, M. Extension of a Monolayer Energy-Budget Degree-Day Model to a Multilayer One. *Water* **2024**, *16*, 1089. https://doi.org/10.3390/w16081089

Academic Editors: Agnieszka Rutkowska and Katarzyna Baran-Gurgul

Received: 6 March 2024
Revised: 3 April 2024
Accepted: 8 April 2024
Published: 10 April 2024

1. Introduction

Understanding the hydrological cycle is a paramount challenge for humanity, as it is essential for protecting against floods, mitigating droughts, meeting water needs for industrial and domestic purposes, and informing weather and climate predictions. Within this cycle, one crucial component is snowfall. Although in the Northern Hemisphere, snow typically constitutes around 6–10% of total precipitation, it can exceed 50% in specific regions [1]. Accumulating as a heat-deficient solid water reservoir, snowpacks experience rapid spring melting, leading to distinctive seasonal flooding patterns. Notably, snowmelt has been found to contribute substantially to annual streamflow in various geographic contexts. For example, in Indian glacier-fed basins, snowmelt accounts for 27–44% [2], and in Czech Republic watersheds 17–42% [3]. Meanwhile, snowmelt can play a pivotal role in groundwater recharge. For example, in the Nelson River Basin, Canada, Jasechko et al. [4] determined that the fraction of precipitation recharging aquifers is 1.3 to 5 times higher during cold months, with negative mean monthly temperatures, than during warmer months. Snow can pose a challenge in mountainous areas like the Andes [5] or Iran [6], and snowmelt remains a concern. Consequently, accurate modeling of snow cover becomes crucial for streamflow modeling.

Snow–water equivalent (or *SWE*) represents one of the key physical characteristics and is defined as the depth of water on the ground if the snow were in a liquid state. For hydrological models simulating water transfers within the hydrological cycle, *SWE* represents an essential variable and is equivalent to the product of snowpack height and snow density (mass of snow per unit volume of snowpack). Another key characteristic, albedo, represents the proportion of solar radiation reflected by the snowpack surface and thus directly affects the amount of absorbed solar energy. In the case of fresh snow, assuming reflectivity is isotropic, its specular component—which entails unidirectional reflection—strengthens as the snowpack ages and undergoes repeated melting and recrystallization events [7], affecting snow metamorphosis and sublimation [8]. Finally, the temperature or calorific deficit of the snowpack assists in determining the snow maturation process and proximity to melting.

Given our physical understanding of snow, multiple snow models have been developed to tackle specific issues related to water resource management, including integrated management, avalanche prediction, climate studies, infrastructure planning, environmental impact, or even scientific research in fields such as ecology or glaciology. As is typical with modeling, the complexity of those models is tailored to the objective they seek to address. Table A1 in Appendix A presents a selection of snow models that differ in their design approach, consideration of the simulated phenomena given the available data, and thus representation of the snowpack structure.

Without delving into the details of each snow model, which would go beyond the scope of this paper, Table A1 highlights a major difference in complexity between monolayer models, which are all daily models and consider only a limited number of phenomena, and multilayer ones that can provide snow cover modeling at 10 min intervals and consider a wider range of phenomena. The snow model of HYDROTEL stands out among monolayer models as the most advanced in terms of physical representation. It encompasses phenomena found in simple models like CEMANEIGE and HBV (i.e., snow accumulation and melting), as well as numerous phenomena typically associated with multilayer modeling approaches (e.g., convective heat, precipitation heat, soil heat, compaction, mixing, radiation heat/melting degree-day, water retention). HYDROTEL [9,10] is the semi-distributed hydrological model at the core of operational hydrologic forecasting systems in the Quebec [11,12], Yukon [13–15], and Southern Québec Hydroclimatic Atlas [16–18], as well as in several other studies, such as on the effect of global warming on environmental flows [19–21] or the role of wetlands on mitigating floods and droughts [22–24]. These applications are made in a Canadian context, where most watersheds are subject to significant snowfall. In these regions, spring freshets often result in annual streamflow peaks, sometimes accompanied by rain-on-snow events [25], which can augment lateral outflows and impede soil infiltration [26]. As a result, accurate simulation of snowmelt becomes critical for effectively predicting streamflow, accounting for both surface runoff and groundwater recharge [27]. One notable feature of HYDROTEL's snow model is its consideration of snow as a monolayer structure [28]. However, the literature suggests that adopting a multilayer representation of the snowpack can significantly improve *SWE* dynamics. For instance, Saha et al. [29] demonstrated substantial enhancements in snowpack height and *SWE* estimations with the use of the six-layer Noah model compared to its conventional monolayer version. In addition, Domine et al. [30] highlighted the significance of accurately modeling the thermal properties of snow for estimating soil water mass balance, suggesting that a multilayer structure can effectively capture density profiles and improve the representation of thermal characteristics. In addition to incorporating a multilayer structure, some models treat snow as a heterogeneous material, accounting for the proportions of air, ice, and water in the snow cover. For example, the SNOWPACK model [31] considers these factors, while the GEOTOP [32] and SeNORGE [33] models represent snow as a mixture of solid and liquid water. Furthermore, the integration of freezing rain enables the direct formation of an ice layer over an existing snow cover, as observed in studies by Henson et al. [34] and Quéno et al. [35].

These different considerations offer potential directions of improvement for snow modeling in HYDROTEL. Given the advancements in modeling sophistication and computational capabilities, this paper focuses on developing a multilayer version of the hybrid energy balance/degree-day snow model of HYDROTEL, assuming snowpack is predominantly composed of ice with interspersed air. As ice exhibits distinct thermal properties compared to those of air, this development impacts heat transfer between layers while creating discontinuities in the physical properties and ensuing temperature and density profiles. This development aligns with the parsimonious structure of HYDROTEL, which has positioned the model as a robust model in Canada. The end goal is not to transform HYDROTEL into a complex and computationally intensive model but rather to assess the potential improvements associated with using a multilayer structure within a relatively simple, physics-based, semi-distributed model.

This paper is organized as follows. First, we describe the original snow model design of HYDROTEL, detailing the modifications from a monolayer model to a multilayer one, and present a sensitivity analysis of the additional parameters. The mono- and multilayer models are then calibrated based on *SWE*, and the resulting differences are highlighted. The modeling is validated using Gamma MONitor (GMON) stations in the Necopastic watershed (Quebec), Lower Fantail, and Wheaton (Yukon) River basins. The effects of the model design on energy balance dynamics, state variables, and characteristic dates of the snowpack are analyzed, followed by a discussion and conclusion.

2. Materials and Methods

2.1. Core Equations of the Monolayer Snow Model

This section presents the governing equations of the monolayer snow model of HYDROTEL [28] focusing on modeling the phenomena introduced in Table A1, namely, snow accumulation, advected heat transfer from precipitation, soil heat transfer, snow compaction, snow water content, and construction of the thermal energy budget (through blending net short-wave radiation and degree-day concepts).

The operation of the model is parsimonious and only requires three input variables, that is, daily total precipitation, and minimum and maximum air temperatures. The model is physics-based, using degree-day equations while building a thermal energy budget based on the heat deficit of a monolayer snowpack. This budget is as follows (see Appendix B for a detailed mathematical description of each term):

$$\frac{\Delta U}{\Delta t} = u_r + u_c + u_{s-s} + u_{a-s} + u_{ac} - u_s \tag{1}$$

where $\frac{\Delta U}{\Delta t}$ is the daily rate of change in the snowpack heat deficit (J.m^{-2}.s^{-1}); u_r, u_c, u_{s-s}, u_{a-s}, and u_{ac} are decreases in heat deficits due to rainfall, conduction, transfer from the soil (at the snow–soil interface), net radiation (at the air–snow interface), and from the water retained on the previous day, respectively; and u_s is the increase in heat deficit due to solid precipitation.

The energy assessment is applied to a snow layer. Liquid and solid precipitations are derived from total precipitation, daily minimum and maximum air temperatures, and a temperature threshold. When the air temperature is sufficiently cold (below the threshold), all precipitation falls as snow (Equation (2a)), whereas when the temperature is warm enough (above the threshold), it falls as rain (Equation (2b)). In between, total precipitation results in a mix of snow and rain (Equation (2c)).

$$R = 0; S = P_t \ if \ T_{max} \leq T_s \tag{2a}$$

$$R = P_t; S = 0 \ if \ T_{min} > T_s, \tag{2b}$$

$$R = P_t \left(\frac{T_{max} - T_s}{T_{max} - T_{min}} \right); S = P_t \left(\frac{T_s - T_{min}}{T_{max} - T_{min}} \right) \ otherwise \tag{2c}$$

where R, S, and P_t are liquid, solid, and total daily precipitation rates (m.s^{-1}), respectively; T_{max} and T_{min} are the maximum and minimum daily air temperatures, respectively; and T_s is the temperature threshold.

The density of falling snow is computed as follows:

$$\rho_s = 151 + 10.63 \left(\frac{T_{max} + T_{min}}{2} \right) + 0.2767 \left(\frac{T_{max} + T_{min}}{2} \right)^2 \text{ if } \frac{T_{max} + T_{min}}{2} \geq -17 \quad (3a)$$

$$\rho_s = 50 \ if \ \frac{T_{max} + T_{min}}{2} < -17 \quad (3b)$$

where ρ_s is the density of fresh snowfall (kg.m^{-3}), and T_{max} and T_{min} are the maximum and minimum daily air temperatures, respectively.

The snowpack is subject to compression, and a reduction in height (*Sett*) is estimated using Equation (4). Thus, *Sett* is subtracted from the current height of the snow layer. When negative, *Sett* is set to 0.

$$Sett = H \ Set_{Coef} \left(1 - \frac{\rho_{snow}}{\rho_{max}} \right) \quad (4)$$

where *Sett* is snowpack height lost to compaction (m), H is the snow height (m), Set_{Coef} is the compaction coefficient ($-$), and ρ_{max} is the maximum achievable density (kg.m^{-3}).

When the total snowpack heat deficit is replenished, a potential snow melt is computed from the excess heat, triggering a phase change as per Equation (5).

$$PM = \frac{\Delta U_{tot}}{C_f \ \rho_w} \quad (5)$$

where PM is the resulting amount of SWE undergoing a phase change (m), ΔU_{tot} is the total heat deficit (J.m^{-2}), ρ_w is the liquid water density (1000 kg.m^{-3}), and C_f is the latent heat of the fusion of water (335,000 J.kg^{-1}).

The maximum water retention capacity (RC_{max}) is computed as follows:

$$RC_{max} = 0.1 \frac{\rho_{snow}}{\rho_w} SWE \quad (6)$$

where RC_{max} is the maximum snow cover capacity of water retention (m), and SWE is the snow water equivalent following the removal of PM (m).

The actual snowmelt (AR) is computed as the difference between potential melt and RC_{max} (Equations (7a) and (7b)).

$$AR = \frac{PM}{\Delta t} \text{ if } PM \leq RC_{max} \text{ then } AM = 0 \quad (7a)$$

$$AR = \frac{RC_{max}}{\Delta t} \text{ and } AM = \frac{PM - RC_{max}}{\Delta t} \text{ otherwise} \quad (7b)$$

where AM is the actual snowmelt (m.s^{-1}), AR is the actual retention, and Δt is the computational time step.

Finally, the snowpack mass balance is the sum of the snowfall and rainfall when there is snow on the ground; otherwise, rainfall either percolates or runs off.

$$\frac{\Delta SWE}{\Delta t} = R + S - AM \quad (8)$$

2.2. Extension of the Monolayer Snow Model

Several modifications to the model are considered, including a change from a monolayer to a multilayer structure. Additionally, some variables are estimated by considering snow as a material composed of both ice and air. Furthermore, freezing rain is made possible given its potential to alter the heat transfer inertia between each layer. Finally, some modifications are introduced to the equations describing snow compression and maximum water retention capacity to account for the changes. These modifications are described in the next subsections.

2.2.1. A Multilayered Structure

Any snowfall in the absence of snow on the ground or above a pure ice layer (layer with a density of 917 kg.m^{-3}) leads to the creation of a new layer with a specific mass and heat deficit. If these criteria are not met, and if the snowfall water equivalent is less than a threshold value St, then the incoming mass and heat deficit are incorporated into the current layer at the air–snow interface. Otherwise, a new layer is established, as illustrated in Figure 1. St serves as a calibration parameter, allowing for concurrent optimization of energy transfers and restricting the number of layers. Some energy transfer processes solely affect specific layers. For instance, heat input from the ground solely influences the layer at the ground–snow interface, while radiation exclusively warms up the layer at the air–snow interface. For the latter layer of the monolayer model, heat loss through conduction and heat gain via radiation are enabled when the air temperature is below or above the melting threshold temperature T_0, respectively. Furthermore, in instances where melting exceeds the water retention capacity, excess water seeps into the underlying layer at a temperature of 0 °C. Excess heat is used for phase change; if the uppermost snow layer has undergone a phase change, any residual heat is then transferred downwards. Consequently, the energy balance can be expressed using Equations (9a)–(9c) for the top layer, any intermediate layers, and the bottom layer, respectively.

$$\frac{\Delta U_k}{\Delta t} = u_r + u_c + u_{a-s} + u_{ac} - u_s \tag{9a}$$

$$\frac{\Delta U_k}{\Delta t} = u_c + u_{ac} + u_{ex,k+1} + u_{perc,k+1} - u_{ex,k} - u_{perc,k} \tag{9b}$$

$$\frac{\Delta U_1}{\Delta t} = u_c + u_{s-s} + u_{ac} + u_{ex,2} + u_{perc,2} \tag{9c}$$

where u_{ex} is the excess heat from melting in the upper layer or the heat transfer due to phase change of freezing rain from the upper layer (more detail below in the article) (J.m^{-2}.s^{-1}), u_{perc} is the heat variation due to infiltration from the upper layer (J.m^{-2}.s^{-1}), and k stands for the kth snow layer from the ground surface.

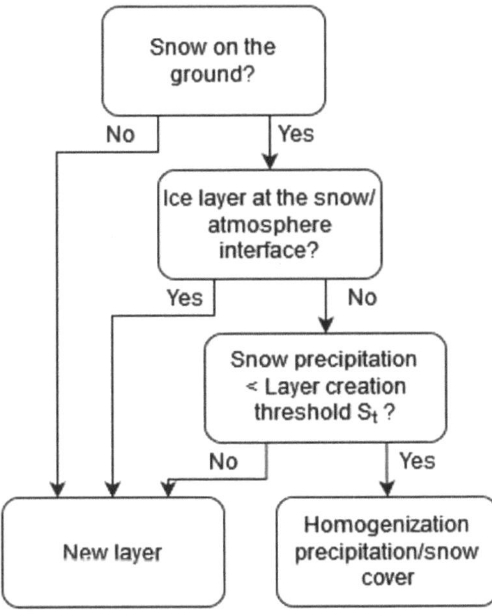

Figure 1. Snow layer creation scheme for the proposed multilayer model.

The heat input from percolation, u_{perc}, is evaluated in a similar manner to u_r. In both cases, the heat input to the snowpack comprises the cumulative sensible heat loss of the liquid water lowered to 0 °C, the ensuing latent heat of fusion (phase change), and the heat released to adjust the new ice crystals to the snowpack temperature. They are described by Equations (10a) and (10b) for the modification of thermal energy from rainfall and percolation, respectively. The rainfall occurred on the top layer, which is noted as k' below.

$$\begin{cases} u_r = \rho_w R \left(C_w T_m + C_f \right) \left(1 - \frac{R}{SWE_{k'} + R} \right) + \frac{R\, U_{k'}}{SWE_{k'} + R} & \text{if } T_m > 0 \\ u_r = \rho_w R \left(C_s T_m + C_f \left(1 - \frac{R}{SWE_{k'} + R} \right) \right) + \frac{R\, U_{k'}}{SWE_{k'} + R} & \text{otherwise} \end{cases} \tag{10a}$$

$$u_{perc,k} = \rho_w Ru_{k+1} C_f \left(1 - \frac{Ru_{k+1}}{SWE_k + Ru_{k+1}} \right) + \frac{Ru_{k+1} U_k}{SWE_k + Ru_{k+1}} \tag{10b}$$

where C_w and C_s are specific heat capacities of water and snow (4184 J.kg^{-1}.°C^{-1} and 2093.4 J.kg^{-1}.°C^{-1}), respectively; C_f is the heat of fusion of water (335,000 J.kg^{-1}); R is the rainfall rate (m.s^{-1}); Ru_{k+1} is the percolation rate of the $k + 1^{th}$ layer (m.s^{-1}); T_m is the mean air temperature (°C); SWE_k is the snow water equivalent (m); and U_k is the heat deficit of the kth layer.

2.2.2. Snow as a Medium of Ice and Air

The snowpack is regarded as a medium comprising different constituents whereby the properties and proportions of each component contributes to the estimation of various snow characteristics. Appendix E describes how the volumetric proportions of air and ice are estimated, assuming liquid water constitutes a non-significant portion of the snowpack during winter. This assumption is based on observations made by Koch et al. [36], where the volumetric liquid water content peaked at a maximum of 8% at the end of the melting phase or during instances of liquid precipitation. This is consistent with the assumption that liquid water in the original snow model is entirely frozen at the daily time step. Leveraging the relationship derived for snow density (Appendix E) and the linear correlation proposed by Evans [37] to gauge the relative dielectric permittivity of snow from those of ice and air, all snow layer characteristics are determined based on the proportions of ice and air. For heat loss by conduction, the thermal diffusivity of snow is computed for each layer using Equation (11).

$$D_{s,k} = \left(\frac{\rho_{s,k} - \rho_{a,k}}{\rho_i - \rho_{a,k}} \right) D_{i,k} + \left(\frac{\rho_i - \rho_{s,k}}{\rho_i - \rho_{a,k}} \right) D_{a,k} \tag{11}$$

where $D_{s,k}$ is the snow diffusivity (m^2.s^{-1}); $\rho_{s,k}$, $\rho_{a,k}$, and ρ_i are the snow, air, and ice densities (kg.m^{-3}), respectively; $D_{i,k}$ and $D_{a,k}$ are the ice and air thermal diffusivities (m^2.s^{-1}), respectively; and k stands for the kth snow layer.

The thermal diffusivities of ice and air are computed using Equation (12):

$$D_{m,k} = \frac{K_{m,k}}{\rho_{m,k} C_{s,m,k}} \tag{12}$$

where $D_{m,k}$ is the thermal diffusivity of the kth snow layer made of a material m (m^2.s^{-1}), $K_{m,k}$ is the thermal conductivity (W.m^{-1}.°C^{-1}), $\rho_{m,k}$ is the density (kg.m^{-3}), and $C_{s,m,k}$ is the specific heat (J.kg^{-1}.°C^{-1}).

Estimates of the thermal conductivities of ice [38] and air [39] are derived from Equations (13) and (14), respectively.

$$K_{i,k} = 1.16 \left(1.91 - 8.66.\,10^{-3}\, T_k + 2.97.10^{-5}\, T_k^2 \right) \tag{13}$$

where T_k is the temperature of the kth layer (°C).

$$K_{a,k} = 1.5207.10^{-11}(273.15 + T_k)^3 - 4.857.10^{-8}(273.15 + T_k)^2 \\ +1.0184.10^{-4}(273.15 + T_k) - 3.9333.10^{-4} \tag{14}$$

T_k is a function of the total heat deficit $\Delta U_{tot,k}$ computed for the kth layer using Equation (15):

$$T_k = \frac{\Delta U_{tot,k}}{SWE_k \, C_s \, \rho_w} \tag{15}$$

For ice, the density and specific heat are deemed constant for any temperature and are set at 917 kg.m^{-3} and 2093.4 J.kg^{-1}.°C^{-1}, respectively. For air, the density (ideal gas law under normal pressure conditions) and specific heat [39] are computed using Equations (16) and (17), respectively:

$$\rho_{a,k} = 1.292 \, \frac{273.15}{273.15 + T_k} \tag{16}$$

$$C_{s,a,k} = 1.9327.10^{-10}(273.15 + T_k)^4 - 7.999.10^{-7}(273.15 + T_k)^3 \\ +1.1407.10^{-3}(273.15 + T_k)^2 - 4.489.10^{-1}(273.15 + T_k) \\ +1.0575.10^3 \tag{17}$$

where T_k stands for the temperature of the kth layer (°C).

Snow albedo is determined by snow grain metamorphism, which also causes the snowpack to become denser. However, our snow model assesses albedo based on snow density since snow grain size and shape are not evaluated. Here, snow albedo is estimated based on the proportion of ice and air in the surface layer. This approach is reminiscent of the optical paths of radiation that are absorbed by ice crystals instead of being reflected or transmitted through them. Nevertheless, since the albedo of air cannot be defined, fresh snow was employed as a surrogate material. Indeed, fresh snow constitutes a blend of ice and air with a very high porosity.

Perovich et al. [40] measured an ice albedo of 0.5 in the Arctic for snow on a frozen pothole. The albedo of fresh snow is 0.9 [41] for a 50 kg.m^{-3} density, which is consistent with that of snowfall computed in the monolayer mode. The albedo of snow as a composite material is thus computed using Equation (18):

$$\alpha_s = \left(\frac{\rho_s - \rho_{fs}}{\rho_i - \rho_{fs}}\right)\alpha_i + \left(\frac{\rho_i - \rho_s}{\rho_i - \rho_{fs}}\right)\alpha_{fs} \tag{18}$$

where α_i and α_{fs} are albedos of ice (0.5) and fresh snow (0.9), respectively, and ρ_{fs} is the fresh snow density (50 kg.m^{-3}).

2.2.3. Freezing Rain

Freezing rain occurs upon contact with surfaces when raindrops become supercooled while passing through a freezing layer of air. It is characterized by a heat deficit due to changes in both phase and air temperature. Like how the monolayer model manages precipitation that freezes within the snow cover, the freezing rain heat deficit from the newly created layer is computed using Equation (19):

$$u_s = \rho_w \left(C_f - C_w \frac{T_{max} + T_{min}}{2}\right) R \tag{19}$$

where ρ_w is the liquid water density (1000 kg.m^{-3}); C_f is the heat of fusion of water (335,000 J.kg^{-1}); C_w is the specific heat capacity of water (4184 J.kg^{-1}.°C^{-1}); T_{max} and T_{min} are the maximum and minimum daily air temperatures, respectively (°C); and R is the daily liquid precipitation rate (m.s^{-1}).

Ice is a better heat conductor than air—about 100 times more, according to Equations (13) and (14). That is why upon freezing, the excess heat from the phase change is transferred to the snowpack (see Equation (20)). The ice-layer temperature subsequently impacts the conduction heat loss of the lower layer.

$$u_{ex} = \rho_w C_f R \tag{20}$$

where ρ_w is the liquid water density (1000 kg.m^{-3}), C_f is the heat of fusion of water (335,000 J.kg^{-1}), and R is the daily liquid precipitation rate (m.s^{-1}).

It is noteworthy that in the original monolayer model, the cooling of ice from 0 °C down to the snow layer temperature was neglected. This oversight stands corrected in the multilayer model.

2.2.4. Compression

Snow is made of ice crystals and can undergo compression due to its own weight. Throughout this process, there is no melting or loss of mass, and the snow is contained within a time-dependent volume, as the bonds between ice crystals strengthen, resulting in a structure that can better withstand gravitational force. For this purpose, compaction is computed using Equation (4), with a distinct maximum density $\rho_{max,l}$.

2.2.5. Maximum Water Retention Capacity

Some snow models, such as MASiN [42], estimate the maximum water retention capacity of a layer as a proportion of the volume of air that can retain the melted snow. Since the volume of air is now a variable in the proposed model (see the Section 2.2), this capacity can be computed as follows:

$$RC_{max,k} = \%air \, \frac{\rho_i - \rho_{s,k}}{\rho_i - \rho_{a,k}} \, H_k \tag{21}$$

where $RC_{max,k}$ is the maximum water retention capacity of the kth layer (m), $\%air$ is the ratio of the volume of air that can be filled in by water (−), and H_k is the height of the kth layer after melting (m).

Table 1 displays the calibration parameters and their respective physical ranges considered for the two versions of the snow model. They align with typical values employed in HYDROTEL. However, the lower limit of parameter T_0 is relatively small, intended for an open vegetation environment. Despite the low probability of reaching this value during the calibration of the hydrological model, it was retained to evaluate the behavior of the snow model should an optimal solution be identified using such a value.

Table 1. Snow model calibration parameters.

Parameter	Model	Meaning	Lower Threshold	Upper Threshold
ρ_{max}	Original	Maximum snow density (kg.m^{-3})	250	550
T_0	Original/multilayer	Temperature threshold for net radiation heat gain (°C)	−8	3
T_s	Original/multilayer	Precipitation separation temperature (°C)	−1	3
Set_{Coef}	Original/multilayer	Settling coefficient (−)	0.0001	0.1
MR_{a-s}	Original/multilayer	Melt rate at air–snow interface (m.day^{-1}.°C^{-1})	0.001	0.04
MR_{s-s}	Original/multilayer	Melt rate at snow–ground interface (m.day^{-1})	0.0001	0.002
S_t	Multilayer	New-layer snow precipitation threshold (m.day^{-1})	0	0.06
$\rho_{max,l}$	Multilayer	Settling maximum snow-layer density (kg.m^{-3})	350	750
$\%air$	Multilayer	Ratio of the volume of air that can be filled in by water (−)	0.05	0.15

It is noteworthy that the multilayer snow model introduces three additional calibration parameters while removing one, keeping it relatively parsimonious while allowing for the integration of one new phenomenon: freezing rain.

2.3. Framework for Evaluating Different Versions of the Snow Model

The models were calibrated using OSTRICH [43], which provides a choice of different deterministic algorithms, such as steepest descent [44] or multi-start GML with trajectory repulsion [45], as well as stochastic algorithms such as dynamically dimensioned search (DDS) [46] or shuffled complex evolution [47,48]. For this study, we used DDS following the guidelines proposed by Tolson et al. [46]. For the mono- and multilayer versions of the snow model, there are six calibration parameters, requiring at least 18 calibration repetitions (trials) of 100 iterations each.

The Kling–Gupta efficiency (KGE) was used as the objective function [49]:

$$\mathrm{KGE} = 1 - \left[(1 - \mu_s/\mu_o)^2 + (1 - \sigma_s/\sigma_o)^2 + (1 - r)^2 \right]^{1/2} \tag{22}$$

where μ_s et μ_o are the simulated and observed *SWE* averages, respectively; σ_X is the standard deviation; and r is the Pearson correlation coefficient.

We conducted a sensitivity analysis using the variogram analysis of response surface (VARS) toolbox from Razavi et al. [50]. Among the various suggested tools, the STAR-VARS method [51], based on a "star based" sampling strategy, was retained because it is an efficient global sensitivity analysis (GSA) technique for analyzing the variograms of the model. A variogram characterizes the model's spatial covariance structure and takes the following form:

$$\gamma(\vec{h}) = \frac{1}{2|N(\vec{h})|} \sum_{(i,j) \in N(\vec{h})} (y(\vec{x}^A) - y(\vec{x}^B))^2 \tag{23}$$

where $\vec{h}$ is the distance (or direction) between the parameter sets $\vec{x}^A$ and $\vec{x}^B$ in the factor space, $N(\vec{h})$ is the number of pairs of points in the factor space with a distance $\vec{h}$ between them, and $y(\vec{x}^A)$ and $y(\vec{x}^B)$ are the response of the model in the parametric space at locations $\vec{x}^A$ and $\vec{x}^B$, respectively.

Therefore, an increase in the variogram in a direction $\vec{h}$ in the factor space implies a greater variation on $\vec{h}$, indicating a higher sensitivity of the model in this direction.

To combine the various variograms for each parameter, a sensitivity index (IVAR) is generated for each one of them, which integrates the variograms over a scale interval from 0 to H_i for a parameter i:

$$IVAR_i(H_i) = \int_0^{H_i} \gamma(h_i)dh_i \tag{24}$$

Based on the recommendation of Razavi and Gupta [52], we calculated the sensitivity index for 50% of the interval ($IVAR_i(0.5)$), corresponding to a scale of $H_i = 0.5$. To facilitate parameter comparison, a relative sensitivity index ($IVAR_{i,50n}$) is estimated for each parameter i as follows:

$$IVAR_{i,50n} = \frac{IVAR_i(0.5)}{\sum_{j=1}^n IVAR_j(0.5)} \tag{25}$$

A temporal sensitivity analysis was performed by estimating the $IVAR_{i,50n}$ for each day using a generalized global sensitivity matrix approach (or GGSM) instead of the previously employed GSA method.

The Latin hypercube sampling method was adopted to generate the parameter sets, using a sampling of parameter sets based on 50 stars with a resolution of 0.1. The time frame aligns with each period of accessible data, which will be elaborated upon in the case study section.

Two calibration strategies were evaluated to optimize the information obtained from the different datasets of the *SWE* gauge stations presented below. The first strategy was to test the prediction ability of both the monolayer model and the multilayer model. Various calibrations were performed by extracting one year of the datasets for validation, while the remaining years were used for calibration. All possible permutations were evaluated. The second strategy involved using the complete dataset to compare the overall performance of each model. The top ten KGE performances, assessed on *SWE* during the calibration period (or as the optimal compromise between calibration and validation periods for the first strategy), were compared for both models at every *SWE* gauge station. A Wilcoxon rank sum test was performed to compare the median of these performances at each station. A *p*-value of less than 0.05 indicated a significant difference at a 5% type-I error rate. The second strategy consisted of using all available data for calibration with the KGE. In addition to the KGE, the root mean squared error (RMSE) and Nash–Sutcliffe Efficiency (NSE) [53] were computed. For the remainder of this paper, the monolayer model is referred to as "Mo", while the multilayer model is referred to as "Multi".

$$RMSE = \sqrt{\frac{\sum_{i=1}^{n}(SWE_{o,i} - SWE_{s,i})^2}{n}} \tag{26}$$

where n is the number of daily time steps, and $SWE_{o,i}$ and $SWE_{s,i}$ are the observed and simulated *SWE* for day i (m), respectively.

$$NSE = 1 - \frac{\sum_{i=1}^{n}(SWE_{s,i} - SWE_{o,i})^2}{\sum_{i=1}^{n}(SWE_{o,i} - \overline{SWE_o})^2} \tag{27}$$

where $\overline{SWE_o}$ is the mean observed *SWE* over the entire dataset.

Finally, to further substantiate differences between the Mo and Multi models, the snowpack onset and end dates as well as the date of maximum *SWE* and height were compared on an annual basis. The results are presented relative to their absolute seasonal deviations for each set of parameters using Equation (28). The median results are then compared between models at each *SWE* station.

$$A_c = \sqrt{(C_{k,m} - C_{k,o})^2} \ or \ A_c = 100\frac{\sqrt{(C_{k,m} - C_{k,o})^2}}{C_{k,o}} \ (SWE \ max, \ in \ \%) \tag{28}$$

where A_c is the mean value of characteristic C, and m stands for the tested model (Mo or Multi) and o the observations for year k.

2.4. Case Study

Three *SWE* stations were selected for this study based on their differences in altitude and climate. As shown in Figure 2, they are in two distinct regions of Canada. The first *SWE* station (a.k.a. GMON station) and meteorological stations are in the Necopastic River watershed, in a subboreal climate. It is in a 50 m-radius forest clearing, surrounded by a 7 to 8 m-tall spruce trees, with vegetation reaching 3 to 4 m beyond 30 m. The exact altitude of the station is uncertain, but the altitudes of the watershed are between 100 and 180 m. The observed data used in this study were taken from Oreiller et al. [54]. The two other *SWE* stations are in the Upper Yukon River watershed, namely, the Lower Fantail and the Wheaton stations [55]. The Lower Fantail stations are located on an outcrop surrounded by a wetland, at the bottom of a river valley, while the Wheaton stations are located on a ridge crest close to a glacier, surrounded partially by subalpine firs and shrubs. These stations are located in the alpine, subalpine, and boreal eco-climatic regions of the Northern and Central Cordillera [56]. The exact station altitude is uncertain, but the altitudes of this watershed fall between 640 and 2010 m.

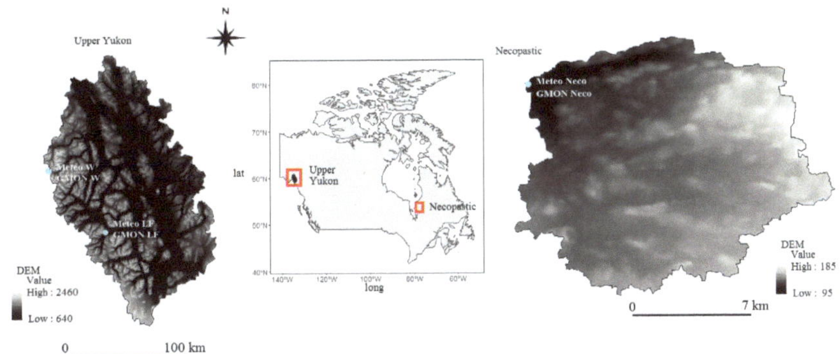

Figure 2. Locations of the upper Yukon (**left**) and Necopastic (**right**) watersheds in Canada. Weather and ground snow stations are in blue circles. LF stands for Lower Fantail, W for Wheaton, and Neco for Necopastic.

The weather and *SWE* station metadata are provided in Table 2. During the study, ground-based precipitation measurements were non-continuous in Yukon. Given these conditions, the precipitation times series for modeling was assumed to be the daily increase in observed water equivalents due to the lack of information about wind-related snow transport. The modeled *SWE* was compared to data from the GMON stations, which measure gamma rays naturally emitted by the Earth and attenuated by the snowpack. The measuring principle, developed by Choquette et al. [57], converts gamma radiation measurements into *SWE* (mm). The station sensors at Necopastic and Upper Yukon are GMON3 [58] and CS275s [59], respectively, with measurement uncertainties ranging from ± 15 mm (for *SWE* less than 300 mm) to $\pm 15\%$ otherwise. Figure 3a–c depict precipitation, average air temperature, and *SWE* time series at the three stations. The evaluation of model performance excluded days without *SWE* data.

Table 2. Weather and GMON station metadata. Data for the Necopastic watershed are from Oreiller et al. [54]; Upper Yukon data were provided by Yukon Energy.

Station	Code	Period	Temporal Resolution	Type	Basin
Necopastic	Meteo_Neco GMON Neco	2006–2011	Daily and hourly 6 h	Auto	Necopastic
Lower Fantail	Meteo_LF GMON LF	2014–2017	Daily and hourly 6 h	Auto	Upper Yukon
Wheaton	Meteo_W GMON W	2014–2017	Daily and hourly 6 h	Auto	Upper Yukon

Table 3 shows the average temperature and cumulative precipitation for each hydrological year. The fifth year of the Necopastic station appears to be an aberration. However, the dataset for that year did not account for the summer temperature or precipitation.

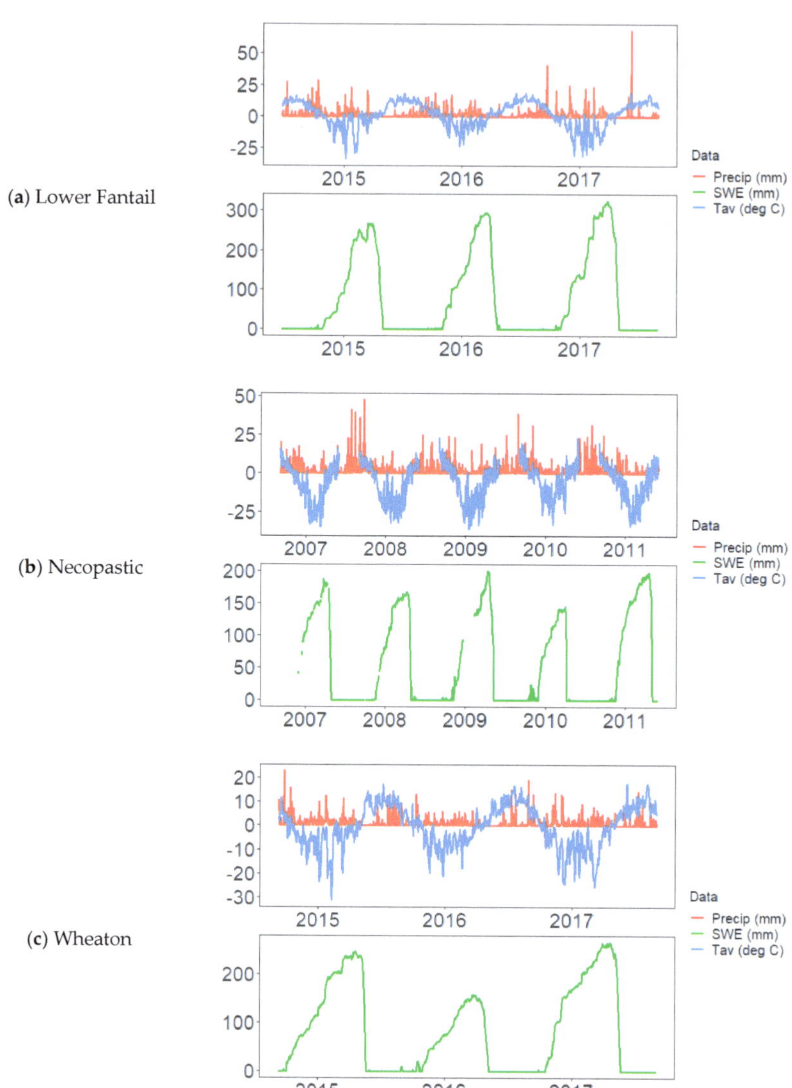

Figure 3. Daily precipitation, average air temperature, and *SWE* at the three stations.

Table 3. Cumulative precipitations and average temperature for each hydrological year.

Station	Data	Y1	Y2	Y3	Y4	Y5
	Years	2014/2015	2015/2016	2016/2017	-	-
Lower Fantail	Precipitation (mm)	643	556	721	-	-
	Average temperature (°C)	1.8	1.5	2.0	-	-
	Years	2006/2007	2007/2008	2008/2009	2009/2010	2010/2011
Necopastic	Precipitation (mm)	803	855	840	819	462
	Average temperature (°C)	2.2	2.3	2.3	2.2	1.7
	Years	2014/2015	2015/2016	2016/2017	-	-
Wheaton	Precipitation (mm)	525	352	489	-	-
	Average temperature (°C)	−0.2	0.6	−1.9	-	-

3. Results

3.1. Sensitivity Analyses

The sensitivity analysis was conducted for the three stations. Figure 4 depicts the relative sensitivity index $IVAR_{i,50n}$ for each parameter for both models.

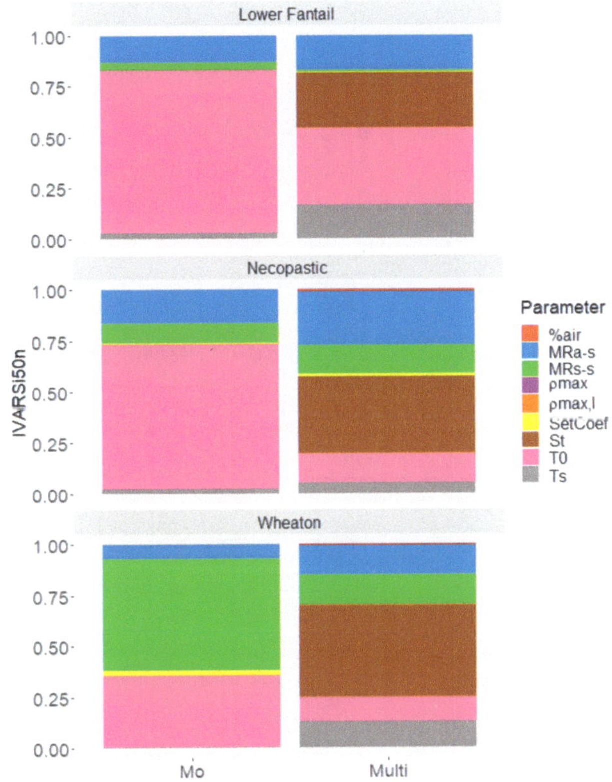

Figure 4. Normalized sensitivity analysis of the monolayer (Mo) and multilayer (Multi) snow model.

Before comparing parameter sensitivity differences between the two models, it is necessary to evaluate those of the three additional parameters of the Multi snow model. Notably, S_t, the new snow layer precipitation threshold, displayed high sensitivity and requires calibration, while $\rho_{max,l}$ and *%air* exhibited minimal sensitivity values and could therefore be set to a constant value prior to calibration. Both Dahe et al. [60] and Nishimura et al. [61] observed a maximum value of 550 kg.m^{-3} for $\rho_{max,l}$. Considering its sensitivity and range of values set at 250–550 kg.m^{-3} in the Mo model, it was set to 550 kg.m^{-3} for the Multi model. As for *%air*, it was established as 10% of the snowpack depth in the Mo model. In the multilayer model MASiN [42], it was set at 8% of the volume of the snowpack not occupied by the *SWE* or the liquid water content, with some allowance possible for values varying between 5 and 10%. Würzer et al. [62] set a value of 3.5% of the snow depth in the SNOWPACK model. Taking these divergent values into account, *%air* was set at 10% of the snowpack height occupied by air in the Multi model.

The most sensitive phenomenon in the Mo model was located at the boundary between the atmosphere and snow. Two parameters, T_0 (the threshold temperature for considering melt due to radiation) and MR_{a-s} (the degree-day rate of melt due to radiation), are crucial in this context. Set_{Coef} (i.e., compression rate) and T_s (threshold temperature for precipitation partitioning into rain and snow) are insensitive. By incorporating a multilayer structure into the model, the significance of T_s is given greater importance while simultaneously minimizing the relative sensitivity of the boundary phenomenon between the atmosphere and snow.

Figure 5 illustrates the daily relative sensitivity $IVAR_{i,50n}$ of both models at the three stations. The discontinuity arose from limited data over few years. The parameter factor space did not allow the Mo model to simulate snow during the summer season, in contrast to the Multi model.

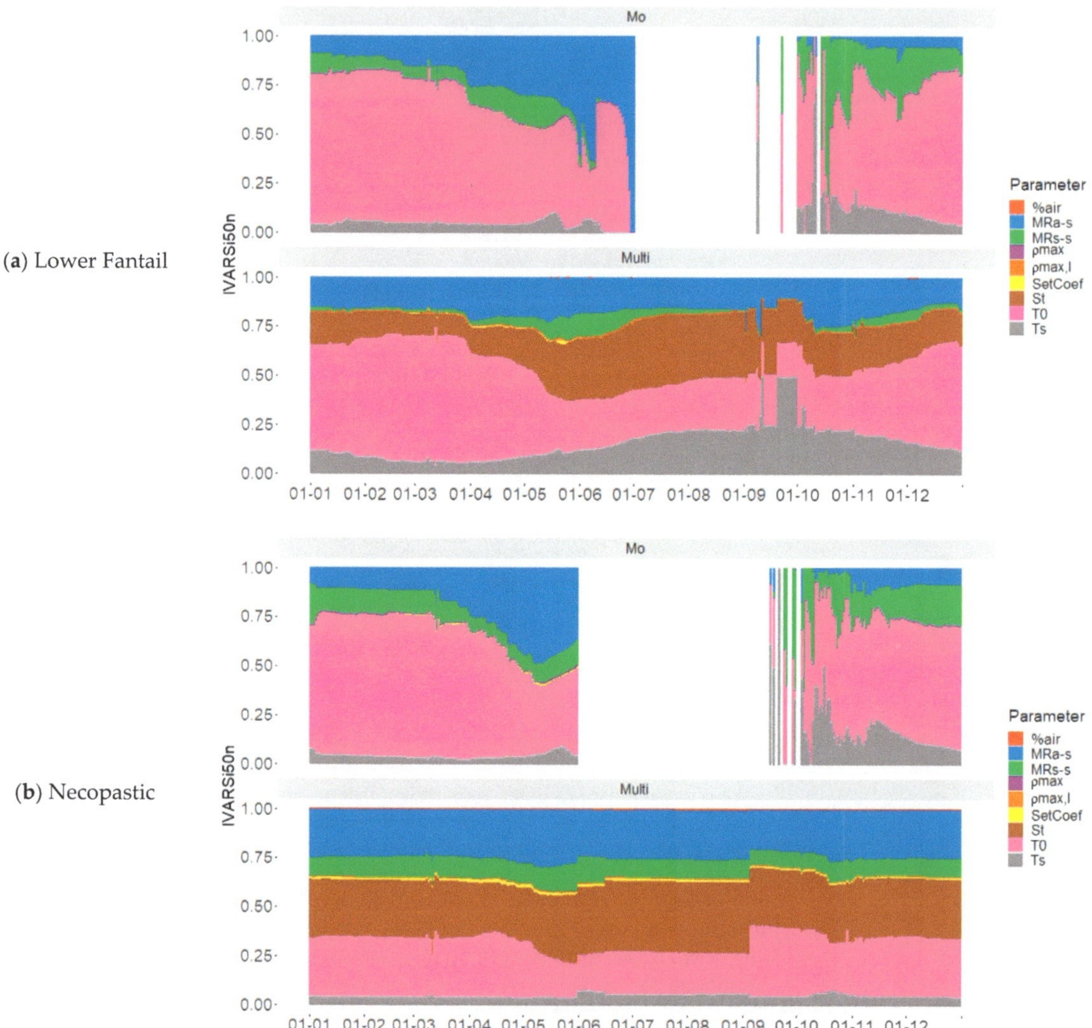

Figure 5. *Cont.*

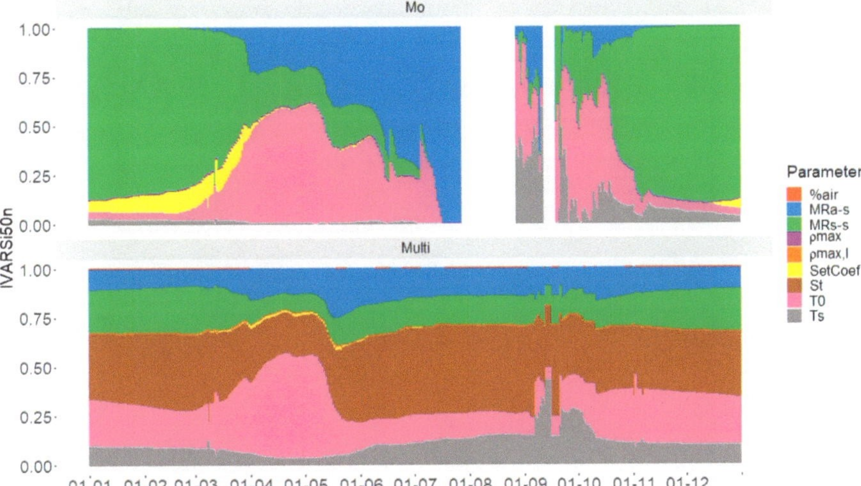

(c) Wheaton

Figure 5. Daily relative sensitivity analysis of the monolayer and multilayer models at the (**a**) Lower Fantail station, (**b**) Necopastic station, and (**c**) Wheaton station.

The seasonal phenomena are highlighted in both models. For the Mo model, the temperature threshold for separating precipitation (T_s) was quite sensitive early in the formation of the snowpack. The most significant phenomenon during spring was melting caused by radiation (MR_{a-s}). As the melting season drew to a close, melting from the soil became increasingly important (MR_{s-s}). The parameter Set_{Coef} (settling rate) was also quite sensitive prior to the melting season, particularly at the Wheaton station.

For the Multi model, variations in sensitivity were less severe but still revealed the same seasonal phenomena as in the Mo model. However, the new snow layer precipitation threshold (S_t) served as a buffer during the melting period.

3.2. Modeling Performances—Validations

Figure 6 depicts the performances of the snow models for the top ten best parameter sets for the calibration and validation periods at the three GMON stations.

For the Lower Fantail station, the Wilcoxon test indicated no significant difference (p-value > 0.05) between the median of the models during the "Y23" combination calibration period, where the first year of data was used for validation, which was the driest and coldest year. For the remaining combinations, the Multi model improved median performances by 0.021 to 0.033 for the calibration period and by 0.125 to 0.223 for the validation period.

The performances of both models for the Necopastic station did not exhibit significant differences over the calibration period for combinations "Y1235" and "Y1245", and over the validation period for the combinations "Y1234" and "Y1245". However, during the calibration period, the Multi model boosted performance by 0.01 to 0.017 of KGE, and during the validation period, it improved by 0.009 to 0.154. The Mo model improved the performance by 0.012 for "Y2345" for the calibration period and by 0.012 for the validation period for "Y1345". Notably, there was no relationship with annual meteorological characteristics.

Conversely, for the Wheaton station, there was no significant difference between the models over the validation period for combination "Y12" or "Y23". However, the Mo model enhanced the performance by 0.03 for combination "Y13", which considered the driest and warmest year for validation. During the calibration period, the Mo model improved the performance by 0.008 to 0.04.

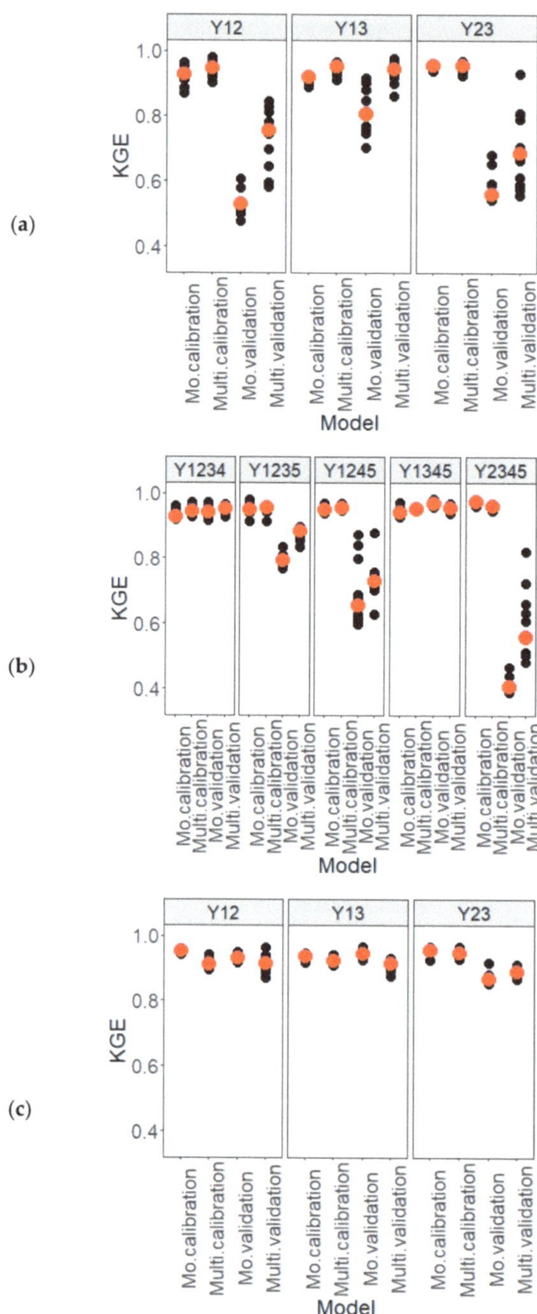

Figure 6. KGE values for Mo and Multi models for (**a**) Lower Fantail, (**b**) Necopastic, and (**c**) Wheaton. In red are the median performances of the top ten best parameter sets. The missing number in each column corresponds to the year used for validation; for example, Y12 means that year 3 was used for validation.

Of the eleven configurations detailed above for the calibration period, there were three cases where the snow models performed equally. It is important to note that this evaluation is objective and based solely on performance. The Mo model performed better in four configurations (with an average gain of 0.019 of KGE), whereas the Multi model performed better in the remaining four configurations (with an average gain of 0.020 of KGE). The gains were comparable across the calibration periods. Out of the configurations for the validation period, there were four cases where the snow models showed no difference. The Mo model performed better on two configurations (average gain: 0.021 of KGE), whereas the Multi model performed better on the remaining five configurations (average gain: 0.146 of KGE). The Multi model demonstrated a clear improvement in result consistency over the validation period.

3.3. Modeling Performances—All Calibrations

In the third part of this paper, calibration was performed using all years. For the Lower Fantail and Wheaton stations, the results of the Wilcoxon tests rejected the median equality hypothesis, yielding p-values of 9.8×10^{-3} and 2×10^{-3}, respectively (with Multi median values of 0.95 and 0.92, respectively, and Mo model median values of 0.93 and 0.95, respectively). Conversely, for the Necopastic station, the medians (Multi: 0.95, and Mo: 0.96) are considered equal, given an 8.4×10^{-2} p-value. Regarding the root mean squared error (RMSE) and the Nash–Sutcliffe efficiency (NSE), the Wilcoxon test failed to reject the null hypothesis that the medians are equal. Figure 7 illustrates the calibration performances (KGE, RMSE, and NSE values) of the top 10 sets of parameters obtained for each model at the three stations, as well as the coefficients derived from linear regression analysis.

It is evident that the slopes obtained from the Mo model had a narrower range than those of the Multi model during the snow accumulation (defined as the observed period between the first day of snow on the ground and the winter peak) and the melt period (defined as the observed period between the winter peak and the day when the snow cover has completely melted). Furthermore, the range increased more during the melting period compared to the accumulation period for each model.

Figure 8 depicts *SWE* simulations based on the top ten parameter sets for each model at the Lower Fantail station. Results for the two other stations can be found in Appendix F.

The results show minimal disparities in the optimal performances, with KGE values consistently exceeding 0.95 for the optimal sets of parameter values. Assessing robustness through the minimum values of the red interval indicated a similarity for both models. However, because of their inherent differences, *SWE* absolute values differed substantially between models. Notably, the Multi model showed more pronounced seasonal variability (red interval width), thereby enabling a more precise representation of the first winter peak at the Lower Fantail station with certain parameter sets, whereas the Mo model failed to represent adequately the observed *SWE* profiles.

Similarly, Figure 9 displays the range of snow height and density modeled by the top ten sets of parameter values for each model. Analyzing the snow height time series is relevant, as this variable is used for the *SWE* estimation in both models. The snow height series was overestimated by the Mo model, whereas the Multi model underestimated them, except for a few sets of parameters. This resulted in underestimated snow densities by the Mo model, as opposed to the output of the Multi model. It is evident that the Multi model overestimated the density during each phase of melting.

The modeled snow height and density time series for the Lower Fantail and Necopastic GMON stations are presented in Appendix F. Figure 10 shows the KGE values for snow height and density time series achieved by the top ten sets of parameter values, calibrated on *SWE* for both models.

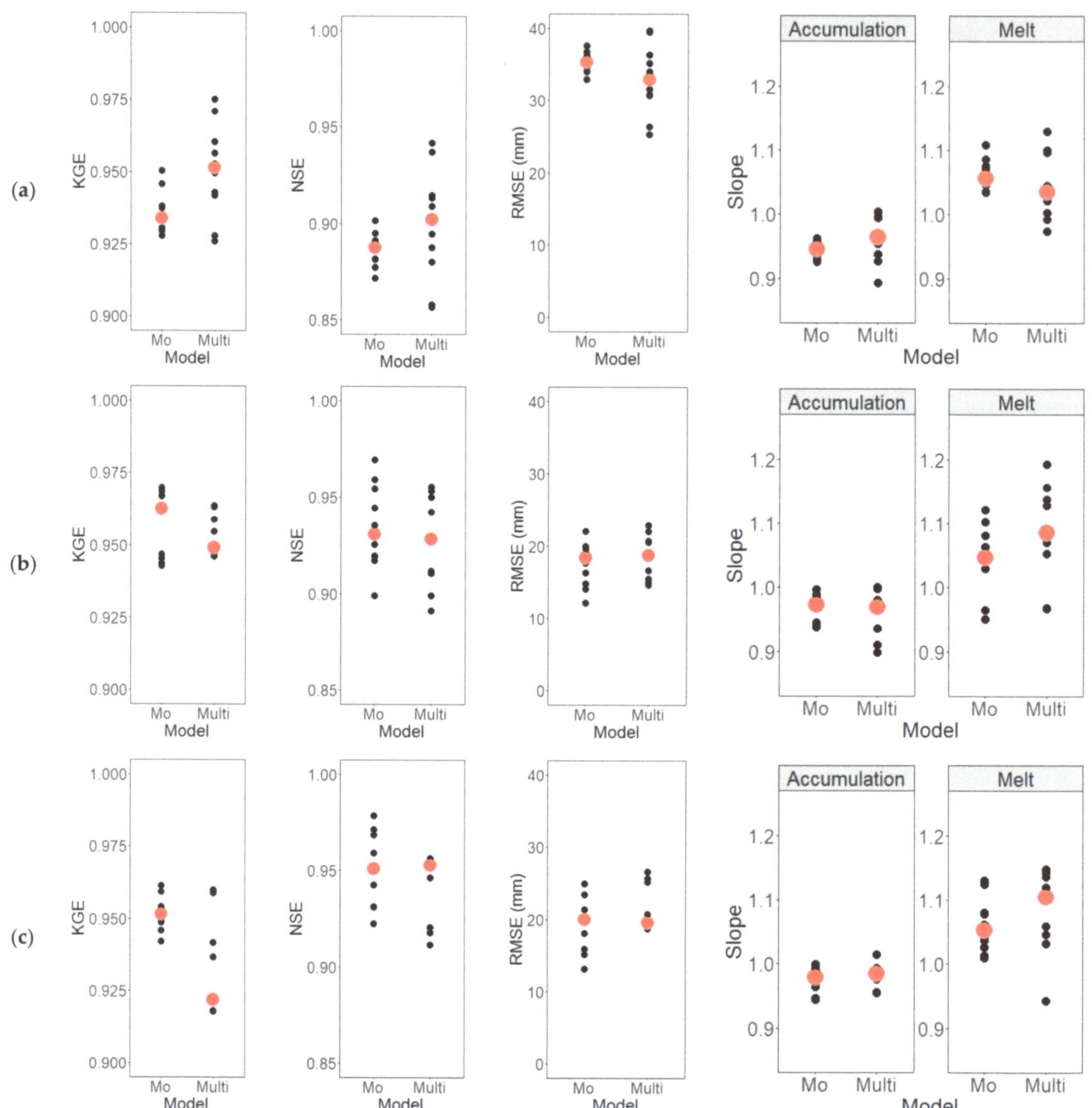

Figure 7. Modeling performances (KGE, RMSE, and NSE) and average rate of change (i.e., slope) of *SWE* during the snow accumulation and melt periods of the top ten sets of parameter values obtained for the multilayer snow model (Multi) and the monolayer model (Mo) for (**a**) the Lower Fantail station (LF), (**b**) the Necopastic station (Neco), and (**c**) the Wheaton GMON station (W). In orange is the median performance. KGE, RMSE, and NSE stand for Kling–Gupta efficiency, root mean squared error, and Nash–Sutcliffe efficiency, respectively.

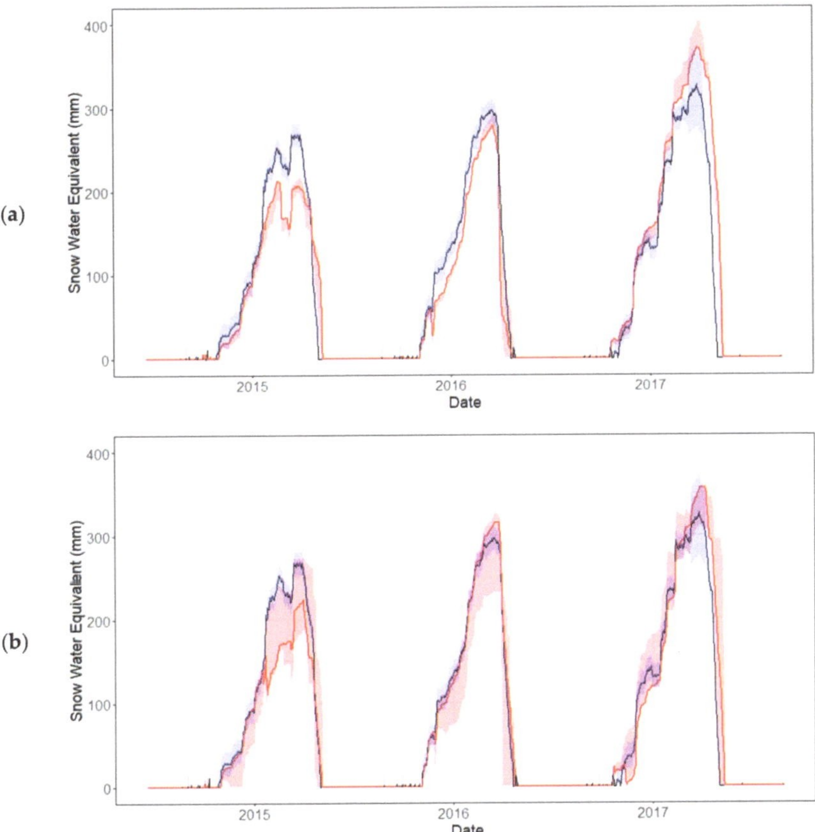

Figure 8. Modeled *SWE* time series at the Lower Fantail station for the (**a**) monolayer (Mo) and (**b**) multilayer (Multi) models. The red shaded interval shows the range of values provided by the top ten sets of parameters, with the red line for the best parameter set. The observed *SWE* time series is shown in black, while the blue interval depicts the measurement uncertainty.

Although the minimum performances can be considered unsatisfactory for each model, the median performances indicate that the Multi model more frequently generated physically accurate simulations (with a KGE around or greater than 0.5), whereas the acceptable results provided by the Mo model were achieved only by a few sets of parameters. Consequently, the Multi model can offer more parameter sets for *SWE*, providing satisfactory performances for snow height, compared to the Mo model. However, it is noteworthy that during the melting period, the densities of the snowpack layers remained high for the Multi model, incorporating layers of ice (density of 917 kg.m^{-3}) with thicknesses exceeding 20 cm.

Furthermore, the modeling of freezing rain was of little impact. Out of the ten best sets of parameter values obtained for each station, only one parameter set modeled this type of rain for Necopastic, and none for Lower Fantail and Wheaton. More importantly, during calibration, only 9.6% of the parameter sets accounted for any freezing rain event for the Necopastic GMON station, and none for the Lower Fantail and Wheaton stations.

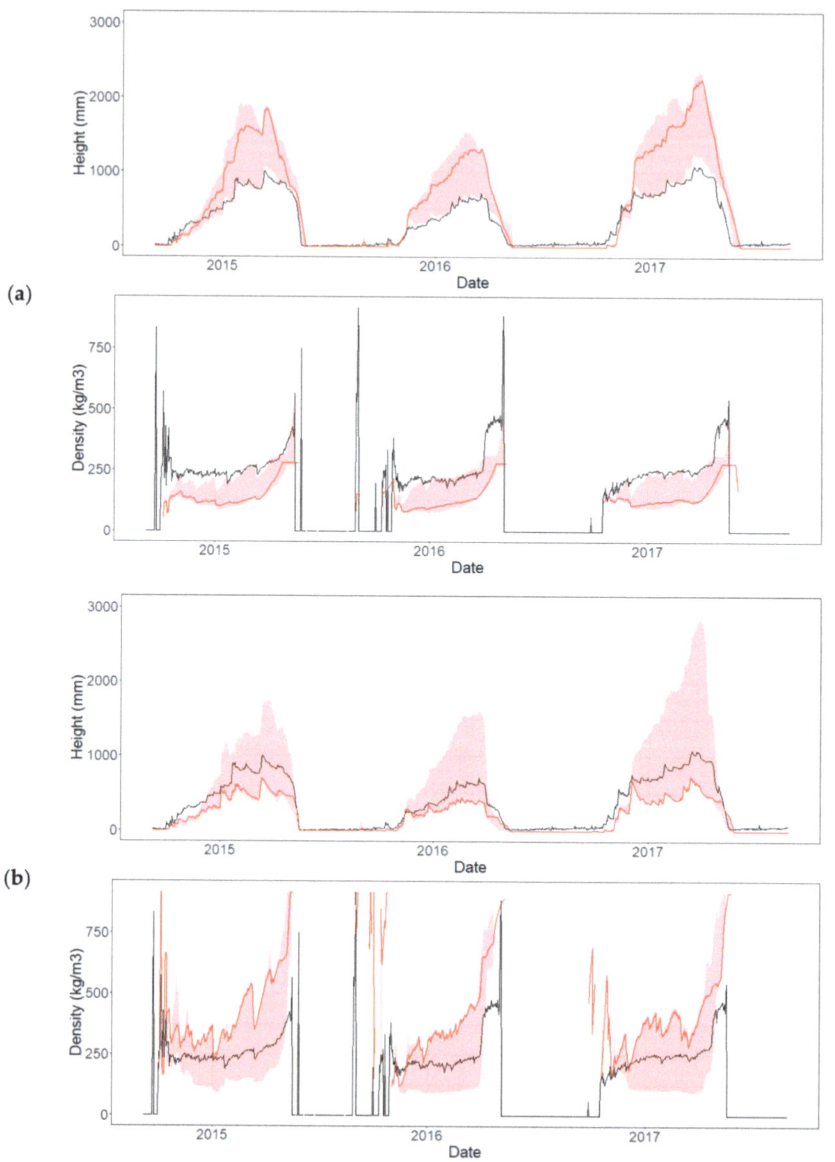

Figure 9. Modeled height and density time series at the Wheaton station for the (**a**) monolayer (Mo) and (**b**) multilayer (Multi) models. The red shaded interval shows the range of values provided by the top ten sets of parameters, with the red line for the best parameter set. The observed height and density time series are shown in black.

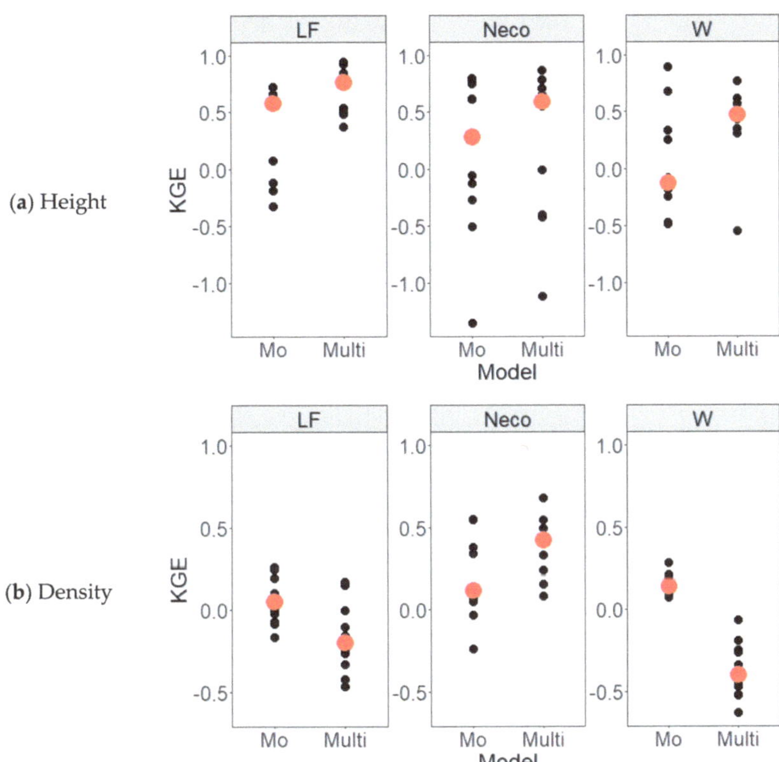

Figure 10. KGE values computed from the ten best parameter sets using time series of observed and modeled snow heights (**a**) and densities (**b**) for the Mo and Multi models. In orange is the median performance.

3.4. Modeling Snowpack Characteristics

Snowpack characteristics derived from the top ten sets of parameter values of each model were compared in terms of the onset and end dates of the snowpack, as well as the maximum *SWE* values and dates. The seasonal discrepancies between the modeled and observed data were analyzed across the top sets of parameter values in Table 4. This assessment provides insights into the equifinality of each feature of interest. For instance, the top 10 parameter sets presented here for each station and model yielded global KGE values greater than 0.9. However, snow peaks or melting periods may be modeled differently given the set of parameter values used.

Table 4. Medians of annual differences between observations and snowpack characteristics from the top ten best sets of parameter values of each model at the three GMON stations.

Station	Lower Fantail		Necopastic		Wheaton	
Models	**Mo**	**Multi**	**Mo**	**Multi**	**Mo**	**Multi**
Onset date (days)	3	4	3	3	3	4
End date (days)	8	7	4.5	4	6	1.5
Maximum *SWE* date (days)	5.5	7	9	11	1	8
Maximum *SWE* relative difference (%)	17	6.7	11	5.9	8.8	13.2

Both Mo and Multi onset dates showed consistent median deviations of 3–4 days from the observed data. The end date deviations were similar, except for the Wheaton station, where the Multi model showed a 4-day improvement over the Mo model. Comparing with the Multi model, the maximum *SWE* dates were better represented with the Mo model by 1.5, 2, and 7 days for the Lower Fantail, Necopastic, and Wheaton stations, respectively. Notably, the Multi model outperformed the Mo model in representing the maximum *SWE*, particularly exhibiting a halved error at the Lower Fantail and Necopastic stations, but with a higher error at the Wheaton station.

As previously introduced, the Multi model uses a different approach to estimate snow albedo compared to the Mo model. Mo assumes that the albedo decays with time as a function of snowpack liquid water content, whereas Multi estimates albedo as a linear function based on the proportion of ice and air in the top snow layer. Figure 11 illustrates the albedo values of the top ten best parameter sets for both models for the Wheaton station (the albedos for the Lower Fantail and Necopastic stations are depicted in Appendix F). It can be observed that the estimated albedo for the Mo remained consistent across each parameter set, whereas more variations were observed for Multi. Although both approaches demonstrate a decreasing albedo over winter, Multi's behavior was consistent throughout the winter, except following a snowfall, which could have temporarily increased the albedo after the new snow blended in the uppermost layer or after adding a new layer. The decreasing albedo of Mo fluctuated within a certain range during winter until the spring melt, when it strongly decreased. Finally, the albedo of Multi was greater than that of Mo because it is calculated for the uppermost snow layer only, whereas Mo considers an equivalent albedo for the entire snow cover.

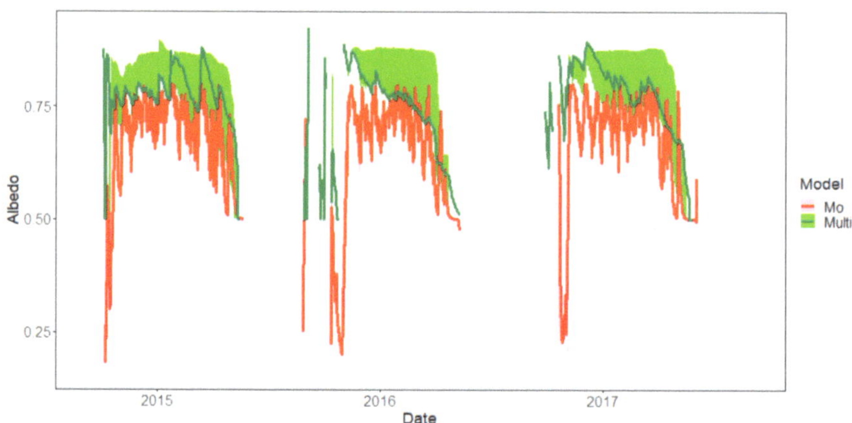

Figure 11. Albedo time series modeled by the top ten best sets of parameter values for the Mo model (pink envelope) and the Multi model (green envelope) for the Wheaton GMON station. The best parameter sets are depicted by the red and green lines for the Mo and Multi models, respectively.

4. Discussion

This paper has proposed a set of modifications to the monolayer snow model of HYDROTEL, including the integration of a multilayer structure, estimation of snowpack properties based on the proportion of ice and air, freezing rain modeling, and changes in compression and maximum water retention capacity. The modeling was assessed with respect to *SWE* modeling and other snow characteristics, such as snow height and density.

The sensitivity analysis indicated that amongst the changes implemented in the Mo model, the precipitation threshold for adding a snow layer (S_t) was highly sensitive, whereas the ratio of the volume of air that can be filled in by water (%*air*) and the settling maximum snow-layer density ($\rho_{max,l}$) were less sensitive. The addition of these parameters changed the hierarchy of sensitivity of the other parameters. For instance, whereas the melting temperature threshold at the air–snow interface became less sensitive (T_o), the melting rate sensitivity at this interface (MR_{a-s}) increased. Similarly, the temperature precipitation threshold temperature (T_s) becoming more sensitive was deemed significant. In addition, the melting rate at the ground–snow interface (MR_{s-s}) became less sensitive, except for a slight increase in sensitivity for the Necopastic GMON station. These modifications rendered the Multi model more sensitive to phenomena at the snow–atmosphere interface. Furthermore, the melt rate at the snow–ground interface (MR_{s-s}) became generally less sensitive, emphasizing the influence of the atmosphere on snow melt rather than at the snow–ground interface. This change in behavior is consistent with observations made by Lackner et al. [63], who showed that temperature variations within the snowpack exhibit amplitudes more akin to those in the atmosphere than those at the ground level.

The calibration/validation strategies were based on 22 combinations when examining their respective periods separately. Among these, seven combinations showed no significant difference in performance between models. During the calibration period, increases in performance did not exceed 0.04. Thus, both models demonstrated similar levels of performance over this period. However, during the validation period, the Mo model's performance did not surpass 0.03, whereas the range of increased performance for the Multi model varied between 0.046 and 0.223. Notably, there was a subset of four combinations that exhibited an increase in performance of more than 0.1. Overall, the Multi snow model demonstrated greater robustness during the validation period compared to the Mo model. When both models were calibrated using the full datasets, with respect to their relative performances in reproducing SWE, the results highlighted some very good performances, with KGE values consistently greater than 0.9. Thus, neither model gained a clear advantage over the other. The reconstruction of precipitation records for the Lower Fantail and Wheaton stations may have contributed to these performances, providing the appropriate amount of water to the snowpack on the correct days until the melting period. However, for the Necopastic station, performances were still good even though precipitation records were not reconstructed. This suggests that the reconstruction of precipitation does not necessarily affect the conclusions of this paper. The modifications introduced in the Multi model made it possible to maintain a level of performance similar to that of the Mo model while also providing more flexibility for the computation of energy transfer within the snowpack, as suggested by the sensitivity of the additional parameter (S_t) on modeled $SWEs$. Furthermore, from a hydrological modeling perspective, snowpack melt rates are crucial for estimating streamflow, especially the maximum SWE, with snowpack heights being a somewhat secondary objective.

Although SWE modeling performances were comparable, model behaviors for snow heights were not. The Mo model tended to overestimate snow heights, whereas the Multi model tended to be consistent with observed heights or even slightly underestimate them. For the Mo model, the height is used solely to estimate compression while affecting thermal diffusivity; it can also be adjusted using the calibrated maximum density. In contrast, for the Multi model, although the height is used for compression, it is also used to compute snowpack density, which is required for computations of thermal diffusivity, albedo, and maximum water retention. Since energy transfer by radiation governs snowmelt, a low albedo increases this transfer. During spring snowmelt, minimizing snowpack height implies high densities—which is not surprising given that SWE is also equivalent to the product of snow height and relative snow density—which in return reduces albedo. An additional indication that simulated densities are larger than what may be observed in general can be inferred through a comparison reported by Keenan et al. [64] between simulated and observed density profiles using the SNOWPACK model. The densities they observed reached values of about 475 kg.m^{-3} at ground level, whereas the Multi model formed snow layers limited to 550 kg.m^{-3}, or 917 kg.m^{-3} for ice layers during the spring melt, with thicknesses exceeding 20 cm, which is unrealistic. Indeed, these densities are more akin to those observed for glaciers [65].

The attempt to model freezing rain indicated that this phenomenon seldom occurred for all the tested parameter sets. Indeed, the required condition that the atmospheric temperature near the ground be negative may be too restrictive, and it is emphasized that atmospheric phenomena must be considered to model this type of precipitation as effectively as possible. However, it was decided to keep this phenomenon in the model, as it is a mean of creating snow layers under conditions of temperatures close to 0 °C.

The results of this study showed that transforming the Mo model into a Multi model improves the simulation of the end date of the snowpack as well as the seasonal maximum *SWE*, albeit at the expense of the occurrence date. Oreiller et al. [54] considered wind-induced snow transport as a plausible explanation for *SWE* discrepancies for the Necopastic station. This could also be a plausible hypothesis for discrepancies at the other GMON stations, but that remains to be validated. The different approaches used by the Multi and Mo models to estimate snow albedo can be interpreted in terms of the location where phenomena are assessed. For the Mo model, the albedo mimics the distribution of the radiative heat flux throughout the snow cover. In contrast, the approach used by the Multi model emphasizes the distribution of this flux throughout the top layer. Furthermore, the albedo of the Mo model varies within a certain range during winter before decreasing during the spring melt, whereas that of the Multi model decreases throughout the winter. Based on observations made by Gray et al. [66] and Stroeve et al. [67], the behavior of the Mo model albedo is more accurate, but the range of values of the Multi model remains coherent (albedo > 0.65 during winter). In other words, (i) the Mo albedo is for the whole snow cover; (ii) the observed albedo is based on upgoing and outgoing radiation measurements, which depends on the depth of snow penetrated by shortwave radiation; and (iii) the albedo of the Multi model is assumed to be that of the top snow layer only, regardless of the thickness.

5. Conclusions

The snow model of HYDROTEL is a daily monolayer (Mo) model combining degree-day and physics-based equations. This paper proposed a multilayer (Multi) alternative, modifying some of the fundamental equations while preserving the overall computational structure and limiting the addition of new calibration parameters. These modifications increased the sensitivity of processes occurring at the atmosphere–snow interface and the subsequent energy balance of each snow layer, improving the realism of the model. Although snow heights were overestimated by the Mo model, the Multi model more accurately depicted them, although some underestimation persisted. These underestimations resulted from the development of excessively dense, thick, and persistent snow layers during melting periods. Nonetheless, the vertical density profiles became consistent, with the densest layers located at ground level. Also, *SWE* modeling performances were very good (KGE consistently above 0.9) for both models, with the Multi snow model demonstrating more robustness during the validation period. By focusing on snowpack characteristics, the Multi model improved estimations of snowpack end dates and maximum *SWE* but compromised the modeled dates of the latter occurrence. These behavioral changes point towards the potential for improving snowmelt runoff and consequently spring peak flows, which are ultimately linked to the maximum *SWE*. As the frequency of the freezing rain events will, in all likelihood, increase in Eastern Canada given global warming [68], it would be relevant to find a parsimonious way to model these events. However, given that it is primarily an atmospheric phenomenon, the challenge remains. As the hydrological science community is becoming increasingly interested in rain-on-snow events [69–73], the suggested modifications can be viewed as a first step toward modeling them using the Multi version of the HYDROTEL snow model. From a structural standpoint, it may be beneficial to include a basal snow layer to emphasize the thermal discontinuity at ground level. Moreover, future work will involve integration of the multilayer snow model into HYDROTEL to evaluate the effect on stream flow modeling.

Author Contributions: Conceptualization, J.A. and A.N.R.; method, J.A., A.N.R. and E.F.; software, J.A.; validation, J.A., A.N.R. and E.F.; interpretation, J.A. and M.B.; data management, J.A.; visualization, J.A. and A.N.R.; drafting and edition, J.A., E.F., A.N.R. and M.B. All authors have read and agreed to the published version of the manuscript.

Funding: The authors wish to gratefully acknowledge the financial support of the Natural Sciences and Engineering Research Council of Canada (NSERC) and Yukon Energy (YE) through Collaborative (#CRDPJ 499954-16) and Applied (#CARD2 500263-16) Research and Development grants.

Data Availability Statement: Data available on request due to restrictions.

Acknowledgments: This project would not have been possible without substantial contributions from staff at the Yukon Research Center, namely, Brian Horton and Maciej Stetkiewicz, and at Yukon Energy, namely, Shannon Mallory, Kevin Maxwell, and Andrew Hall, as well as Mathieu Oreiller.

Conflicts of Interest: The authors declare no conflicts of interest.

Appendix A. Snow Model Characteristics

Table A1. Snow model characteristics.

Model	Design	Input Data	Considered Phenomena	Structure/Time Step	Reference
CEMANEIGE	Conceptual	- Atmospheric temperature - Precipitation	- Accumulation - Melt	Monolayer/ day	[74]
HDV	Physics-based Degree-day	- Atmospheric temperature - Precipitation	- Accumulation - Degree-day melt - Latent heat flux	Monolayer/ day	[75–77]
SWAT	Physics-based Degree-day	- Atmospheric temperature (min and max) - Precipitation	- Accumulation - Degree-day melt - Sublimation	Monolayer/ day	[78,79]
HYDROTEL	Physics-based Degree-day	- Atmospheric temperature (min and max) - Precipitation	- Accumulation - Compression - Mixed (radiation and degree-day melt) - Precipitation heat - Soil heat - Sensible heat flux - Water retention	Monolayer/ day	[28]
VIC	Physics-based Complex	- Atmospheric temperature - Precipitation - Relative humidity (may be estimated) - Short- to long-wave radiations (may be estimated) - Wind speed	- Accumulation - Compression - Precipitation heat - Turbulent heat flux (Sensible and latent) - Radiation - Water retention	Bilayer/ hourly to daily	[80]
CROCUS	Physics-based complex	- Atmospheric temperature - Precipitation - Relative humidity - Short- and long-wave radiations - Wind speed	- Accumulation - Compression - Heat conduction - Metamorphism - Precipitation heat - Radiations - Runoff and intra-snow cooling - Soil heat - Sublimation - Turbulent heat flux (sensible and latent) - Wind transport	Multilayer/ hour	[81,82]

Table A1. *Cont.*

Model	Design	Input Data	Considered Phenomena	Structure/Time Step	Reference
MASiN	Physics-based Complex	- Atmospheric temperature - Precipitation - Relative humidity - Wind speed	- Accumulation - Soil heat - Cloud cover - Compression - Conduction - Radiation (estimation) - Turbulent heat flux (sensible and latent) - Water retention	Multilayer/ hour	[42]
SnowPack	Physics-based Complex	- Atmospheric temperature - Precipitation - Relative humidity - Wind speed and direction	- Accumulation - Compression - Conduction - Turbulent heat flux (sensible and latent) - Radiation (estimation) - Water retention	Multilayer/ 10 min to day	[83,84]
SNOWPACK	Physics-based Complex	- Atmospheric temperature - Precipitation - Relative humidity - Short- and long-wave radiations - Wind speed	- Accumulation - Compression - Microstructure - Precipitation heat - Radiation - Runoff - Subsurface melt - Surface haze - Surface melt - Turbulent heat flux (sensible and latent) - Wind erosion - Wind transport	Multilayer/ hour	[31,85,86]

Appendix B. Energy Balance Terms of the HYDROTEL Monolayer Snow Model

The different terms of the energy balance equation of HYDROTEL's monolayer snow model are described below.

The heat input from rain, u_r, is computed as follows:

$$u_r = \rho_w \left(C_f + C_w \frac{T_{max} + T_{min}}{2} \right) R \tag{A1}$$

where ρ_w is the density of water (1000 kg.m^{-3}); C_f is the latent heat of fusion of water (335,000 J.kg^{-1}); C_w is the specific heat capacity of water (4184 J.kg^{-1}.°C^{-1}); T_{min} and T_{max} are the minimum and maximum air temperatures (°C), respectively; and R is the daily rainfall rate (m.s^{-1}).

The heat input from the ground, u_{s-s}, is computed as follows:

$$u_{s-s} = \rho_w C_f \frac{MR_{s-s}}{86400} \tag{A2}$$

where MR_{s-s} is the melting rate at the snow–ground interface (m.day^{-1}), and 86,400 is the conversion from day to seconds.

The snow heat deficit, u_s, is computed as follows:

$$u_s = \rho_w C_s \frac{T_{max} + T_{min}}{2} S \tag{A3}$$

where C_s is the specific heat capacity of snow (2093.4 J.kg^{-1}.°C^{-1}), and S is the daily snowfall rate (m.s^{-1}).

Heat loss by conduction and heat gain by radiation are enabled depending on the temperature threshold for radiation heat gain T_0. Indeed, if the daily average air temperature is lower than T_0, the conduction heat losses are estimated; otherwise, the heat gain estimation by radiation is enabled. Heat loss by conduction is estimated using the solution for heat transfer in a semi-infinite material with air temperature as a Dirichlet boundary condition. Thermal diffusivity is computed using estimations of the conductivity and depth of snow. The heat deficit is then updated using the snow temperature resulting from the conductive heat loss.

The radiation heat input, u_{a-s}, is computed as follows:

$$u_{a-s} = \rho_w C_f \frac{M_{pot}}{86400} \tag{A4}$$

where M_{pot} is the potential melting rate due to radiation (m.day^{-1}), computed as follows:

$$M_{pot} = I\, MR_{a-s} \left(\frac{T_{max} + T_{min}}{2} - T_0 \right) (1 - \alpha) \tag{A5}$$

where I is a radiation index, MR_{a-s} is the melting rate at the air–snow interface (m.day^{-1}.$^{\circ}$C^{-1}), and α is the snow albedo.

The radiation index is the ratio of the index for a sloped surface to that of a flat surface [87]. The snow albedo is computed using the snowpack and fresh snowfall albedos, accounting for the exponential decay of radiation penetration within the snowpack [28]. The equations are presented in Appendices B and C.

When the snowpack melts, water is retained within the medium and is considered frozen at the next computational time step. The phase change then warms up the snowpack as follows:

$$u_{ac} = \rho_w C_f \frac{AR}{86400} \tag{A6}$$

where AR is the water retained within the snowpack of the previous day (m.day^{-1}). It is computed using Equation (7) from the maximum water retention capacity estimated in Equation (6).

Appendix C. Radiation Index Equations of the Monolayer Snow Model

θ is the GMON station latitude in radians:

$$\theta = \frac{lat}{rad1} \tag{A7}$$

where lat is the GMON station latitude ($^{\circ}$), and $rad1$ is the conversion factor from radians to degrees ($\approx 57.295779513^{\circ}$.rad^{-1} = $(180^{\circ})/\pi$.rad^{-1}).

k is the slope angle (rad):

$$k = arctan(slope) \tag{A8}$$

where slope is the ground slope (rad).

h is the surface azimuth angle (rad):

$$h = \frac{(495 - 45ori)360}{rad1} \tag{A9}$$

where ori is the ground orientation (1 for east, 2 for north/east, 3 for north, ..., and 8 for south/east). Detailed information is available in Rousseau et al. [88].

θ_1 is the equivalent slope latitude (rad):

$$\theta_1 = arcsin(sin(k)\,\cos(h)\,cos(\theta) + cos(k)\,sin(\theta)) \tag{A10}$$

α is the longitude variation between the slope and its horizontal surface:

$$\alpha = arctan\left(\frac{sin(k)\ sin(h)}{cos(k)\ cos(\theta) - cos(h)\ sin(k)\ sin(\theta)}\right) \tag{A11}$$

$e2$ is the Sun's/Earth's distance to its average on a specific day:

$$e2 = \left(1 - exc\ cos\left(\frac{day - 4}{deg1}\right)\right)^2 \tag{A12}$$

where exc is the Earth's orbit eccentricity (=0.01673), day is the Julian day, and $deg1$ ($\approx$58.1313429644 day.rad^{-1} = $2\pi/365.25$), 4 January, is the Earth at its perihelion.

i_{e2} is the solar constant as a function of the Earth–Sun distance (W.m^{-2}):

$$i_{e2} = \frac{i0}{e2} \tag{A13}$$

where $i0$ is the solar constant (=1361 W.m^{-2}).

$decli$ is the solar declination (rad), which is the angle between solar rays and the plane of the equator:

$$decli = 0.410152374218 sin\left(\frac{day - 80.25}{deg1}\right) \tag{A14}$$

$tampon$ and $tampon1$ are the angles (rad) that correspond to the sunshine duration on a flat surface and on a sloped surface, respectively:

$$tampon = -tan(\theta)\ tan(decli) \tag{A15}$$

$$tampon1 = -tan(\theta_1)\ tan(decli) \tag{A16}$$

dur_{hor} is the sunshine duration on a flat surface:

$$dur_{hor} = 0\ if\ tampom > 1 \tag{A17a}$$

$$dur_{hor} = 12\ if\ tampon < -1 \tag{A17b}$$

$$dur_{hor} = \frac{arccos(tampon)}{w}\ otherwise \tag{A17c}$$

where w is the Earth's angular speed (15°.h^{-1} =15/rad1 rad.h^{-1}).

dur_{slp} is the sunshine duration on a sloped surface:

$$dur_{slp} = 0\ if\ tampon1 > 1 \tag{A18a}$$

$$dur_{slp} = 0\ if\ tampon1 < -1 \tag{A18b}$$

$$dur_{slp} = \frac{arccos(tampon1)}{w}\ otherwise \tag{A18c}$$

$t1_{slp}$ and $t2_{slp}$ are the irradiation starting and end times on a sloped ground, respectively.

$$t1_{slp} = -dur_{slp} - \frac{\alpha}{w} \tag{A19a}$$

$$t1_{slp} = -dur_{hor}\ if\ t1_{pte} < -dur_{hor} \tag{A19b}$$

$$t2_{slp} = dur_{slp} - \frac{\alpha}{w} \tag{A20a}$$

$$t2_{slp} = dur_{hor}\ if\ t2_{slp} > dur_{hor} \tag{A20b}$$

$t1_{hor}$ and $t2_{hor}$ are the irradiation starting and end times on flat ground, respectively.

$$t1_{hor} = -dur_{hor} \tag{A21a}$$

$$t2_{hor} = dur_{hor} \tag{A21b}$$

i_{j1} and i_{j2} are the radiation for a flat and a sloped surface, respectively.

$$i_{j1} = 0 \text{ if } t1_{hor} > t2_{hor} \tag{A22a}$$

$$i_{j1} = 3600\, i_{e2} \left((t2_{hor} - t1_{hor})sin(\theta)sin(decli) + \frac{cos(\theta)cos(decli)(sin(w\, t2_{hor}) - sin(w\, t1_{hor}))}{w} \right) \text{ otherwise} \tag{A22b}$$

$$i_{j2} = 0 \text{ if } t1_{sip} > t2_{sip} \tag{A23a}$$

$$i_{j2} = 3600\, i_{e2} \left(\left(t2_{slp} - t1_{slp}\right)sin(\theta_1)sin(decli) + \frac{cos(\theta_1)cos(decli)\left(sin\left(w\, t2_{slp} + \alpha\right) - sin\left(w\, t1_{slp} + \alpha\right)\right)}{w} \right) \text{ otherwise} \tag{A23b}$$

I is the radiation index.

$$I = \left| \frac{i_{j2}}{i_{j1}} \right| \text{ if } i_{j1} \neq 0 \tag{A24a}$$

$$I = 1 \text{ otherwise} \tag{A24b}$$

Appendix D. Albedo Equations of the Monolayer Snow Model

wet stands for a wet snowpack:

$$wet = 1 \text{ if } R > 0 \text{ or } T > 0 \tag{A25a}$$

$$wet = 0 \text{ otherwise} \tag{A25b}$$

where R is rainfall, and T is the snow temperature (relative to the heat deficit).

With snow on the ground:

A maximum snowpack albedo alb_{t+1} is computed relative to the snowfall's or snowpack's state of humidity.

$$alb_{t+1} = (1 - exp(-0.5\, eq_{snow}))0.8 + (1 - (1 - exp(-0.5\, eq_{snow})))\left(0.5 + (alb - 0.5)exp\left(-0.2\frac{pdth}{24}(1 + wet) \right) \right) \tag{A26}$$

where eq_{snow} is the snowfall water equivalent (mm), alb is the snowpack albedo of the previous time step, and $pdth$ is the time step's number of hours.

$beta2$ is the snowpack radiation penetration exponential decay coefficient.

$$beta2 = 0.2 \text{ if } alb < 0.5 \tag{A27a}$$

$$beta2 = 0.2 + (alb - 0.5)\text{otherwise} \tag{A27b}$$

$$alb = (1 - exp(-beta2\, st_{snow}))alb_{t+1} + (1 - (1 - exp(-beta2\, st_{snow})))0.15 \tag{A28}$$

where st_{snow} is the snowpack water equivalent (mm).

Without snow on the ground:

$$alb = (1 - exp(-0.5\, eq_{snow}))0.8 + (1 - (1 - exp(-0.5\, eq_{snow})))0.15 \tag{A29}$$

Appendix E. Relationships between the Densities of Snow, Ice, and Air

The mass of a composite material is that of its constituent elements. The mass of snow as a mixture of ice and air is computed as follows:

$$W_s = W_i + W_a \tag{A30}$$

where W_s, W_i, and W_a are the snow, ice, and air weights (kg), respectively.

The snow density is estimated for a snow volume that is the sum of the ice and air volumes.

$$\rho_s = \frac{W_s}{V_i + V_a} = \frac{W_i}{V_i + V_a} + \frac{W_a}{V_i + V_a} \tag{A31}$$

where ρ_s is the snow density (kg.m^{-3}), and V_i and V_a are the ice and air volumes (m^3), respectively.

Per the definition of density, $W_i = V_i \rho_i$ and $W_a = V_a \rho_a$:

$$\rho_s = \frac{V_i}{V_i + V_a}\rho_i + \frac{V_a}{V_i + V_a}\rho_a \tag{A32}$$

where ρ_i and ρ_a are the ice and air densities (kg.m^{-3}), respectively.

This equation then shows that by considering snow a composite material, its density can be related to the densities of ice and air, with coefficients corresponding to the respective proportions. In general, this amounts to considering that there is the following relationship:

$$\rho_s = A\rho_i + B\rho_a \ (avec \ A + B = 1) \tag{A33}$$

Since the volumes of ice and air are not explicitly estimated in the snow models proposed in this paper, and knowledge of the volumetric proportions A and B is necessary, an alternative method must be used:

$$\rho_s = A\rho_i + (1 - A)\rho_a \tag{A34}$$

Thus, the volumetric proportion of ice A in the snow can be estimated from equation $A + B = 1$ as follows:

$$A = \frac{\rho_s - \rho_a}{\rho_i - \rho_a} \tag{A35}$$

Thus, the volumetric proportion of air B in the snow can be estimated from equation $A + B = 1$; that is:

$$B = \frac{\rho_i - \rho_s}{\rho_i - \rho_a} \tag{A36}$$

Thus, the knowledge or estimation of the densities of ice, air, and snow enables the derivation of the volumetric proportions of ice and air within the snow from Equations (A35) and (A36), respectively.

Appendix F. Results

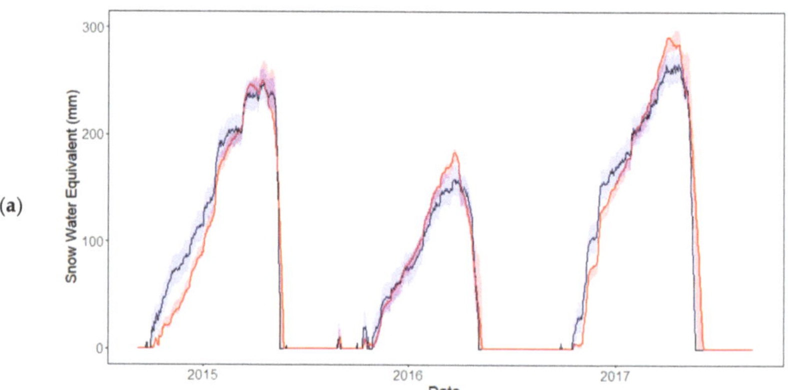

Figure A1. *Cont.*

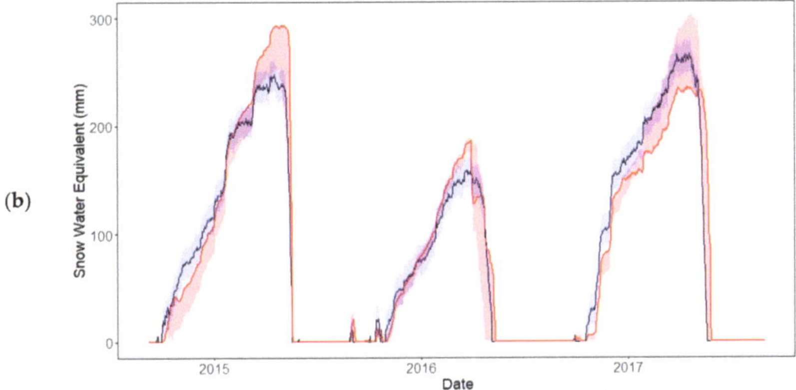

Figure A1. Modeled *SWE* series at the Wheaton station (W) for the (**a**) monolayer (Mo) and (**b**) multilayer (Multi) models. The red shaded interval shows the range of values provided by the top ten sets of parameters values. The observed *SWE* time series is shown in black, while the blue interval depicts the measurement uncertainty.

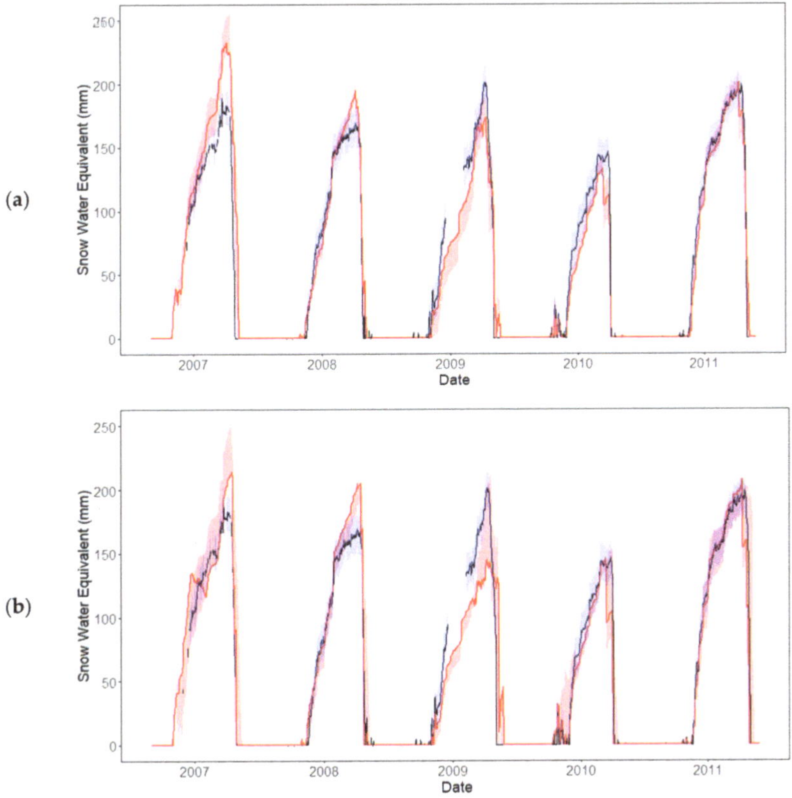

Figure A2. Modeled *SWE* series at the Necopastic station for the (**a**) monolayer (Mo) and (**b**) multilayer (Multi) models. The red shaded interval shows the range of values provided by the top ten sets of parameters values. The observed *SWE* time series is shown in black, while the blue interval depicts the measurement uncertainty.

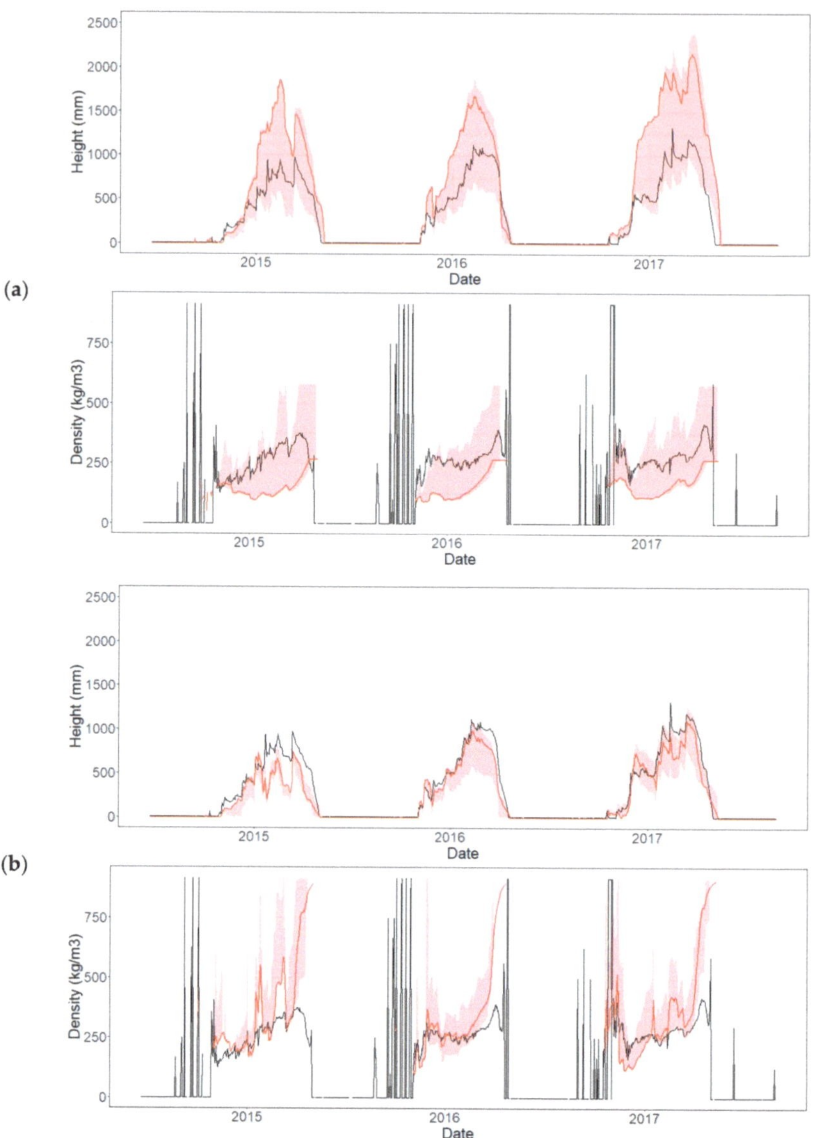

Figure A3. Modeled height and density series at the Lower Fantail station (LF) for the (**a**) monolayer (Mo) and (**b**) multilayer (Multi) models. The red shaded interval shows the range of values provided by the top ten sets of parameters values. The observed height and density time series is shown in black.

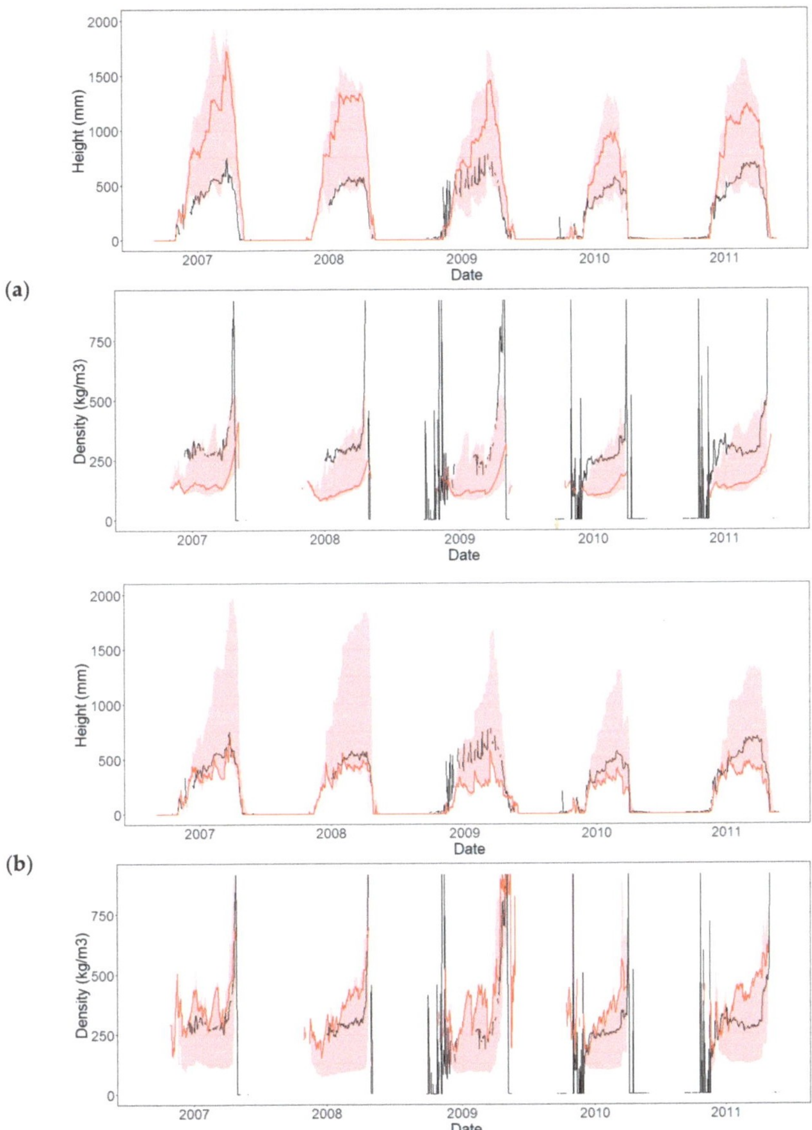

Figure A4. Modeled height and density series at the Necopastic station for the (**a**) monolayer (Mo) and (**b**) multilayer (Multi) models. The red shaded interval shows the range of values provided by the top ten sets of parameters values. The observed height time and density series is shown in black.

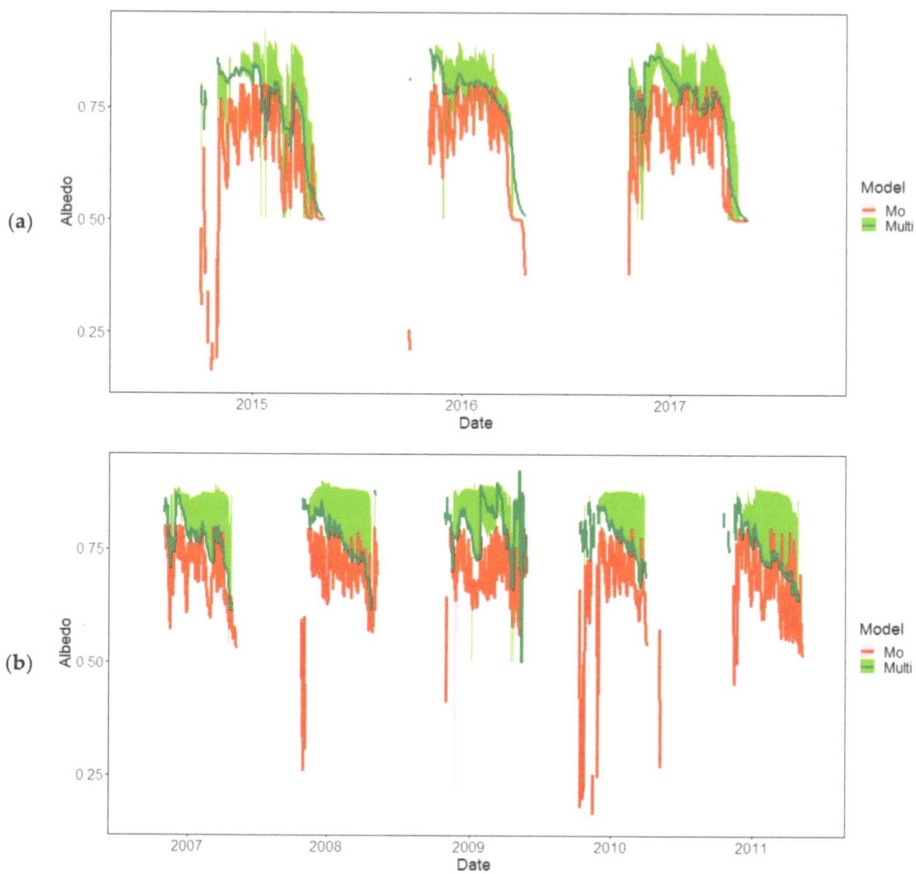

Figure A5. Albedo time series modeled by the top ten best sets of parameter values for the Mo model (pink envelop) and the Multi model (green envelop) for the (**a**) Lower Fantail and (**b**) Necopastic GMON stations. The best parameter sets are depicted by the red and green lines for the Mo and Multi models, respectively.

References

1. Adhikari, A.; Liu, C.; Kulie, M.S. Global Distribution of Snow Precipitation Features and Their Properties from 3 Years of GPM Observations. *J. Clim.* **2018**, *31*, 3731–3754. [CrossRef]
2. Mukhopadhyay, B.; Khan, A. A Reevaluation of the Snowmelt and Glacial Melt in River Flows within Upper Indus Basin and Its Significance in a Changing Climate. *J. Hydrol.* **2015**, *527*, 119–132. [CrossRef]
3. Jenicek, M.; Ledvinka, O. Importance of Snowmelt Contribution to Seasonal Runoff and Summer Low Flows in Czechia. *Hydrol. Earth Syst. Sci.* **2020**, *24*, 3475–3491. [CrossRef]
4. Jasechko, S.; Wassenaar, L.I.; Mayer, B. Isotopic Evidence for Widespread Cold-season-biased Groundwater Recharge and Young Streamflow across Central Canada. *Hydrol. Process.* **2017**, *31*, 2196–2209. [CrossRef]
5. Mernild, S.H.; Liston, G.E.; Hiemstra, C.A.; Malmros, J.K.; Yde, J.C.; McPhee, J. The Andes Cordillera. Part I: Snow Distribution, Properties, and Trends (1979–2014). *Int. J. Climatol.* **2017**, *37*, 1680–1698. [CrossRef]
6. Nouri, M.; Homaee, M. Spatiotemporal Changes of Snow Metrics in Mountainous Data-Scarce Areas Using Reanalyses. *J. Hydrol.* **2021**, *603*, 126858. [CrossRef]
7. Dirmhirn, I.; Eaton, F.D. Some Characteristics of the Albedo of Snow. *J. Appl. Meteorol. Climatol.* **1975**, *14*, 375–379. [CrossRef]
8. Calleja, J.F.; Muniz, R.; Fernandez, S.; Corbea-Perez, A.; Peon, J.; Otero, J.; Navarro, F. Snow Albedo Seasonal Decay and Its Relation with Shortwave Radiation, Surface Temperature and Topography Over an Antarctic Ice Cap. *IEEE J. Sel. Top. Appl. Earth Obs. Remote Sens.* **2021**, *14*, 2162–2172. [CrossRef]

9. Fortin, J.-P.; Turcotte, R.; Massicotte, S.; Moussa, R.; Fitzback, J.; Villeneuve, J.-P. Distributed Watershed Model Compatible with Remote Sensing and GIS Data. I: Description of Model. *J. Hydrol. Eng.* **2001**, *6*, 91–99. [CrossRef]

10. Turcotte, R.; Rousseau, A.N.; Fortin, J.-P.; Villeneuve, J.-P. A Process-Oriented, Multiple-Objective Calibration Strategy Accounting for Model Structure. In *Water Science and Application*; Duan, Q., Gupta, H.V., Sorooshian, S., Rousseau, A.N., Turcotte, R., Eds.; American Geophysical Union: Washington, DC, USA, 2003; Volume 6, pp. 153–163, ISBN 978-0-87590-355-2.

11. Turcotte, R.; Fortier Filion, T.-C.; Lacombe, P.; Fortin, V.; Roy, A.; Royer, A. Simulation Hydrologique Des Derniers Jours de La Crue de Printemps: Le Problème de La Neige Manquante. *Hydrol. Sci. J.* **2010**, *55*, 872–882. [CrossRef]

12. Turcotte, R.; Lacombe, P.; Dimnik, C.; Villeneuve, J.-P. Prévision Hydrologique Distribuée Pour La Gestion Des Barrages Publics Du Québec. *Can. J. Civ. Eng.* **2004**, *31*, 308–320. [CrossRef]

13. Samuel, J.; Rousseau, A.N.; Abbasnezhadi, K.; Savary, S. Development and Evaluation of a Hydrologic Data-Assimilation Scheme for Short-Range Flow and Inflow Forecasts in a Data-Sparse High-Latitude Region Using a Distributed Model and Ensemble Kalman Filtering. *Adv. Water Resour.* **2019**, *130*, 198–220. [CrossRef]

14. Rousseau, A.N.; Savary, S.; Tremblay, S.; Caillouet, L.; Doumbia, C.; Augas, J.; Foulon, É.; Abbasnezhadi, K. *A Distributed Hydrological Modelling System to Support Hydroelectric Production in Northern Environment under Current and Changing Climate Conditions*; INRS-ETE: Québec, QC, Canada, 2020.

15. Abbasnezhadi, K.; Rousseau, A.N.; Foulon, É.; Savary, S. Verification of Regional Deterministic Precipitation Analysis Products Using Snow Data Assimilation for Application in Meteorological Network Assessment in Sparsely Gauged Nordic Basins. *J. Hydrometeorol.* **2021**, *22*, 859–876. [CrossRef]

16. Lachance-Cloutier, S.; Turcotte, R.; Ricard, S. *Atlas Hydroclimatique du Québec Méridional: Impact des Changements Climatiques sur les Régimes de Crue, D'étiage et D'hydraulicité à L'horizon 2050*; Centre d'expertise hydrique du Québec: Québec, QC, Canada, 2013; 51p, Available online: https://www.researchgate.net/publication/267451351_Atlas_hydroclimatique_du_Quebec_meridional (accessed on 16 April 2022).

17. Centre d'expertise hydrique du Québec. *Atlas Hydroclimatique du Québec Méridional: Impact des Changements Climatiques sur les Régimes de Crue, D'étiage et D'hydraulicité à L'horizon 2050*; Centre d'expertise hydrique du Québec: Québec, QC, Canada, 2015; ISBN 978-2-550-72964-8.

18. Direction de l'expertise hydrique. *Document D'accompagnement de L'atlas Hydroclimatique*; Ministère du Développement durable, de l'Environnement et de la Lutte contre les changements climatiques; Direction de l'expertise hydrique: Québec, QC, Canada, 2018; ISBN 978-2-550-82571-5.

19. Berthot, L.; St-Hilaire, A.; Caissie, D.; El-Jabi, N.; Kirby, J.; Ouellet-Proulx, S. Environmental Flow Assessment in the Context of Climate Change: A Case Study in Southern Quebec (Canada). *J. Water Clim. Chang.* **2021**, *12*, 3617–3633. [CrossRef]

20. Poulin, A.; Brissette, F.; Leconte, R.; Arsenault, R.; Malo, J.-S. Uncertainty of Hydrological Modelling in Climate Change Impact Studies in a Canadian, Snow-Dominated River Basin. *J. Hydrol.* **2011**, *409*, 626–636. [CrossRef]

21. Quilbé, R.; Rousseau, A.N.; Moquet, J.-S.; Trinh, N.B.; Dibike, Y.; Gachon, P.; Chaumont, D. Assessing the Effect of Climate Change on River Flow Using General Circulation Models and Hydrological Modelling—Application to the Chaudière River, Québec, Canada. *Can. Water Resour. J.* **2008**, *33*, 73–94. [CrossRef]

22. Blanchette, M.; Rousseau, A.N.; Foulon, É.; Savary, S.; Poulin, M. What Would Have Been the Impacts of Wetlands on Low Flow Support and High Flow Attenuation under Steady State Land Cover Conditions? *J. Environ. Manag.* **2019**, *234*, 448–457. [CrossRef] [PubMed]

23. Rousseau, A.N.; Savary, S.; Bazinet, M.-L. *Flood Water Storage Using Active and Passive Approaches—Assessing Flood Control Attributes of Wetlands and Riparian Agricultural Land in the Lake Champlain-Richelieu River Watershed. A Report to the International Lake Champlain—Richelieu River Study Board*; INRS—Centre Eau Terre Environnement: Québec, QC, Canada, 2022.

24. Wu, Y.; Zhang, G.; Rousseau, A.N.; Xu, Y.J. Quantifying Streamflow Regulation Services of Wetlands with an Emphasis on Quickflow and Baseflow Responses in the Upper Nenjiang River Basin, Northeast China. *J. Hydrol.* **2020**, *583*, 124565. [CrossRef]

25. Sui, J.; Koehler, G. Rain-on-Snow Induced Flood Events in Southern Germany. *J. Hydrol.* **2001**, *252*, 205–220. [CrossRef]

26. Paquotte, A.; Baraer, M. Hydrological Behavior of an Ice-layered Snowpack in a Non-mountainous Environment. *Hydrol. Process.* **2022**, *36*, e14433. [CrossRef]

27. Mohammed, A.A.; Pavlovskii, I.; Cey, E.E.; Hayashi, M. Effects of Preferential Flow on Snowmelt Partitioning and Groundwater Recharge in Frozen Soils. *Hydrol. Earth Syst. Sci.* **2019**, *23*, 5017–5031. [CrossRef]

28. Turcotte, R.; Fortin, L.G.; Fortin, V.; Fortin, J.P.; Villeneuve, J.-P. Operational Analysis of the Spatial Distribution and the Temporal Evolution of the Snowpack Water Equivalent in Southern Québec, Canada. *Hydrol. Res.* **2007**, *38*, 211–234. [CrossRef]

29. Saha, S.K.; Sujith, K.; Pokhrel, S.; Chaudhari, H.S.; Hazra, A. Effects of Multilayer Snow Scheme on the Simulation of Snow: Offline Noah and Coupled with NCEPCFSv2. *J. Adv. Model. Earth Syst.* **2017**, *9*, 271–290. [CrossRef]

30. Domine, F.; Barrere, M.; Sarrazin, D. Seasonal Evolution of the Effective Thermal Conductivity of the Snow and the Soil in High Arctic Herb Tundra at Bylot Island, Canada. *Cryosphere* **2016**, *10*, 2573–2588. [CrossRef]

31. Bartelt, P.; Lehning, M. A Physical SNOWPACK Model for the Swiss Avalanche Warning Part I: Numerical Model. *Cold Reg. Sci. Technol.* **2002**, *35*, 123–145. [CrossRef]

32. Zanotti, F.; Endrizzi, S.; Bertoldi, G.; Rigon, R. The GEOTOP Snow Module. *Hydrol. Process.* **2004**, *18*, 3667–3679. [CrossRef]

33. Saloranta, T.M. Simulating Snow Maps for Norway: Description and Statistical Evaluation of the seNorge Snow Model. *Cryosphere* **2012**, *6*, 1323–1337. [CrossRef]

34. Henson, W.; Stewart, R.; Kochtubajda, B. On the Precipitation and Related Features of the 1998 Ice Storm in the Montréal Area. *Atmos. Res.* **2007**, *83*, 36–54. [CrossRef]

35. Quéno, L.; Vionnet, V.; Cabot, F.; Vrécourt, D.; Dombrowski-Etchevers, I. Forecasting and Modelling Ice Layer Formation on the Snowpack Due to Freezing Precipitation in the Pyrenees. *Cold Reg. Sci. Technol.* **2018**, *146*, 19–31. [CrossRef]

36. Koch, F.; Henkel, P.; Appel, F.; Schmid, L.; Bach, H.; Lamm, M.; Prasch, M.; Schweizer, J.; Mauser, W. Retrieval of Snow Water Equivalent, Liquid Water Content, and Snow Height of Dry and Wet Snow by Combining GPS Signal Attenuation and Time Delay. *Water Resour. Res.* **2019**, *55*, 4465–4487. [CrossRef]

37. Evans, S. Dielectric Properties of Ice and Snow–a Review. *J. Glaciol.* **1965**, *5*, 773–792. [CrossRef]

38. Sakazume, S.; Seki, N. Thermal Properties of Ice and Snow at Low Temperature Region. *Bull. Jpn. Soc. Mech. Eng.* **1978**, *44*, 2059–2069.

39. Reid, R.C.; Prausnitz, J.M.; Poling, B.E. *The Properties of Gases and Liquids*, 4th ed.; McGraw-Hill: New York, NY, USA, 1987; ISBN 978-0-07-051799-8.

40. Perovich, D.K.; Grenfell, T.C.; Light, B.; Hobbs, P.V. Seasonal Evolution of the Albedo of Multiyear Arctic Sea Ice. *J. Geophys. Res.* **2002**, *107*, 8044. [CrossRef]

41. Hartmann, D.L. *Global Physical Climatology*; International geophysics; Academic Press: San Diego, CA, USA, 1994; ISBN 978-0-12-328530-0.

42. Mas, A.; Baraer, M.; Arsenault, R.; Poulin, A.; Préfontaine, J. Targeting High Robustness in Snowpack Modeling for Nordic Hydrological Applications in Limited Data Conditions. *J. Hydrol.* **2018**, *564*, 1008–1021. [CrossRef]

43. Matott, L.S. *OSTRICH—An Optimization Software Toolkit for Research Involving Computational Heuristics Documentation and User's Guide Version 17.12.19*. 79; University at Buffalo Center for Computational Research: New York, NY, USA, 2017; p. 79.

44. Bertsekas, D.P. *Constrained Optimization and Lagrange Multiplier Methods*; Elsevier: Amsterdam, The Netherlands, 2014.

45. Skahill, B.E.; Doherty, J. Efficient Accommodation of Local Minima in Watershed Model Calibration. *J. Hydrol.* **2006**, *329*, 122–139. [CrossRef]

46. Tolson, B.A.; Shoemaker, C.A. Dynamically Dimensioned Search Algorithm for Computationally Efficient Watershed Model Calibration. *Water Resour. Res.* **2007**, *43*, W01413. [CrossRef]

47. Duan, Q.; Sorooshian, S.; Gupta, V. Effective and Efficient Global Optimization for Conceptual Rainfall-Runoff Models. *Water Resour. Res.* **1992**, *28*, 1015–1031. [CrossRef]

48. Duan, Q.Y.; Gupta, V.K.; Sorooshian, S. Shuffled Complex Evolution Approach for Effective and Efficient Global Minimization. *J. Optim. Theory Appl.* **1993**, *76*, 501–521. [CrossRef]

49. Gupta, H.V.; Kling, H.; Yilmaz, K.K.; Martinez, G.F. Decomposition of the Mean Squared Error and NSE Performance Criteria: Implications for Improving Hydrological Modelling. *J. Hydrol.* **2009**, *377*, 80–91. [CrossRef]

50. Razavi, S.; Sheikholeslami, R.; Gupta, H.V.; Haghnegahdar, A. VARS-TOOL: A Toolbox for Comprehensive, Efficient, and Robust Sensitivity and Uncertainty Analysis. *Environ. Model. Softw.* **2019**, *112*, 95–107. [CrossRef]

51. Razavi, S.; Gupta, H.V. A New Framework for Comprehensive, Robust, and Efficient Global Sensitivity Analysis: 2. Application. *Water Resour. Res.* **2016**, *52*, 440–455. [CrossRef]

52. Razavi, S.; Gupta, H.V. A New Framework for Comprehensive, Robust, and Efficient Global Sensitivity Analysis: 1. Theory. *Water Resour. Res.* **2016**, *52*, 423–439. [CrossRef]

53. Nash, J.E.; Sutcliffe, J.V. River Flow Forecasting through Conceptual Models Part I—A Discussion of Principles. *J. Hydrol.* **1970**, *10*, 282–290. [CrossRef]

54. Oreiller, M.; Nadeau, D.F.; Minville, M.; Rousseau, A.N. Modelling Snow Water Equivalent and Spring Runoff in a Boreal Watershed, James Bay, Canada. *Hydrol. Process.* **2014**, *28*, 5991–6005. [CrossRef]

55. Samuel, J.; Kavanaugh, J.; Benkert, B.; Samolczyck, M.; Laxton, S.; Evans, R.; Saal, S.; Gonet, J.; Horton, B.; Clague, J.; et al. *Evaluating Climate Change Impacts on the Upper Yukon River Basin: Projecting Future Conditions Using Glacier, Climate and Hydrological Models*; Northern Climate ExChange, Yukon Research Centre: Whitehorse, YT, Canada, 2016.

56. Strong, W.L. Ecoclimatic Zonation of Yukon (Canada) and Ecoclinal Variation in Vegetation. *Arctic* **2013**, *66*, 52–67. [CrossRef]

57. Choquette, Y.; Lavigne, P.; Nadeau, M. GMON, a New Sensor for Snow Water Equivalent via Gamma Monitoring. In Proceedings of the Whistler 2008 International Snow Science Workshop, Whistler, BC, Canada, 21–27 September 2008; p. 802.

58. Campbell Scientific Instruction Manual: GMON3 Snow Water Equivalency Sensor. Available online: https://s.campbellsci.com/documents/es/manuals/gmon3.pdf (accessed on 16 April 2022).

59. Campbell Scientific Product Manual: CS275 Snow Water Equivalency Sensor. Available online: https://s.campbellsci.com/documents/us/manuals/cs725.pdf (accessed on 16 April 2022).

60. Dahe, Q.; Young, N.W. Characteristics of the Initial Densification of Snow/Firn in Wilkes Land, East Antarctica (Abstract). *Ann. Glaciol.* **1988**, *11*, 209. [CrossRef]

61. Nishimura, H.; Maeno, N.; Satow, K. Initial Stage of Densification of Snow in Mizuho Plateau, Antarctica. *Mem. Natl. Inst. Polar Res.* **1983**, *29*, 149–158.

62. Würzer, S.; Jonas, T.; Wever, N.; Lehning, M. Influence of Initial Snowpack Properties on Runoff Formation during Rain-on-Snow Events. *J. Hydrometeorol.* **2016**, *17*, 1801–1815. [CrossRef]

63. Lackner, G.; Domine, F.; Nadeau, D.F.; Parent, A.-C.; Anctil, F.; Lafaysse, M.; Dumont, M. On the Energy Budget of a Low-Arctic Snowpack. *Cryosphere* **2022**, *16*, 127–142. [CrossRef]

64. Keenan, E.; Wever, N.; Dattler, M.; Lenaerts, J.T.M.; Medley, B.; Kuipers Munneke, P.; Reijmer, C. Physics-Based SNOWPACK Model Improves Representation of near-Surface Antarctic Snow and Firn Density. *Cryosphere* **2021**, *15*, 1065–1085. [CrossRef]
65. Cuffey, K.; Paterson, W.S.B. *The Physics of Glaciers*, 4th ed.; Butterworth-Heinemann/Elsevier: Burlington, MA, USA, 2010; ISBN 978-0-12-369461-4.
66. Gray, D.M.; Landine, P.G. Albedo Model for Shallow Prairie Snow Covers. *Can. J. Earth Sci.* **1987**, *24*, 1760–1768. [CrossRef]
67. Stroeve, J.C.; Box, J.E.; Haran, T. Evaluation of the MODIS (MOD10A1) Daily Snow Albedo Product over the Greenland Ice Sheet. *Remote Sens. Environ.* **2006**, *105*, 155–171. [CrossRef]
68. Cheng, C.S.; Li, G.; Auld, H. Possible Impacts of Climate Change on Freezing Rain Using Downscaled Future Climate Scenarios: Updated for Eastern Canada. *Atmos.-Ocean* **2011**, *49*, 8–21. [CrossRef]
69. Vickers, H.M.S.; Mooney, P.; Malnes, E.; Lee, H. Comparing Rain-on-Snow Representation across Different Observational Methods and a Regional Climate Model. The Cryosphere Discuss. 2022. Available online: https://tc.copernicus.org/preprints/tc-2022-57/ (accessed on 16 April 2022).
70. Hotovy, O.; Jenicek, M. *Changes in the Frequency and Extremity of Rain-on-Snow Events in the Warming Climate*; EGU General Assembly: Vienna, Austria, 2022.
71. Bouchard, B.; Nadeau, D.F.; Domine, F. Comparison of Snowpack Structure in Gaps and under the Canopy in a Humid Boreal Forest. *Hydrol. Process.* **2022**, *36*, e14681. [CrossRef]
72. Yang, T.; Li, Q.; Hamdi, R.; Chen, X.; Zou, Q.; Cui, F.; De Maeyer, P.; Li, L. Trends and Spatial Variations of Rain-on-Snow Events over the High Mountain Asia. *J. Hydrol.* **2022**, *614*, 128593. [CrossRef]
73. Hotovy, O.; Nedelcev, O.; Jenicek, M. Changes in Rain-on-Snow Events in Mountain Catchments in the Rain–Snow Transition Zone. *Hydrol. Sci. J.* **2023**, *68*, 572–584. [CrossRef]
74. Valery, A. Modélisation Precipitations—Débit sous Influence Nivale. Elaboration d'un Module Neige et Évaluation sur 380 Bassins Versants. Ph.D. Thesis (Hydrobiologie), Cemagref, Antony, France, AgroParisTech, Paris, France, 2010.
75. Bergström, S. *Development and Application of a Conceptual Runoff Model for Scandinavian Catchments*; Reports RHO 7; SMHI: Norrköping, Sweden, 1976; Volume 134.
76. Girons Lopez, M.; Vis, M.J.P.; Jenicek, M.; Griessinger, N.; Seibert, J. Complexity and Performance of Temperature-Based Snow Routinesfor Runoff Modelling in Mountainous Areas in Central Europe. *Hydrol. Earth Syst. Sci.* **2020**, *24*, 4441–4461. [CrossRef]
77. Lindström, G.; Johansson, B.; Persson, M.; Gardelin, M.; Bergström, S. Development and Test of the Distributed HBV-96 Hydrological Model. *J. Hydrol.* **1997**, *201*, 272–288. [CrossRef]
78. Pradhanang, S.M.; Anandhi, A.; Mukundan, R.; Zion, M.S.; Pierson, D.C.; Schneiderman, E.M.; Matonse, A.; Frei, A. Application of SWAT Model to Assess Snowpack Development and Streamflow in the Cannonsville Watershed, New York, USA. *Hydrol. Process.* **2011**, *25*, 3268–3277. [CrossRef]
79. Tuo, Y.; Marcolini, G.; Disse, M.; Chiogna, G. Calibration of Snow Parameters in SWAT: Comparison of Three Approaches in the Upper Adige River Basin (Italy). *Hydrol. Sci. J.* **2018**, *63*, 657–678. [CrossRef]
80. Andreadis, K.M.; Storck, P.; Lettenmaier, D.P. Modeling Snow Accumulation and Ablation Processes in Forested Environments. *Water Resour. Res.* **2009**, *45*, 1–13. [CrossRef]
81. Brun, E.; Martin, E.; Simon, V.; Gendre, C.; Coleou, C. An Energy and Mass Model of Snow Cover Suitable for Operational Avalanche Forecasting. *J. Glaciol.* **1989**, *35*, 333–342. [CrossRef]
82. Brun, E.; David, P.; Sudul, M.; Brunot, G. A Numerical Model to Simulate Snow-Cover Stratigraphy for Operational Avalanche Forecasting. *J. Glaciol.* **1992**, *38*, 13–22. [CrossRef]
83. Liston, G.E.; Hall, D.K. An Energy-Balance Model of Lake-Ice Evolution. *J. Glaciol.* **1995**, *41*, 373–382. [CrossRef]
84. Liston, G.E.; Mernild, S.H. Greenland Freshwater Runoff. Part I: A Runoff Routing Model for Glaciated and Nonglaciated Landscapes (HydroFlow). *J. Clim.* **2012**, *25*, 5997–6014. [CrossRef]
85. Lehning, M.; Bartelt, P.; Brown, B.; Fierz, C.; Satyawali, P. A Physical SNOWPACK Model for the Swiss Avalanche Warning: Part II. Snow Microstructure. *Cold Reg. Sci. Technol.* **2002**, *35*, 147–167. [CrossRef]
86. Lehning, M.; Bartelt, P.; Brown, B.; Fierz, C. A Physical SNOWPACK Model for the Swiss Avalanche Warning: Part III: Meteorological Forcing, Thin Layer Formation and Evaluation. *Cold Reg. Sci. Technol.* **2002**, *35*, 169–184. [CrossRef]
87. Franck, E.C.; Lee, R. *Potential Solar Beam Irradiation on Slopes: Tables for 300 to 500 Latitude*; US Rocky Mountain Forest and Range Experiment Station: Fort Collins, CO, USA, 1966; Volume 18.
88. Rousseau, A.N.; Savary, S.; Tremblay, S. *Développement de PHYSITEL 64 Bits Avec Interface Graphique Pour Supporter les Applications d'HYDROTEL sur des Bassins Versants de Grande Envergure: (Incluant des Compléments D'aide et des Développements pour HYDROTEL): Travaux 2016. Rapport Final*; (INRS Centre Eau Terre Environnement Documents scientifiques et techniques; R1724); Institut National de la Recherche Scientifique—Centre Eau Terre Environnement: Québec, QC, Canada, 2017; ISBN 978-2-89146-878-7.

Article

Study of the River Discharge Alteration

Alina Bărbulescu [1,*] and **Nayeemuddin Mohammed** [1,2]

[1] Faculty of Civil Engineering, Transilvania University of Brașov, 900152 Brașov, Romania; mdnayeem0811@gmail.com

[2] Faculty of Civil Engineering, Bauhaus-Universität Weimar, 99423 Weimar, Germany

* Correspondence: alina.barbulescu@unitbv.ro

Abstract: This article aims to analyze the alteration in water discharge due to the building of one of the largest dams in Romania. Modifications in the hydrological patterns of the studied river were emphasized by a complex technique that includes decomposition models of the series into trends, seasonal indices, and random components, as well as into Intrinsic Mode Functions (IMFs). The Mann–Kendall trend test indicates the existence of different positive slopes for the subseries S1 and S2 (before and after the inception of the Siriu dam, respectively) built from the raw series, S. The stationarity hypothesis was rejected for all series. The multifractal analysis shows two different patterns of the data series. After decomposing the subseries S1 and S2, it resulted that the seasonality indices are not the same. Moreover, the seasonal variations decreased after building the dam. Empirical Mode Decomposition (EMD) unveils different short- and long-term patterns of the series before and after building the dam, concluding that there is a significant alteration in the river discharge after the dam's inception.

Keywords: monthly water discharge series; decomposition model; seasonal components; empirical mode decomposition

Citation: Bărbulescu, A.; Mohammed, N. Study of the River Discharge Alteration. *Water* **2024**, *16*, 808. https://doi.org/10.3390/w16060808

Academic Editors: Agnieszka Rutkowska and Katarzyna Baran-Gurgul

Received: 21 January 2024
Revised: 5 March 2024
Accepted: 7 March 2024
Published: 8 March 2024

1. Introduction

River systems are lifelines that sustain ecological balance, support civilizations, and reflect the complex interplay between natural processes and human interventions. Ancient civilizations thrived along riverbanks, leveraging the waterways for sustenance, irrigation, and transportation. The river discharge dynamic is influenced by diverse factors, from climatic variations to anthropogenic alterations [1]. Infrastructural developments like dam-building can profoundly impact the natural river flows [2–4]. When dams interfere with flow regimes, it often results in significant environmental damage and biodiversity deterioration [5–7]. China's Yangtze River's Three Gorges Dam stands out as a testament to this influence, causing extensive alterations in river flow patterns. These changes have ripple effects, challenging ecosystems and posing notable dilemmas for local communities reliant on the river [8]. Significant river flow alterations have trapped sediments, modified the natural flow pattern, and disrupted the nutrient balances, affecting delta, estuarine, and marine ecosystems [9–11]. These ecological concerns emphasize the need for a balanced approach to infrastructural developments [12].

Despite the extensive studies on the dam's impact on river flow and the environment, some gaps persist. One concerns the study of the cumulative impacts of smaller dams [13]. Although large dams have been the target of most research, the role of smaller dams has been emphasized recently [14]. Unlike mega-projects like the Three Gorges Dam which received extensive coverage, smaller dam projects need to be scrutinized more, leaving potential knowledge unknown [15]. Even if small dams might appear insignificant individually, their collective impact can be at least as potent as bigger dams, especially when it comes to flows that matter ecologically [14,15]. While the scientific literature contains numerous studies on this subject, a comprehensive critique is essential. There is a pressing need to examine how much dam constructions deviate river flows from their natural state, preserving intact ecological functions [16–18].

In pursuing a more holistic understanding, the research community has seen a convergence of conventional wisdom and innovative techniques for analyzing the rivers' discharge. Given the urgent global imperative for sustainable water resource management, these research endeavors take on heightened significance [14]. As time evolved, the methodologies to study these dynamics expanded, including statistical analysis [19–22], wavelets decomposition [23], artificial intelligence methods—neural networks [24–26], support vector regression [27], time series models [28], hydrological simulation [29], etc.

IHA (Indicator of Hydrologic Alteration) [30] represents another tool to assess the hydrological alteration that is widely used by researchers [31–38]. The IHA software [30] can compute 33 flow statistical parameters divided into three classes—low, medium, and high—containing values less than or equal to the 33rd percentile, between the 33rd and the 67th percentiles, and above the 67th percentile, respectively. The software allows for the determination of the frequency of each annual post-impact value belonging to each category [37]. One main drawback is that many of these indices are intercorrelated [37], so the question is how many indicators are necessary to describe the river flow alteration. Other shortcomings are that no IHA directly quantifies the amplitude of high flow conditions, and no seasonality indices are provided. The latter are essential for understanding the seasons with high flows corresponding to possible floodings.

Within this complex realm, our research combines different approaches to cross-validate the results of the hydrological alteration in the Buzău River flow after operating the Siriu dam, Romania's second biggest accumulation lake. Very few articles [38–40] have approached the impact of building the Siriu dam on the river flow, one of them using IHA [38] and the other [39] using statistical methods to test the existence of specific trends in seasonal series (winter, spring, summer, and autumn) before and after the dam's inception. Another attempt was made by modeling the daily mean flow series using general regression neural networks (GRNNs) [40]. None of these studies have provided a complete analysis given that (1) assessing the tendency by the Mann–Kendall test followed by utilizing Sen's slope (which computes a linear trend) [39] cannot capture the complex behavior of the data series (such as seasonality), (2) the linear parametric regression [38,40] was not satisfactory from the viewpoint of the extracted information, and (3) the GRNN model [40] was not accurate enough. In the GRNN, the correlation between the actual and predicted values on the test set was not very high, and the time taken to run the algorithm was very long (a few hours).

Our approach complements these procedures, providing seasonality indicators and emphasizing the flow pattern at short and long scales through Empirical Mode Decomposition (EMD), which offers granularity by breaking down intricate oscillatory patterns in river data [32]. The procedures utilized in the present study are easy to use and do not involve the computation and comparison of a high number of indicators, which can sometimes lead to confusing interpretations (as in IHA methods). Moreover, the time necessary to run the algorithms is extremely low (a few seconds) compared to the GRNN model.

The novelty of this research is that it introduces a unique framework that amalgamates the strengths of multifractal analysis with time series decomposition and EMD. The first one emphasizes the existence of periods with different behaviors of monthly river discharge series, the second provides the seasonality factors, and the last underscores the differences between the river's short- and long-term variations before and after the dam's construction. This combination gives a more nuanced view of the dam's impact on Buzău discharge dynamics and clarifies the extent of the streamflow alteration.

2. Materials and Methods

2.1. Study Area

The Buzău River's catchment, with a surface of 5264 km^2 and an average elevation of 1043 m (Figure 1), is situated in the Romanian Curvature Carpathians. The climate of the study area is temperate–continental. From 1955 to 2010, the average (minimum) temperature was 6 °C (1 °C), the multiannual mean precipitation varied between 500 mm and 1000 mm [41], and the mean monthly precipitation (%) varied from 4.4 (in January) to 15 (in June and July).

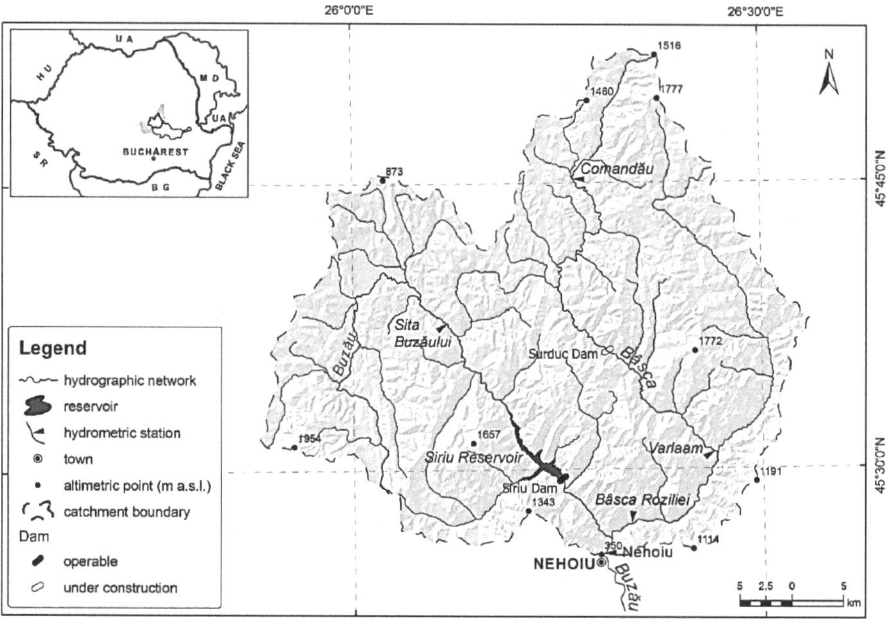

Figure 1. The study area [38].

The sub-basin catchment from where the River's water is drained at Nehoiu is 1567 km^2. The average river discharge ranges from 0.76 to 5000 m^3/s, while the specific and multiannual averages are 17 L/s.km^2 and 25.2 m^3/s [42]. The Siriu dam, which can store up to 125 million m^3 of water, was put in operation on 1 January 1984, on the Buzău River, upstream of Nehoiu. It drains about 56% of the water of this river and its tributaries [43] and supplies water to settlements and industrial plants downstream and for irrigating 50,000 ha [41].

Many catastrophic floods were recorded on the Buzău catchment after 1948, the biggest one in 1975, recording a maximum discharge of 2100 m^3/s. One of the reasons that the dam was built was to avoid or at least diminish the effects of such events. Still, it does not protect the settlements downstream from floods produced by the Buzău's tributaries (as in the case of flooding events in July 2004 and May 2005) [43]. In June–July 2010, a great flood affected many villages downstream of Nehoiu, damaging 14 pedestrian bridges and 1.2 km of water supplies and interrupting the communication between different villages.

Details about the geography and geomorphology of the Buzău River catchment and the hydrological constructions can be found in [41,44,45] and on the flooding events and damages in [43].

2.2. Data Series

The studied series is formed by the monthly mean flow of the Buzău River measured at the Nehoiu station (45°25′29″ latitude and 26°18′27″ longitude) during January 1955–December 2010 (Figure 2).

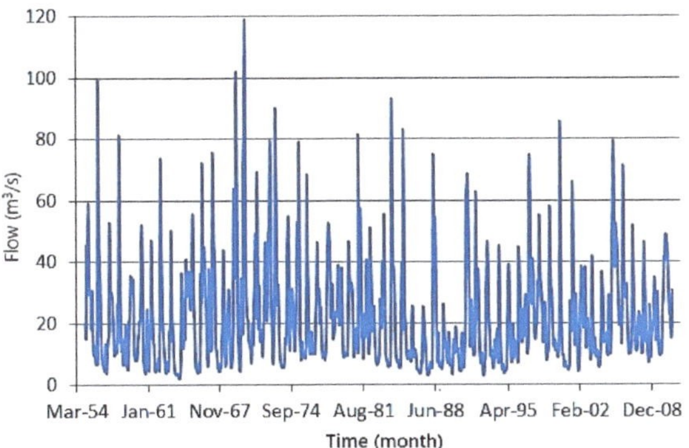

Figure 2. Monthly series of the Buzău River discharge.

The values of river discharge were measured daily at 7 a.m. and 7 p.m. and provided to the National Institute of Hydrology and Water Management (INGHA), where specialists checked them. The daily series were aggregated to obtain the monthly mean discharge. The dataset had no missing values.

To analyze the changes in water discharge after building the Siriu dam, the entire series, denoted by S, in the following, was split into two parts: S1—the series from January 1955 to December 1983 (before starting operating the Siriu dam), and S2—the series from January 1984 to December 2010 (after starting operating the Siriu dam). This split was performed to compare S1 and S2 and determine if the river discharge was altered due to the dam's operation.

The mean values for S, S1, and S2 are 21.83, 23.15, and 20.41 m^3/s, respectively. The maximum decreases significantly from 117.29—for S and S1—to 92.79 m^3/s for S2. The highest variance is noticed for S1 (347.58) and the lowest for S2 (259.14). There is no significant difference between the series skewness, but the kurtosis decreases from 3.93 (for S and S1) to 3.43 for S2. So, all distributions are right-skewed and leptokurtic. The outliers of S and S1 are in the same range, but those of S2 have values lower than 100 m^3/s, mostly under 70 m^3/s. The histograms and boxplots are shown in Figure 3.

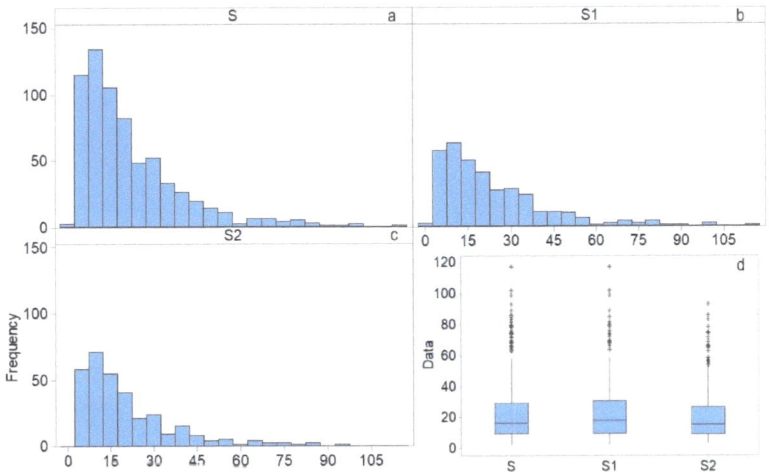

Figure 3. The histograms of S (**a**), S1 (**b**), and S2 (**c**). The boxplots of S, S1, and S2 (**d**).

2.3. The Study Stages

This study introduces a complex framework to clarify the extent of Buzău River water discharge modifications after building the Siriu dam. The study's flowchart is presented in Figure 4 and the detailed steps of the methodology are introduced below.

1. Perform the statistical analysis to determine if there are common futures (trend or stationary) of S, S1, and S2. This includes the following:

    - We rest the null hypothesis that there is no trend in the data series against the alternative of a monotonic trend existence using the Mann–Kendall (MK) [46] and seasonal Mann–Kendall (SMK) trend test [47]. Since there are n seasons, with m observations each, the null hypothesis in SMK is that observations are independent and identically distributed, and the alternative is that a monotonic trend is presented in the data series. First, a test statistic similar to the MK test's is built for each season. Then, the MK statistic for the seasonal test is obtained by summing the n statistics. The decision is made to reject the null as in the MK test [48]. When the null hypothesis is rejected, the trend is determined by the non-parametric Sen's method [49] as the median of the slopes of all of the pairs of ordinal time points.
    - We test the series stationarity by following the KPSS [50] procedure. The null hypothesis is the series level or trend stationarity; the alternative is its non-stationarity. Testing this hypothesis is important for building forecast models for the studied series.

2. Assess the existence of the different behaviors of the studied series on subintervals and emphasize their scaling characteristics through the multifractal analysis [51,52]. It was shown [47,48] that if a time series exhibits scaling properties, its behavior can be expressed in an exponential form (1), with the mass exponent $\tau(q)$, which can be estimated by fitting a linear regression of $\log Z_q(\lambda)$ vs. $\log\lambda$ ($\lambda > 0$) [53]. So,

$$Z_q(\lambda) \sim \lambda^{\tau(q)} \text{ for } \lambda \to 0, \tag{1}$$

where $Z_q(\lambda)$ is the partition function, whose values can be computed by covering the series chart with a certain number of boxes of size λ and summing up the probabilities of the appearance of a gray value in a box [54]. The multifractality can also be assessed by computing Renyi's dimension, $D(q)$ [55], whose relationship with $\tau(q)$ is [56] as follows:

$$D(q) = \tau(q)/(q-1). \tag{2}$$

The series presents multifractality if $D(q)$ decreases when q increases. When $D(q)$ has a constant value (equal to $D(0)$), the series is monofractal. An alternative way to characterize the multifractality is by using the $f(\alpha)$ spectrum, which is defined by the equation

$$f(\alpha(q)) = q\alpha(q) - \tau(q) \tag{3}$$

and can be computed by using a Legendre transform of $\tau(q)$ in the following form:

$$\alpha(q) = \frac{d\tau(q)}{dq} \tag{4}$$

where α is the Hölder exponent [57]. In the multifractality case, the $f(\alpha)$'s chart shape is single-humped. The steps in the MFDFA are [53] as follows:

- Compute the series cumulative sum, F.
- Split F into N_s subseries (each containing s values).
- Apply the least squared method for fitting an n-th order polynomial.
- Build the detrended F series by subtracting the polynomial values from the subseries' values.
- Build the fluctuation function, F_q, by taking the q-th root of the mean of the square functions from the previous stage.

- Fit the function $F_q(\mathrm{s}) \sim s^{h(q)}$, with $h(q)$ as the generalized Hurst exponent.

3. Perform the time series decomposition to analyze the changes in the seasonality factors before and after the dam's inauguration. The series (y_t) is decomposed into a trend (T_t), seasonality (S_t), and random noise component (ε_t) using an additive model or multiplicative model. The best model is selected based on the smallest mean standard error (MSE), mean absolute error (MAE), or mean absolute percentage error (MAPE). So, the additive decomposition model—ADM (the multiplicative decomposition model—MDM)—will be calculated as follows:

$$y_t = T_t + S_t + \varepsilon_t \quad (y_t = T_t \times S_t \times \varepsilon_t). \tag{5}$$

In this approach, the trend is computed by a moving average of the 12th order, and then, the deseasonalized series $(y_t - T_t)$ is determined. The row seasonality indices are computed as averages of the values of each month. These final seasonality indices are calculated by adjusting the raw ones to add up to zero. The random component (residual) is the difference between the detrended series and the series obtained by replicating the 12 seasonality indexes for each year. In the case of the MDM, the deseasonalized series is computed by y_t/T_t. The seasonality indices are obtained similarly to the ADM case, and the residual is obtained by dividing the deseasonalized series by the seasonality indices [58].

4. Perform EMD to determine the short and long-term variations in S1 and S2 and detect the differences in their patterns. EMD is an adaptive data analysis technique of non-linear and non-stationary time series aiming to decompose the series into a collection of oscillatory components called IMFs [59,60]. The importance of this technique is given by the following characteristics [60–62]:

- *Adaptability and Flexibility:* Unlike other decomposition methods, which often impose predetermined basis functions (like sine and cosine functions in Fourier analysis), EMD does not rely on any a priori basis. This means it can adapt to the nature of the data and make the decomposition more accurate and meaningful.
- *Local Characterization:* EMD provides a local representation of data. Each IMF captures oscillations occurring over a specific time scale, crucial in analyzing data where different periodic components might overlap or where transient features (like spikes or dips) are of interest.
- *Versatility:* While initially developed for time series analysis, EMD has shown remarkable performance in various fields, including climatology, biomedical signal analysis, and financial market research.
- *Handling Non-Linearity and Non-Stationarity:* Most traditional methods fail or require stringent preprocessing when dealing with non-linear or non-stationary data. EMD is inherently suited for such data types, providing a robust decomposition even in challenging conditions.

The steps of the EMD algorithm are [60–63] as follows:

(a) Initialization: with the entire dataset, *data(t)*, as the input, identify the sets of local maxima and minima and denote them by Max(*t*) and Min(*t*), respectively.

(b) Building the envelope through interpolation:

- Use a cubic spline interpolation or any suitable method to generate the upper envelope by connecting all of the Max(*t*) points.
- Similarly, connect all of the Min(*t*) points to create the lower envelope.

(c) Compute the mean envelope, *m(t)*, by using the following equation:

$$m(t) = (\text{upper envelope} + \text{lower envelope})/2. \tag{6}$$

(d) Extract the detail, *h(t)*, by using the following equation:

$$h(t) = data(t) - m(t). \tag{7}$$

(e) Verify the Intrinsic Mode Function (IMF) criteria:

- Check if $h(t)$ is an IMF, that is, by the following:
  - ✓ The number of extrema and zero-crossings must be equal or differ at most by one.
  - ✓ For any t, the mean value of the envelope defined by the local maxima (minima, respectively) is null.
- If $h(t)$ is an IMF, go to (f); otherwise repeat the procedure from (b) using $h(t)$ as input.

(f) Perform iterations and compute the residual:

- For an identified IMF, $h(t)$, subtract it from the original data and rename the resultant as the new data.
- Continue the new data-sifting process, extracting further IMFs.
- Continue the iteration until the residue becomes monotonic, showing no more IMFs can be extracted.

(g) Compile the results:

- Collate all extracted IMFs.
- Record the final residue left after all possible IMFs have been derived. It should be a monotonic function.

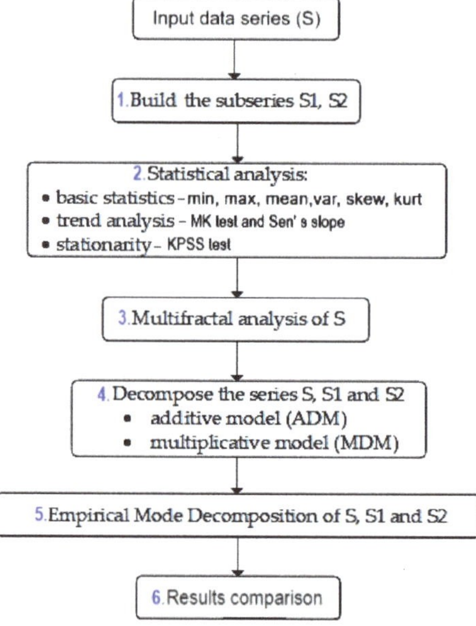

Figure 4. The study's flowchart.

This approach will underscore the differences in trend and seasonality of S1 and S2 and clarify the extent of the river flow alteration after operating the dam.

3. Results

3.1. Results of Statistical Analysis, Multifractal Analysis and Series Decomposition

Table 1 contains the results of the statistical tests on the data series performed at a significance level of 0.05. A p-value (computed in a test) less than 0.05 leads to the rejection of the respective null hypothesis.

Table 1. The *p*-values in the MK, SMK, and KPSS test, and Sen's slopes (marked with * inside the brackets).

Series	MK	SMK	KPSS Trend	KPSS Level
S	0.2851	0.2850	0.0571	0.1000
S1	0.0290 (0.0139 *)	0.0289 (0.0139 *)	0.0891	0.1000
S2	0.0000 (0.0311 *)	0.0000 (0.0311 *)	0.1000	0.6550

The MK and SMK tests could not reject the randomness hypothesis for S. They rejected it when applied to S1 and S2 (Table 1, column 2, row 3). The trend values, computed by Sen's method, are indicated inside the brackets and marked with * in Table 1. They are 0.0139 for S1 and 0.0311 for S2, so there is an increasing trend in the river discharge for both subseries, but not for the entire series. The result indicates that the trends of the subseries are different, indicating a different behavior of S1 and S2. Still, the slopes of the trends of S1 and S2 are very small, and putting together S1 and S2 (resulting in S) will not guarantee a significant trend for S. Indeed, the rejection of the null hypothesis for S shows that no significant trend was detected for it.

The stationarity hypothesis could not be rejected for all series. The results of the multifractal analysis are displayed in Figure 5.

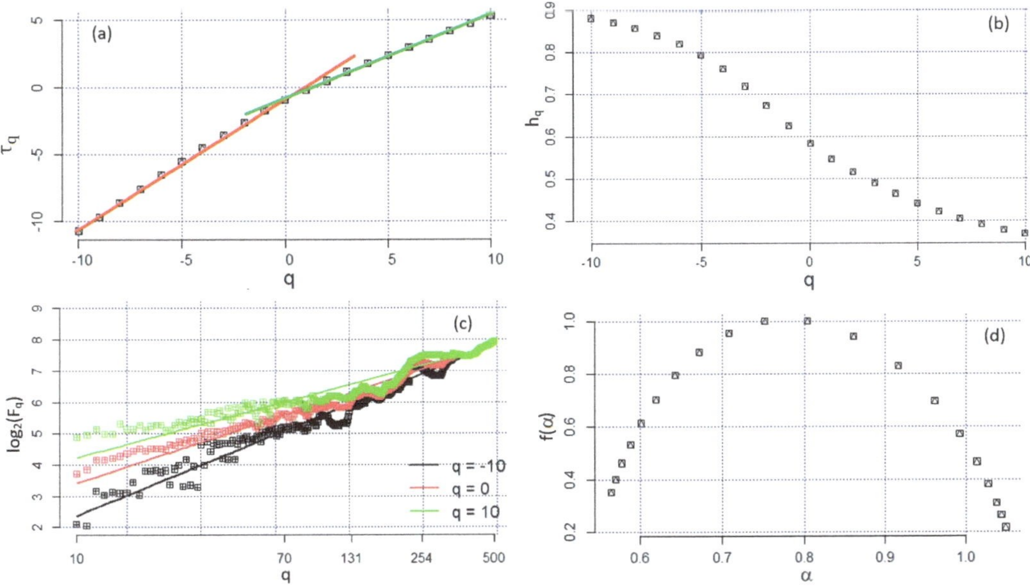

Figure 5. Results of the multifractal analysis: (**a**) the mass exponent; (**b**) the chart of the generalized Hurst exponent; (**c**) the segmentation function; (**d**) the $f(\alpha)$ spectrum.

The chart of the mass exponent (Figure 5a) presents two subseries for which two linear trend lines can be fitted—for $q \in [-10, 0]$ and for $q \in [0, 10]$. The slope change is at zero, so the series has a multifractal character. The shape of the generalized Hurst's exponent chart (h_q vs. q, Figure 5b) is a damped sine shape, with an inflection point at $q = 0$. In Figure 5c, one may notice deviations in the estimated segmentation function values from the linear trend (represented by straight lines in black, blue, and green, respectively). This variability is higher for $q = 10$, as the right-hand side of the chart shows the distribution of the function's values (represented by rectangles) with a higher slope than its values situated at the chart's right hand. The $f(\alpha)$ spectrum (Figure 5d) has a parabolic shape, indicating the series multifractality. The previous remark shows that the Buzău River's flow series has a multifractal character. We may assume that it is the effect of the pattern change in the rivers' discharge after building the dam.

The decomposition of the S series was performed to emphasize this change and find the seasonality components in particular. The elements in the ADM and MDM are presented in Figures 6 and 7.

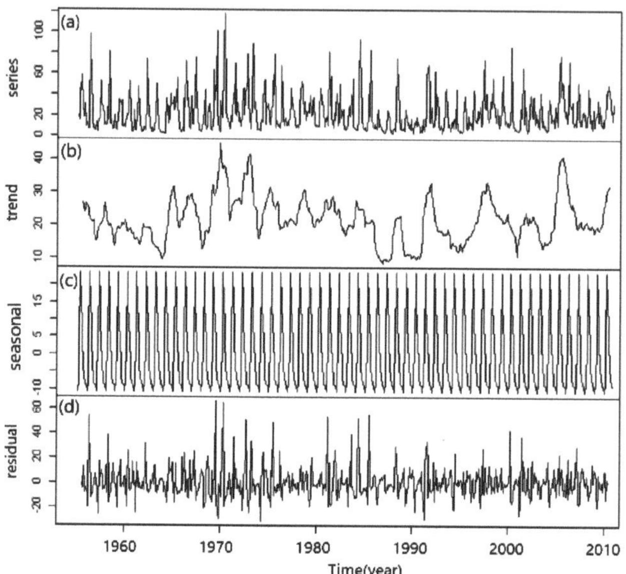

Figure 6. The ADM for S: (**a**) the initial series (series), (**b**) the trend obtained by the moving average of the 12-th order, (**c**) the seasonality indices, and (**d**) the residual.

The raw series and the trend are the same but the seasonal and the residual components are different. The values of the seasonality indices are presented in Table 2. In both cases, the highest indices were registered in April and May and the lowest in January. In the case of the MDM, all indices are positive and vary in the interval [0.4986, 2.1326]. In the ADM, seven indices are negative, only five are positive, and the variation interval is [−10.700, 23.5093].

Table 2. Seasonality indices in the ADM and MDM for the S series.

Index	January	February	March	April	May	June	July	August	September	October	November	December
ADM	−10.7800	−8.3037	3.4178	23.5093	15.5835	7.7747	4.6647	−3.2790	−6.2985	−8.6096	−8.8271	−8.8521
MDM	0.4986	0.6410	1.2014	2.1329	1.6923	1.3937	1.1742	0.834	0.6874	0.5811	0.5816	0.5820

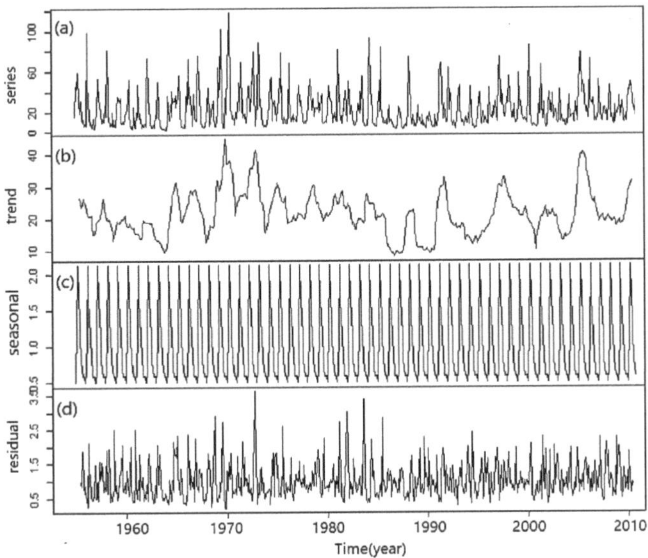

Figure 7. The MDM for S: (**a**) the initial series (series), (**b**) the trend obtained by the moving average of the 12-th order, (**c**) the seasonality indices, and (**d**) the residual.

The positive indices correspond to the spring and summer months (March–July), indicating a higher impact of the seasonal variations on the water flow (more precipitation, and thus a higher flow, is recorded in spring and the beginning of summer). The residuals comparison shows a smaller amplitude for the MDM compared to the ADM, with a lower autocorrelation order (Figure 8a,c) and lower skewness and kurtosis (Figure 8b,d), respectively: 1.4351 compared to 1.5316, and 2.7742 compared to 5.2396.

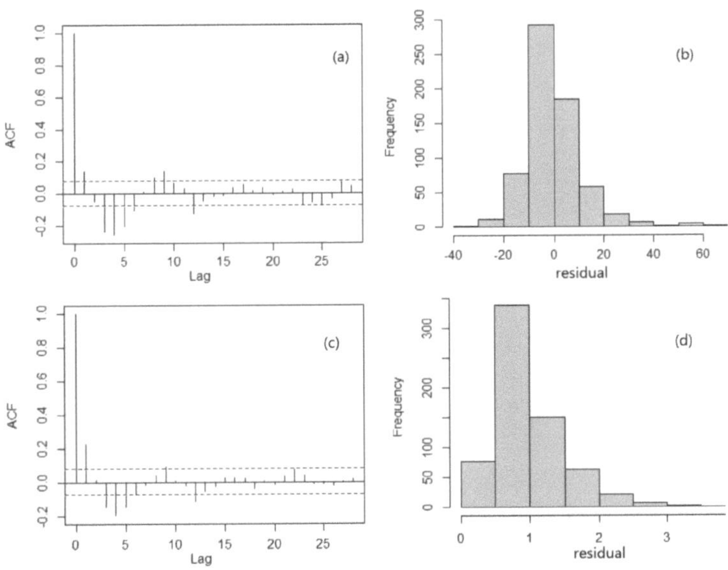

Figure 8. (**a**) The residual correlogram in the ADM, (**b**) the residual histogram in the ADM, (**c**) the residual correlogram in the MDM, and (**d**) the residual histogram in the MDM.

The residuals' MAE, MSE, and MAPE are 8.3273, 144.7643, and 50.5265 in the ADM and 0.9917, 1.2364, and 6.3449 in the MDM. Therefore, the best decomposition is provided by the second method. So, based on the MDM, similar seasonal variations are recorded in the last month of autumn and the first month of winter, being slightly higher in February and September and lower in January. The seasonal factor of 1.2012 in March might be explained by the snowmelt at the beginning of spring, and that from April to June is due to the high precipitation in spring.

The same analysis was performed for the subseries S1 and S2 to examine if there are significant differences between the seasonality factors. These indices in the MDMs are given in Table 3.

Table 3. Seasonality indices in the MDM for S1 and S2.

Series	January	February	March	April	May	June	July	August	September	October	November	December
S1	0.4751	0.6289	1.1361	2.1553	1.7975	1.3180	1.2582	0.8205	0.6734	0.5467	0.5985	0.5918
S2	0.5380	0.6739	1.3000	2.1215	1.5102	1.4778	1.0937	0.7758	0.7134	0.6280	0.5785	0.5893

The seasonality indices' pattern is similar to that in the MDM for S, with the highest values in April, May, June, and March or July and the lowest in January. The amplitude of the seasonality indices' decreased from $1.6902 = 2.1553 - 0.4751$ to $1.5835 = 2.1215 - 0.538$, indicating an attenuation of the extreme events.

3.2. Cross-Validation of the Results Using EMD

EMD was finally applied to cross-validate the above findings. A set of plots was created to represent and analyze the data effectively. The initial step was to display the original data over time as scattered points, with the amplitude of monthly averages depicted through a color map. Subsequent plots showcased each computed IMF, with enhanced subplots for improved readability. Cubic spline interpolation was utilized to visualize each IMF in greater detail. Furthermore, residuals were also visualized individually and in combination with the IMFs to provide an all-encompassing view.

In data analysis, accurate and clear visualization is essential. Therefore, two methods were used to draw the IMFs: fine indexing and cubic spline interpolation. Fine indexing is similar to zooming into an image to see details. It increases the number of examined points, allowing for a more detailed view of data trends. Cubic spline interpolation uses polynomial functions to connect these points, creating a smoother curve. It ensures that the curve passes through the data points and transitions smoothly between them.

These two techniques provide a more detailed and smooth visualization, making data trends more evident. With more data points from fine indexing and smoother curves from interpolation, subtle patterns in the data can be more easily identified. Without these methods, data visualizations can appear disjointed, and some subtler patterns might be overlooked or misinterpreted.

Figure 9 contains details from visualizing IMF 3 and IMF 4 for S using the abovementioned techniques. In the first case, the interpolated IMF 3 passes through the original points of IMF 3. In the second one, the original IMF 4 is formed by points on parallel lines, and the interpolated IMF 4 provides a smooth image of the original one. Upon decomposing the data with EMD, several IMFs are obtained. Figure 10 shows eight of them in EMD for the S series.

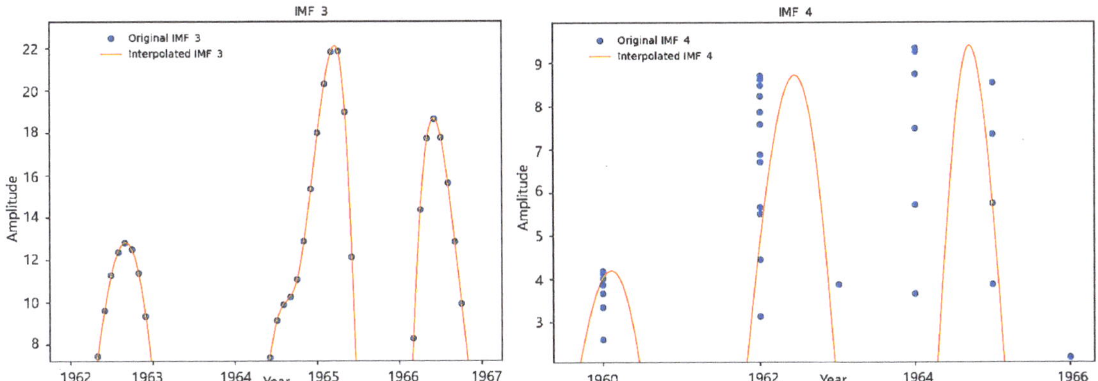

Figure 9. Details for building IMFs with spline interpolation for IMF 3 (**left**) and by fine tuning for IMF 4 (**right**) of the monthly series.

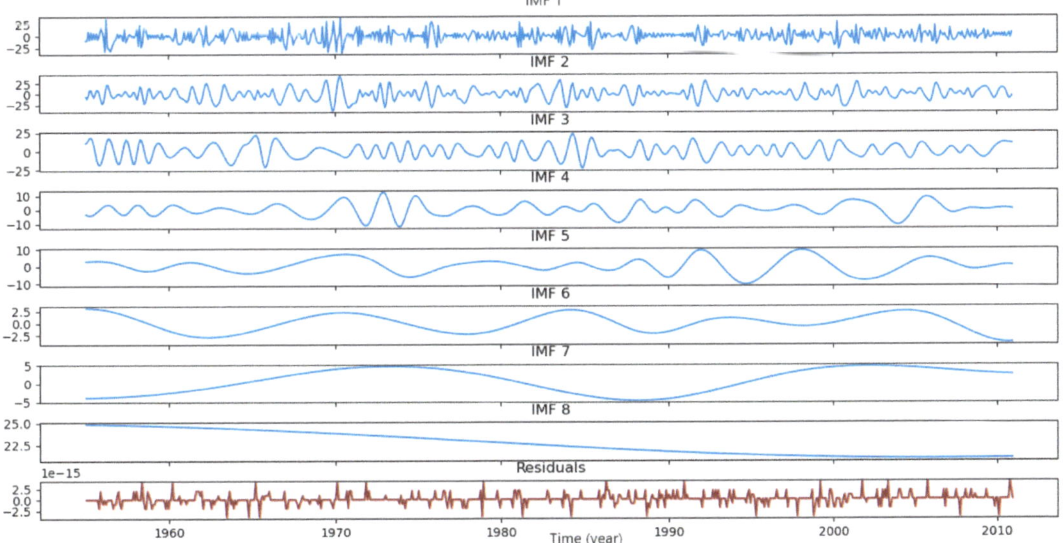

Figure 10. EMD of the S series: IMF 1, IMF8, and the residuals.

The first IMF presents the highest frequency oscillations observed in the dataset, typically corresponding to short-term changes. The periods with high oscillations might correspond to the months when floods were recorded. These oscillations are still kept by IMF 2. IMFs 3 and 4 show long periods with similar behavior, followed by some peaks. IMFs 3–5 reveal a slightly lower frequency, indicating seasonal fluctuations. IMFs 6 and 7 show an almost perfect sine behavior, whereas IMF 8 has a decreasing trend. They display patterns that might be correlated with multiyear climate cycles or longer-term environmental changes. The remaining values are very low (of the order 10–15), signifying good series decomposition.

An additional investigation was carried out for S1 and S2 (Figure 11) to analyze any potential differences or anomalies in the discharge pattern before and after 1984. The EMD process was repeated for each dataset, followed by respective visualizations. Six significant IMFs came from EMD of the series S1 and S2 (Figure 12).

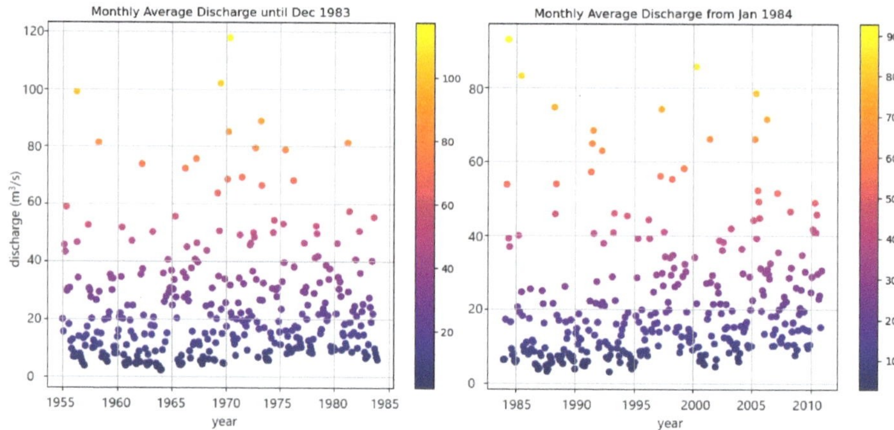

Figure 11. S1 (**left**) and S2 (**right**) series.

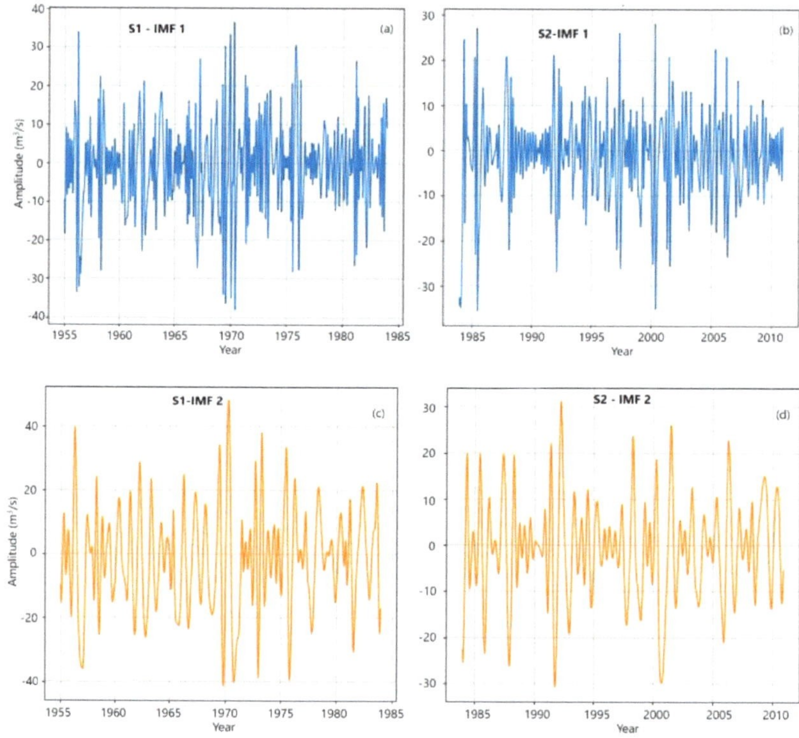

Figure 12. *Cont.*

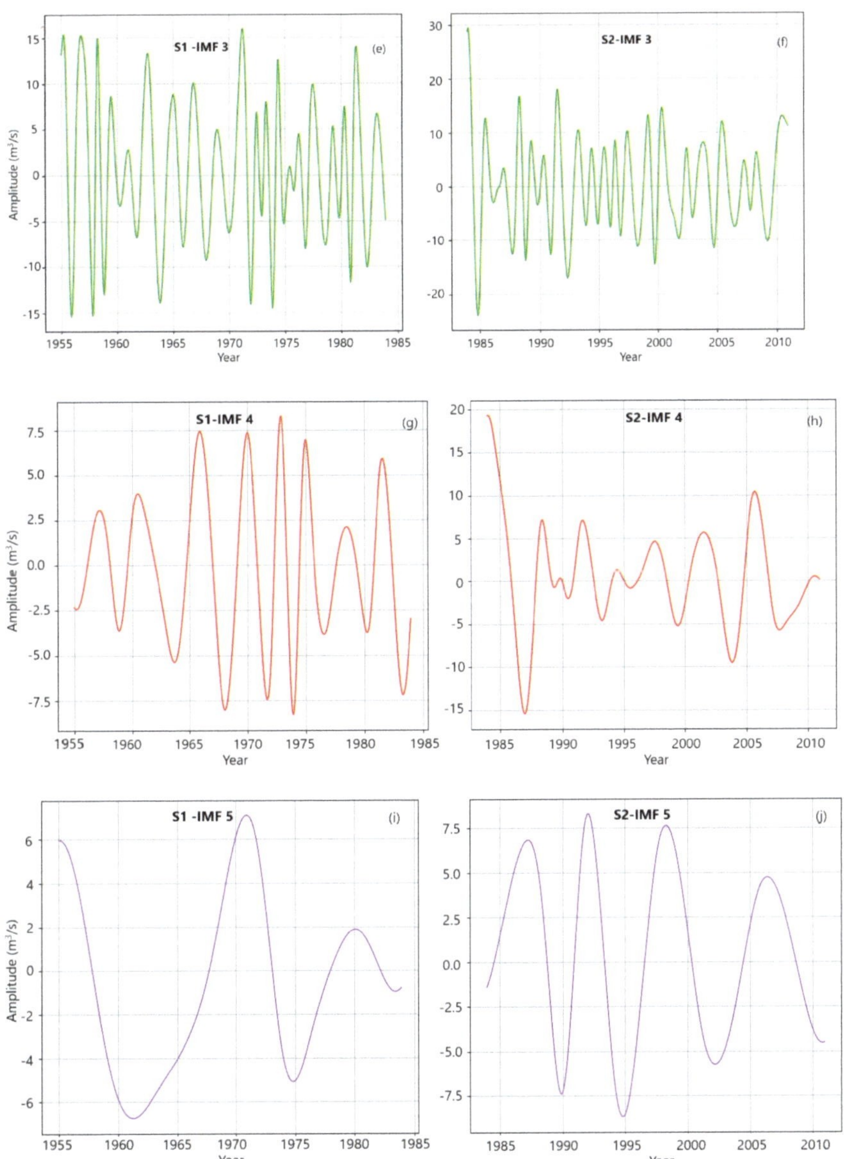

Figure 12. *Cont.*

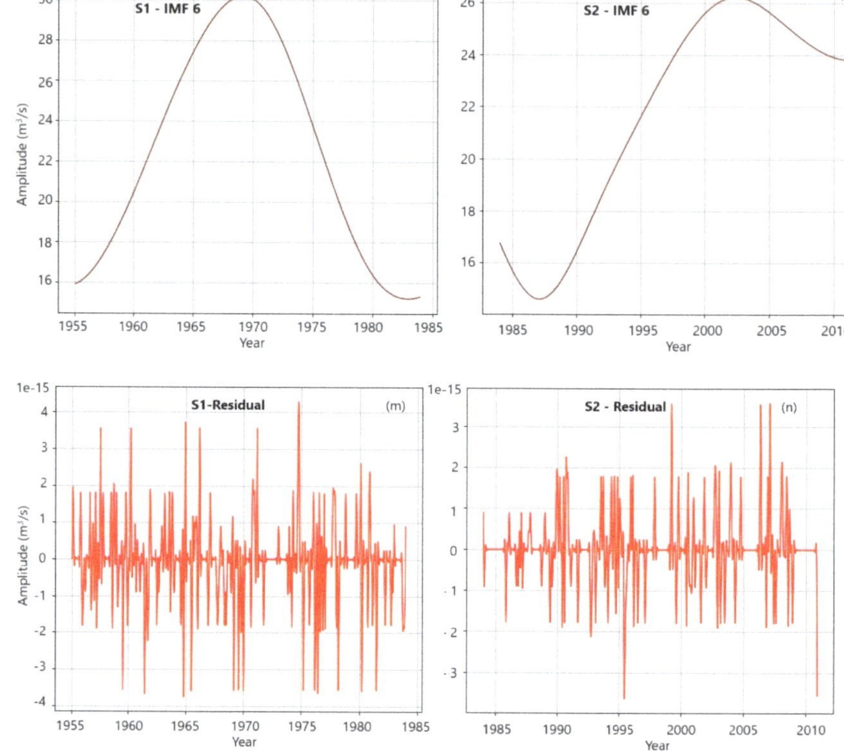

Figure 12. (**a**) IMF1 for S1, (**b**) IMF1 for S2, (**c**) IMF2 for S1, (**d**) IMF2 for S2, (**e**) IMF 3 for S1, (**f**) IMF 3 for S2, (**g**) IMF 4 for S1, (**h**) IMF 4 for S2, (**i**) IMF5 for S1, (**j**) IMF5 for S2, (**k**) IMF6 for S1, (**l**) IMF6 for S2, (**m**) residual for S1, and (**n**) residual for S2.

Analyzing the frequency domain in the IMFs (Figure 12) provides key insights, as follows.

- In all cases, the amplitudes of the IMFs of S1 are higher than those of S2.
- IMF1 reveals high-frequency oscillations, highlighting probable monthly anomalies. Unequal variances in different periods are also recorded, emphasizing inhomogeneities in the data series. At the end of the study period (after 2005), a decreasing amplitude of oscillations is observed.
- IMF2 and IMF3 predominantly illustrate seasonal cycles, hinting at the spring snowmelt and autumnal rain, aligning with the temperate–continental climate attributes. IMF3 for S2 is almost uniformly distributed, with low variations in the amplitudes with respect to IMF3, whose chart presents more accentuated variations in subperiods.
- IMF4–IMF6 are tied to longer temporal scales, suggesting multiyear, possibly decadal trends. These could be attributed to broader climatic shifts, land-use changes, or long-term anthropogenic impacts on the river basin. We can observe different shapes of IMFS 5 and 6 for S1 and S2, suggesting a different distribution of the subsequent series.
- In both cases, residuals have values close to zero (order 10^{-15}), indicating good decomposition.

Comparing these findings with prior studies [29,63], the oscillatory patterns observed in the Buzău River seem consistent with those of other rivers in a temperate–continental climate. The seasonal fluctuations, as captured by the second IMF, align well with the expected regional snowmelt and rainfall patterns.

While the high-frequency oscillations indicated by the first couple of IMFs can be associated with shorter-term events like storms or immediate snowmelt responses, the subsequent IMFs' longer-term patterns might hint at broader climatic or anthropogenic influences on the river flow. It would be interesting to juxtapose these results with other datasets from surrounding river basins to identify if these patterns are localized to Buzău or represent a regional trend.

4. Discussion

In short, the multifaceted approach used in this study provides a more complete understanding of the dam's impact on Buzău discharge dynamics.

The current analysis provides a comprehensive and nuanced view of the dam's impact on Buzău discharge dynamics, validating and complementing the findings from the articles [38,39]. While the research conducted in [39] only tested for the existence of a trend in the annual and quarterly flow series and built an AR(5) model to forecast the mean quarterly series, the present study has a larger scope. With a higher-resolution data series, it can detect monthly seasonality factors. This approach offers increased flexibility and accuracy for forecasting, unlike the AR(5) model which only captures the shape of the series but fails to account for extremes. The EMD analysis highlights differences between the river's short- and long-term variations before and after the dam's construction, emphasizing the need for a more nuanced approach.

Compared to the results obtained in [38] by utilizing the IHA methodology, the present article provides new insights into the series variability by building IMFs. The output of this study is in concordance with the finding of [39], indicating the significant alteration in the river flow after building the dam, emphasized by the differences between the IMFs' amplitudes and frequencies corresponding to S1 and S2. Meanwhile, in [38], it was shown that a decrease of 12.6% in the maximum monthly discharge was noticed after 1984 with respect to 1955–1983, and in the same period, the minimum and the maximum of the 90-day maximum flow decreased from 28.6 to 16.4 m^3/s, and from 92.6 to 69.8 m^3/s, respectively, but no chart of the time series components is provided.

Another advantage of the proposed methodology over the deterministic ones is that no initial conditions on the data series are required, and no validation of certain coefficients (using the least-squares or similar methods) is required, as in the econometric models. The same advantage is shared by building a linear trend (when it exists) by Sen's non-parametric method.

Utilizing the Thomas–Fiering (TF) equation might be another possibility for modeling the S, S1, and S2 series. It relies on the following equation:

$$Q_{i+1} = \overline{Q_{k+1}} + b_k\left(Q_i - \overline{Q_k}\right) + t_i\overline{\sigma_{k+1}}\left(1 - r_k^2\right)^{1/2} \tag{8}$$

where:

Q_{i+1}—flow in month i +1;
$\overline{Q_{k+1}}$—average flow in month k + 1;
b_k—gradient of line between the flow in month j + 1 and the flow in month j;
Q_i—flow in month i;
$\overline{Q_k}$—average of flows in month k;
t_i —normal random function;
$\overline{\sigma_{k+1}}$—standard deviation for flows in month k + 1;
r_k^2—correlation coefficient between flows of months k and k + 1.

The difference between the decomposition models used here and the TF model is that the latter takes into account the correlations between the flows in successive months and the average flow in a month i to estimate the flow in the next month, whereas the ADM and MDM focus on the values recorded in the same months to compute the seasonality indices which are further subtracted from the detrended series to estimate the residual. The TF model fits the series but does not indicate the seasonality indices separately. Utilizing the TF model will add value to the knowledge related to the discharge series, especially when the river flow presents high variations (as in spring when snowmelt or in months with high precipitation).

5. Conclusions

This article aimed to address the Buzau River flow alteration due to the building of the Siriu dam. The multifractal analysis rejected the monofractality, sustaining different scaling behavior of the monthly data series. The decrement in the seasonality indices indicates attenuation of the extreme events (high flows and flooding).

Since the river basin is situated in a temperate continental zone and the climate variation during the study period was not significant, one cannot attribute the modification of the water flow regimen to the change but to the hydrotechnical works on the Buzău River catchment.

When interested in different time horizons for the series evolution, the use of the EMD technique is recommended. The existence of IMF1 to IMF3′s high-frequency oscillations indicates the need for a deeper insight into short-term disturbances, possibly from industrial effluents, rapid urbanization, or land use. The slow, evolving changes in IMFs 4–6 suggest that Buzău River's flow is influenced by more than just its immediate environment. A multivariate analysis must further investigate these correlations to clarify the anthropic impact on the river and the river flow variation on land use.

The findings of this article provide scientific background for flood modeling, which is necessary for building hazard maps. Specific software can be used for this purpose, and different hypotheses on water discharge can be adopted. No matter the underlying modeling technique, the resulting models are calibrated using the existing data series. With knowledge of the existence of two different river flow patterns, the calibration must be performed with the most recent data series (which are those after 1984 for the flood risk evaluation). Using the entire S series for such a purpose would introduce significant bias in the flood forecast. Studying the correlation between precipitation and discharge would be necessary to better understand the impact of torrential rain on the prediction of river discharge and floods. Given that in Romanian, hazard and flood risk maps are essential for establishing the National Plan for Risk Management, comprehensive knowledge of river discharge is essential for molding water management policies and influencing land-use decisions.

Author Contributions: Conceptualization, A.B.; methodology, A.B. and N.M.; software, A.B. and N.M.; validation, A.B. and N.M.; formal analysis, A.B.; investigation, A.B. and N.M.; resources, A.B.; data curation, A.B.; writing—original draft preparation, A.B. and N.M.; writing—review and editing, A.B.; visualization, A.B. and N.M.; supervision, A.B.; project administration, A.B.; funding acquisition, A.B. All authors have read and agreed to the published version of the manuscript.

Funding: The APC was funded for the first author by Transilvania University of Brașov Romania by means of publication grants.

Data Availability Statement: Data will be available on request from the authors.

Conflicts of Interest: The authors declare no conflicts of interest.

References

1. Li, S.-L.; Zhang, H.; Yi, Y.; Zhang, Y.; Qi, Y.; Mostofa, K.M.G.; Guo, L.; He, D.; Fu, P.; Liu, C.-Q. Potential impacts of climate and anthropogenic-induced changes on DOM dynamics among the major Chinese rivers. *Geogr. Sustain.* **2023**, *4*, 329–339. [CrossRef]
2. Hoque, M.M.; Islam, A.; Ghosh, S. Environmental flow in the context of dams and development with special reference to the Damodar Valley Project, India: A review. *Sustain. Water Resour. Manag.* **2022**, *8*, 62. [CrossRef] [PubMed]
3. Habel, M.; Nowak, B.; Szadek, P. Evaluating indicators of hydrologic alteration to demonstrate the impact of open-pit lignite mining on the flow regimes of small and medium-sized rivers. *Ecol. Ind.* **2023**, *157*, 111295. [CrossRef]
4. Mei, X.; Van Gelder, P.H.A.J.M.; Dai, Z.; Tang, Z. Impact of dams on flood occurrence of selected rivers in the United States. *Front. Earth Sci.* **2007**, *11*, 268–282. [CrossRef]

5. Soomro, S.; Guo, J.; Shi, X.; Ke, S.; Li, Y.; Hu, C.; Zwain, H.M.; Gu, J.; Chunyun, Z.; Li, A.; et al. Climate Change Critique on Dams and Anthropogenic Impact to Mediterranean Mountains for Freshwater Ecosystem—A Review. *Pol. J. Environ. Stud.* **2023**, *32*, 2981–2992. [CrossRef]

6. Słowik, M.; Kiss, K.; Czigány, S.; Gradwohl-Valkay, A.; Dezső, J.; Halmai, A.; Tritt, R.; Pirkhoffer, E. The influence of changes in flow regime caused by dam closure on channel planform evolution: Insights from flume experiments. *Environ. Earth. Sci.* **2021**, *80*, 165. [CrossRef]

7. Castello, L.; Mceda, L.M. Large-scale degradation of Amazonian freshwater ecosystems. *Glob. Change Biol.* **2016**, *22*, 990–1007. [CrossRef]

8. Gao, B.; Li, J.; Wang, X. Analyzing Changes in the Flow Regime of the Yangtze River Using the Eco-Flow Metrics and IHA Metrics. *Water* **2018**, *10*, 1552. [CrossRef]

9. Kondolf, G.M.; Rubin, Z.K.; Minear, J.T. Dams on the Mekong: Cumulative sediment starvation. *Water Resour. Res.* **2014**, *50*, 5158–5169. [CrossRef]

10. Li, D.L.; Long, D.; Zhao, J.; Lu, H.; Hong, Y. Observed changes in flow regimes in the Mekong River basin. *J. Hydrol.* **2017**, *551*, 217–232. [CrossRef]

11. Ward, P.J.; de Ruiter, M.C.; Mård, J.; Schröter, K.; Van Loon, A.; Veldkamp, T.; von Uexkull, N.; Wanders, N.; AghaKouchak, A.; Arnbjerg-Nielsen, K.; et al. The need to integrate flood and drought disaster risk reduction strategies. *Water Secur.* **2020**, *11*, 100070. [CrossRef]

12. Deitch, M.J.; Merenlender, A.M.; Feirer, S. Cumulative effects of small reservoirs on streamflow in northern coastal California catchments. *Water Resour. Manag.* **2013**, *27*, 5101–5118. [CrossRef]

13. Habets, F.; Molénat, J.; Carluer, N.; Douez, O.; Leenhardt, D. The cumulative impacts of small reservoirs on hydrology: A review. *Sci. Total Environ.* **2018**, *643*, 850–867. [CrossRef] [PubMed]

14. Mayor, B.; Rodríguez-Muñoz, I.; Villarroya, F.; Montero, E.; López-Gunn, E. The Role of Large and Small Scale Hydropower for Energy and Water Security in the Spanish Duero Basin. *Sustainability* **2017**, *9*, 1807. [CrossRef]

15. Wang, Y.; Wang, D.; Lewis, Q.W.; Wu, J.; Huang, F. A framework to assess the cumulative impacts of dams on hydrological regime: A case study of the Yangtze river. *Hydrol. Process.* **2017**, *31*, 3045–3055. [CrossRef]

16. Zhang, X.; Fang, C.; Wang, Y.; Lou, X.; Su, Y.; Huang, D. Review of Effects of Dam Construction on the Ecosystems of River Estuary and Nearby Marine Areas. *Sustainability* **2022**, *14*, 5974. [CrossRef]

17. Dang, T.D.; Cochrane, T.A.; Arias, M.E.; Van, P.D.T.; de Vries, T.T. Hydrological alterations from water infrastructure development in the Mekong floodplains. *Hydrol. Process.* **2016**, *30*, 3824–3838. [CrossRef]

18. Ekka, A.; Pande, S.; Jiang, Y.; der Zaag, P.V. Anthropogenic Modifications and River Ecosystem Services: A Landscape Perspective. *Water* **2020**, *12*, 2706. [CrossRef]

19. Elesbon, A.A.A.; da Silva, D.D.; Sediyama, G.S.; Guedes, H.A.S.; Ribeiro, C.A.A.S.; Ribeiro, C.B.d.M. Multivariate statistical analysis to support the minimum streamflow regionalization. *Eng. Agríc.* **2015**, *35*, 838–851. [CrossRef]

20. Bărbulescu, A.; Maftei, C. Statistical approach of the behavior of Hamcearca River (Romania). *Rom. Rep. Phys.* **2021**, *73*, 703.

21. Bărbulescu, A.; Maftei, C.E. Evaluating of the Probable Maximum Precipitation. Case study from the Dobrogea region, Romania. *Rom. Rep. Phys.* **2023**, *75*, 704. [CrossRef]

22. Bărbulescu, A.; Dumitriu, C.S.; Maftei, C. On the Probable Maximum Precipitation Method. *Rom. J. Phys.* **2022**, *67*, 801.

23. Gürsoy, Ö.; Engin, S.N. A wavelet neural network approach to predict daily river discharge using meteorological data. *Meas. Control* **2019**, *52*, 599–607. [CrossRef]

24. Tang, D.; Jin, X.; Chen, W.; Pu, N. A combined rotated general regression neural network method for river flow forecasting. *Hydrol. Sci. J.* **2016**, *61*, 669–682.

25. Chakraborty, S.; Biswas, S. River discharge prediction using wavelet-based artificial neural network and long short-term memory models: A case study of Teesta River Basin, India. *Stoch. Environ. Res. Risk A* **2023**, *37*, 3163–3184. [CrossRef]

26. Mehedi, M.A.A.; Khosravi, M.; Yazdan, M.M.S.; Shabanian, H. Exploring Temporal Dynamics of River Discharge Using Univariate Long Short-Term Memory (LSTM) Recurrent Neural Network at East Branch of Delaware River. *Hydrology* **2022**, *9*, 202. [CrossRef]

27. Essam, Y.; Huang, Y.F.; Ng, J.L.; El-Safie, A. Predicting streamflow in Peninsular Malaysia using support vector machine and deep learning algorithms. *Sci. Rep.* **2022**, *12*, 3883. [CrossRef] [PubMed]

28. Bărbulescu, A.; Dumitriu, C.S.; Dragomir, F.-L. Detecting Aberrant Values and Their Influence on the Time Series Forecast. In Proceedings of the 2021 International Conference on Electrical, Computer, Communications and Mechatronics Engineering (ICECCME), Mauritius, 7–8 October 2021; pp. 1–5.

29. Popescu, N.C.; Bărbulescu, A. On the flash flood susceptibility and accessibility in the Vărbilău catchment (Romania). *Rom. J. Phys.* **2022**, *67*, 811.

30. Conservation GATEWAY. Available online: https://www.conservationgateway.org/ConservationPractices/Freshwater/EnvironmentalFlows/MethodsandTools/IndicatorsofHydrologicAlteration/Pages/IHA-Software-Download.aspx (accessed on 12 December 2013).

31. Barbalić, D.; Kuspilić, N. Trends of indicators of hydrological alterations. *Građevinar* **2014**, *66*, 613–624.

32. Macnaughton, C.J.; Mclaughlin, F.; Bourque, G.; Senay, C.; Lanthier, G. The effects of regional hydrologicalteration on fish community structure in regulated rivers. *River Res. Appl.* **2017**, *33*, 249–257. [CrossRef]

33. Eum, H.I.; Dibike, Y.; Prowse, T. Climate-induced alteration of hydrologic indicators in the Athabasca River Basin, Alberta, Canada. *J. Hydrol.* **2017**, *544*, 327–342. [CrossRef]

34. Lin, K.; Lin, Y.; Xu, Y.; Chen, X.; Chen, L.; Singh, V. Inter- and intraannual environmental fow alteration and its implication in the Pearl River Delta South China. *J. Hydro-Environ. Res.* **2017**, *15*, 27–40. [CrossRef]

35. Ge, J.; Peng, W.; Huang, W.; Qu, X.; Singh, S. Quantitative assessment of fow regime alteration using a revised range of variability methods. *Water* **2018**, *10*, 597. [CrossRef]

36. Kumar, A.; Tripathi, V.K.; Kumar, P.; Rakshit, A. Assessment of hydrologic impact on fow regime due to dam inception using IHA framework. *Environ. Sci. Pollut. Res.* **2023**, *30*, 37821–37844. [CrossRef]

37. Olden, J.D.; Poff, N.L. Redundancy and the choice of hydrologic indices for characterizing streamflow regimes. *River Res. Appl.* **2003**, *19*, 101–121. [CrossRef]

38. Minea, G.; Bărbulescu, A. Statistical assessing of hydrological alteration of Buzău River induced by Siriu dam (Romania). *Forum Geogr.* **2014**, *13*, 50–58. [CrossRef]

39. Mocanu-Vargancsik, C.A.; Bărbulescu, A. On the variability of a river water flow, under seasonal conditions. Case study. *IOP Conf. Ser. Earth Environ. Sci.* **2019**, *344*, 012028. [CrossRef]

40. Bărbulescu, A. Statistical Assessment and Model for a River Flow under Variable Conditions. In Proceedings of the 15th International Conference on Environmental Science and Technology, Rhodes, Greece, 31 August–2 September 2017; Available online: https://cest2017.gnest.org/sites/default/files/presentation_file_list/cest2017_00715_poster_paper.pdf (accessed on 21 February 2024).

41. Chendeş, V. *Water Resources in Curvature Subcarpathians. Geospatial Assessments*; Editura Academiei Române: Bucureşti, Romania, 2011; (In Romanian with English abstract).

42. The Arrangement of the Buzău River. Available online: https://www.hidroconstructia.com/dyn/2pub/proiecte_det.php?id=110&pg=1 (accessed on 30 November 2023). (In Romanian).

43. Updated management plan of the buzău-ialomiţa hydrographic area. Available online: http://buzau-ialomita.rowater.ro/wp-content/uploads/2021/02/PMB_ABABI_Text_actualizat.pdf (accessed on 16 December 2023). (In Romanian).

44. Diaconu, D. *Water Resources from of the Buzău River Catchment*; Editura Universitară: Bucureşti, Romania, 2008. (In Romanian)

45. Minea, S.I. *The Rivers from Buzău River Catchment. Hydrographical and Hydrological Constructions*; Editura Alpha MDN: Buzău, Romania, 2011. (In Romanian)

46. Kendall, M.G. *Rank Correlation Methods*, 4th ed.; Charles Griffin: London, UK, 1975.

47. Hirsch, R.M.; Slack, J.R.; Smith, R.A. Techniques of trend analysis for monthly water quality data. *Water Resour. Res.* **1982**, *18*, 107–121. [CrossRef]

48. Hippel, K.; McLeod, A.I. *Time Series Modelling of Water Resources and Environmental Systems*; Elsevier Science B.V.: Amsterdam, The Netherlands, 1994.

49. Sen, P.K. Estimates of the Regression Coefficient Based on Kendall's Tau. *J. Am. Stat. Assoc.* **1969**, *63*, 1379–1389. [CrossRef]

50. Kwiatkowski, D.; Phillips, P.C.B.; Schmidt, P.; Shin, Y. Testing the null hypothesis of stationarity against the alternative of a unit root. *J. Econ.* **1992**, *54*, 159–178. [CrossRef]

51. Bărbulescu, A.; Dumitriu, C.S. Fractal characterization of brass corrosion in cavitation field in seawater. *Sustainability* **2023**, *15*, 3816. [CrossRef]

52. Bărbulescu, A.; Dumitriu, C.S. Assessing the Fractal Characteristics of Signals in Ultrasound Cavitation. In Proceedings of the 25th International Conference on System Theory, Control and Computing (ICSTCC 2021), Iasi, Romania, 20–23 October 2021. [CrossRef]

53. Kantelhardt, J.W.; Zschiegner, S.A.; Koscielny-Bunde, E.; Havlin, S.; Bunde, A.; Stanley, H. Multifractal detrended fluctuation analysis of nonstationary time series. *Phys. A Stat. Mech. Its Appl.* **2002**, *316*, 87–114. [CrossRef]

54. Ferraris, L.; Gabellani, S.; Parodi, U.; Rebora, N. Revisiting Multifractality in Rainfall Fields. *J. Hydrometeorol.* **2003**, *4*, 544–551. [CrossRef]

55. Renyi, A. On a new axiomatic theory of probability. *Acta Math. Hung.* **1955**, *6*, 285–335. [CrossRef]

56. Hentschel, H.G.E.; Procaccia, I. The infinite number of generalized dimensions of fractals and strange attractors. *Phys. D* **1983**, *8*, 435–444. [CrossRef]

57. Halsey, T.; Jensen, M.; Kadanoff, L.; Procaccia, I.; Shraiman, B.I. Fractal measures and their singularities: The characterization of strange sets. *Phys. Rev. A* **1986**, *33*, 1141–1151. [CrossRef] [PubMed]

58. Classical Decomposition. Available online: https://otexts.com/fpp2/classical-decomposition.html (accessed on 2 September 2023).

59. Huang, N.E.; Shen, Z.; Long, S.R.; Wu, M.C.; Shih, H.H.; Zheng, Q.; Yen, N.-C.; Tung, C.C.; Liu, H.H. The empirical mode decomposition and the Hilbert spectrum for nonlinear and non-stationary time series analysis. *Proc. R. Soc. Lond. A* **1998**, *454*, 903–995. [CrossRef]

60. Huang, N.E.; Wu, M.C.S.; Long, R.; Shen, S.; Qu, W.; Gloerson, P.; Fan, K.L. A Confidence Limit for the Empirical Mode Decomposition and Hilbert Spectral Analysis. *Proc. R. Soc. Lond. A* **2003**, *31*, 417–457. [CrossRef]

61. Rilling, G.; Flandrin, P.; Goncalves, P. On empirical mode decomposition and its algorithms. In Proceedings of the 6th IEEE/EURASIP Workshop on Nonlinear Signal and Image Processing (NSIP '03), Grado, Italy, 8–11 June 2003.

62. Flandrin, P. Empirical mode decompositions as data-driven wavelet like expansions. *Int. J. Wavelets Multires. Infor. Proc.* **2004**, *2*, 1–20. [CrossRef]
63. Zhang, Z.; Huang, J.; Huang, Y.; Hong, H. Streamflow variability response to climate change and cascade dams development in a coastal watershed. *Estuar. Coast. Shelf Sci.* **2015**, *166*, 209–217. [CrossRef]

Article

Assessment of the Impact of Spatial Variability on Streamflow Predictions Using High-Resolution Modeling and Parameter Estimation: Case Study of Geumho River Catchment, South Korea

Bomi Kim, **Garim Lee, Yaewon Lee, Sohyun Kim and Seong Jin Noh ***

Department of Civil Engineering, Kumoh National Institute of Technology, Gumi 39177, Republic of Korea; kimbom3835@gmail.com (B.K.)
* Correspondence: seongjin.noh@kumoh.ac.kr

Abstract: In this study, we analyzed the impact of model spatial resolution on streamflow predictions, focusing on high-resolution scenarios (<1 km) and flooding conditions at catchment scale. Simulation experiments were implemented for the Geumho River catchment in South Korea using Weather Research and the Forecasting Hydrological Modeling System (WRF-Hydro) with spatial resolutions of 100 m, 250 m, and 500 m. For the estimation of parameters, an automatic calibration tool based on the Model-Independent Parameter Estimation and Uncertainty Analysis (PEST) method was utilized. We assessed the hydrological predictions across different spatial resolutions considering calibrated parameters, calibration runtime, and accuracy of streamflow before and after calibration. For both Rainfall Events 1 and 2, significant improvements were observed after event-specific calibration in all resolutions. Particularly for 250 m resolution, *NSE* values of 0.8 or higher were demonstrated at lower gauging locations. Also, at a 250 m resolution, the changes in the calibrated parameter values (*REFKDT*) were minimized between Rainfall Events 1 and 2, implicating more effective calibration compared to the other resolutions. At resolutions of 100 m and 500 m, the optimal parameter values for the two events were distinctively different while more computational resources were required for calibration in Event 2 with drier antecedent conditions.

Keywords: spatial resolution; distributed modeling; WRF-Hydro; PEST; parameter estimation; streamflow prediction

Citation: Kim, B.; Lee, G.; Lee, Y.; Kim, S.; Noh, S.J. Assessment of the Impact of Spatial Variability on Streamflow Predictions Using High-Resolution Modeling and Parameter Estimation: Case Study of Geumho River Catchment, South Korea. *Water* **2024**, *16*, 591. https://doi.org/10.3390/w16040591

Academic Editors: Agnieszka Rutkowska and Katarzyna Baran-Gurgul

Received: 8 December 2023
Revised: 15 February 2024
Accepted: 16 February 2024
Published: 17 February 2024

1. Introduction

Hydrological models are simplified representations of real-world hydrological processes with different levels of approximation, and a wide range of such models have been developed and implemented to improve understanding of these processes and to provide better support for making decisions. Among the various models with different spatial abstractions, distributed hydrological modeling has been gaining increased attention in various fields, including streamflow and drought forecasting, climate and land-use change assessment, and water resources management, due to its ability to handle spatial variations of hydrological variables based on the rapidly increasing spatial information. Some recent studies reported that the performance of distributed hydrological modeling could be improved through the use of high-resolution input (Maxwell et al. [1], Abbazadeh et al. [2]). For instance, Maxwell et al. [1] simulated surface and subsurface flows at a high spatial resolution (1 km) in illustrating the feasibility of continental-scale integrated modeling for enhancing our understanding of large-scale hydrological systems. In addition, Abbazadeh et al. [2] investigated the impact of soil moisture on streamflow prediction in the Houston catchment area in Texas, USA at varying spatial resolutions. Their simulation results indicated the improved streamflow with a finer spatial resolution of 1 km compared to coarser resolutions such as 36 km.

Regardless of the degree of input spatial detail, however, hydrological models inherently contain various uncertainties stemming from the parameters, model structure, input data, and initial and boundary conditions (Moges et al. [3]). Calibration and uncertainty estimation techniques, once developed for lumped models with simple model structures, have been modified and employed to reduce gaps between distributed modeling and reality. Chen et al. [4] proposed optimal parameters for a Physically Based Distributed Hydrological Model (PBDHM) using the Particle Swarm Optimization (PSO) technique in the southern region of China. Verri et al. [5] used the Weather Research and Forecasting Hydrological Modeling System (WRF-Hydro) model to reduce uncertainties, analyzing parameter sensitivity and developing an interpolation method for mitigating rainfall uncertainties. Tolson et al. [6] analyzed the sensitivity of model equations and the Nash–Sutcliffe coefficient of efficiency for daily flows in relation to parameters using the Soil and Water Assessment Tool (SWAT 2000) model. Setegn et al. [7] applied the SWAT model in flow analysis of the Tana Lake catchment in Ethiopia, implementing parameter estimation using the SUFI-2, GLUE, and ParaSol algorithms.

The WRF-Hydro adopted in this study is a distributed hydrological model developed by the U.S. National Center for Atmospheric Research (NCAR) that consists of land surface model, terrain routing, river, and reservoir components. This model offers flexible coupling between its components and has been used for diverse applications in various research studies. Most of the research using WRF-Hydro can be categorized into analyses of its integration with the Weather Research and Forecasting (WRF) atmospheric model and proposals for model improvements, including advances achieved through artificial intelligence (AI) and post-processing research. Examples of research on utilizing the integration between WRF and WRF-Hydro include a study by Senatore et al. [8] in which observations with simulations from standalone WRF and combined WRF/WRF-Hydro models were compared at Land Surface Model (LSM) 2.5 km-Routing 250 m resolution in the Crati river catchment in southern Italy. In their study, Naabil et al. [9] demonstrated the improved performance of the WRF/WRF-Hydro coupled model over standalone WRF in rainfall estimation for the Tono catchment. Moreover, Wang et al. [10] evaluated the simulated results of the WRF standalone model and the coupled WRF/WRF-Hydro model as hydrological elements for six different 24 h storm events with varying spatiotemporal homogeneity in rainfall distribution. Lee et al. [11] applied standalone WRF-Hydro to assess drought characteristics in South Korea. Several studies suggest that use of a hybrid model combining Long Short-Term Memory (LSTM) and the WRF-Hydro model enhances the accuracy of streamflow predictions. Cho et al. [12] proposed a hybrid model that combines LSTM and the WRF-Hydro model for improved streamflow prediction. They demonstrated that the predictive capabilities were enhanced using the proposed approach and various targeted model sensitivity analyses. Liu et al. [13] applied four traditional statistical post-processing methods based on Quantile Mapping (QM) and two proposed machine learning methods (SVR, CNN) to a distributed model (WRF-Hydro), aiming to reduce systematic biases in streamflow simulation. Zhang et al. [14] evaluated the impact of soil infiltration processes on simulation by calibrating the WRF-Hydro model using the Dynamically Dimensioned Search (DDS) technique and analyzing the importance of infiltration effects in urban areas. These studies collectively contribute to understanding the uncertainties within distributed hydrological models and their calibration, enabling more accurate interpretations of hydrological responses. Kim et al. [15] investigated the impact of the modeling resolution of WRF-Hydro on the land surface and streamflow in the built environment of Dallas–Fort Worth area, USA. Their findings showed that by increasing the spatial resolution from 100 m to 10 m in surface flow and river routing models, the duration of simulation was extended by over 100 times. As simulation time increased, a corresponding increase in the calibration time of automatic parameter estimation techniques became evident. Therefore, achieving higher accuracy through high-resolution input data requires additional computational resources for both simulation and calibration.

The Model-Independent Parameter Estimation (PEST) method applied in this study is a nonlinear parameter optimization technique that operates independently of the model itself. It requires only the preparation of necessary files for PEST without the need for additional programs. As an example of applying the PEST method to a hydrological model, Abbas et al. [16] applied the SWAT+ model and PEST based on the gwflow module combined with the Morris screening method to evaluate the effect of parameters on streamflow and groundwater head prediction. de Wit et al. [17] presented a dynamic modeling approach with combined SWAP and PEST to simulate groundwater levels and soil moisture. As some examples of the applications of PEST to WRF-Hydro modeling, Fersch et al. [18] implemented parameter optimization for six parameters of the WRF-Hydro across six sub-catchments utilizing PEST. Sofokleous et al. [19] investigated the influence of WRF-Hydro model parameters of soil moisture, groundwater, and vegetation on hydrological balance, employing grid-based calibration methods within PEST to mitigate overestimation and validate its applicability. Furthermore, Wang et al. [20] implemented PEST in a parallelized fashion for High-Performance Computing (HPC), integrating its functionality alongside WRF-Hydro, and assessed the suitability and computational advantages of parallel PEST techniques in a 2013 flooding case study in the Midwestern United States.

The main objective of this investigation is to evaluate the influence of model spatial resolution on parameter estimation in distributed hydrological modeling. To achieve this, we utilize WRF-Hydro models with model spatial resolutions of 100 m, 250 m, and 500 m for the Geumho River catchment, a medium-sized catchment in South Korea. First, we assess the impact of spatial resolution on the hydrological processes under the same default parameter conditions. Then, we evaluate how various resolutions of model impact parameter estimation using the PEST technique, taking into account streamflow prediction results and computational time. In addition, we investigate how the adjusted model affects the prediction results of streamflow at multiple observation gauges within the catchment. The paper is organized as follows: Section 2 describes the materials and methods in detail, including the study area, data, model configurations, simulations, and calibrations; the results with discussions are presented in Section 3; Section 4 presents the conclusions and directions for future research.

2. Materials and Methods

2.1. Study Area and Data

The study area is the Geumho River catchment (area: 2087.9 km^2; river length: 69.3 km). Within the catchment, the Geumho River traverses Daegu Metropolitan City, the fourth largest city in South Korea in terms of population, and merges into the mainstream of the Nakdong River (Figure 1). There are multiple dam reservoirs in the study catchment, including the Yeongcheon Dam (235 km^2), whose artificial controls are not explicitly considered in the hydrological simulation. Land use in the Geumho River catchment comprises 66.4% forested land and 28.8% agricultural land. The average annual temperature is at 13 °C with an annual rainfall of 1007 mm, less than the average annual precipitation of 1306.3 mm in South Korea where about 56% of the precipitation is concentrated during summer.

There are a total of 11 ground weather observation gauges within the Geumho River catchment, including 9 Automatic Weather Stations (AWS) and 2 Automated Synoptic Observing Systems (ASOS). The Inverse Distance Weighted (IDW) method was employed to build meteorological forcing data of WRF-Hydro with eight components (i.e., incoming shortwave radiation, incoming longwave radiation, specific humidity, air temperature, surface pressure, near-surface wind in the u- and v-components, and liquid water precipitation rate). IDW is one of the most widely selected techniques for interpolating spatial data in estimating the value of a point without data using the value of a point with data. The reciprocal of the distance between points is used as a weight to allow the values of closer points greater influence. The two selected events (Table 1) are flood events, each with distinct characteristics. The rainfall event in 2020, referred to as Rainfall Event 1, was a concentrated heavy rainfall event characterized by antecedent rainfall before the event. In contrast, in the year 2022, during Rainfall Event 2, dry conditions were experienced from the beginning of the year until the event, with minimal antecedent rainfall. This event was triggered by the sudden landfall of Typhoon Hinnamnor, which led to heavy rainfall.

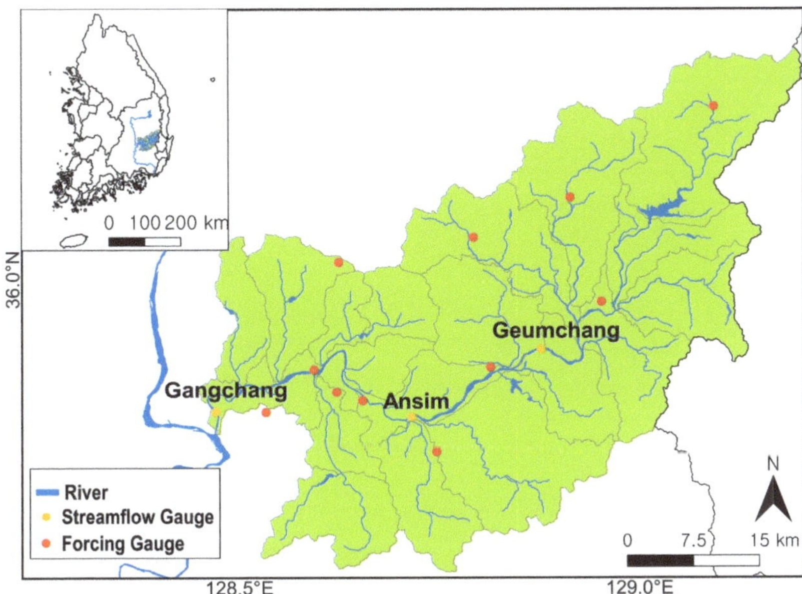

Figure 1. Map of the Geumho River catchment: the three yellow circles represent the streamflow gauges at Geumchang, Ansim, and Gangchang, respectively.

Table 1. Two selected Rainfall Events for simulation with warm-up periods to minimize the impact of initial conditions.

Rainfall Event	Warm-Up Period	Simulation Period
Event 1	1 July–5 August 2020	5–13 August 2020
Event 2	1 July–1 August 2022	1–9 September 2022

The Digital Elevation Model (DEM) data with resolutions of 100 m, 250 m, and 500 m were built based on the National Aeronautics and Space Administration (NASA) Shuttle Radar Topography Mission (SRTM) 1-Arc second dataset through resampling. Table 2 provides detailed information regarding spatial dimensions, with varying resolution. We generated land cover and soil data in a WRF binary format by resampling based on datasets from the Ministry of Environment and the National Institute of Agricultural Sciences in Korea and according to United States Geological Survey (USGS) standards (Figure 2).

Table 2. Comparison of grid numbers by resolution, including details on the number of model input data grids determined by resolutions of 100 m, 250 m, and 500 m for comparison in this study.

Resolution	No. LSM Grids	No. Routing Grids	No. Channel Grids
100 m	459,441 (639 × 719)	459,441	9804
250 m	72,865 (247 × 295)	72,865	1484
500 m	18,081 (123 × 147)	18,081	378

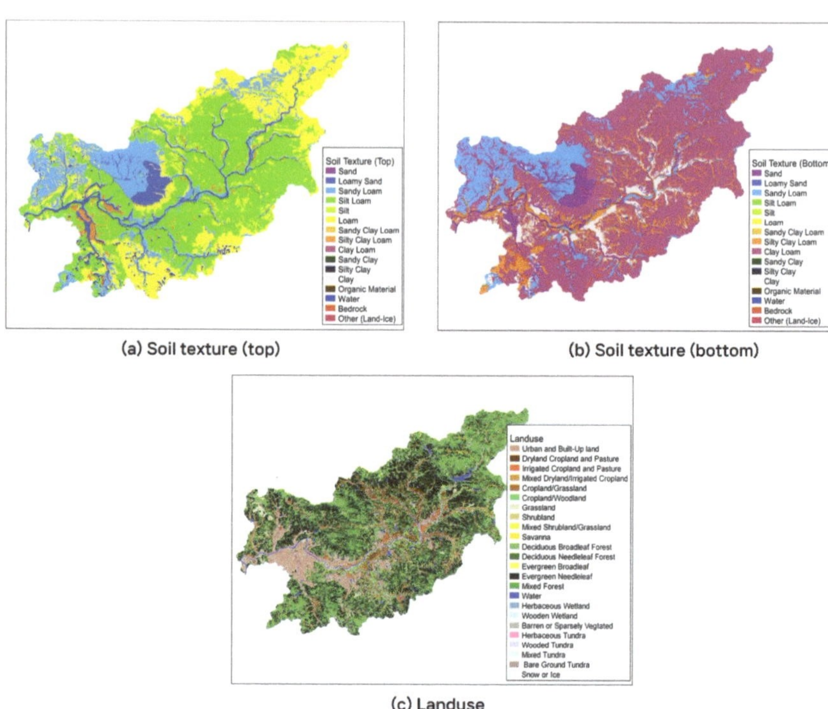

(a) Soil texture (top)

(b) Soil texture (bottom)

(c) Landuse

Figure 2. Soil and land use distribution maps (100 m resolution).

2.2. Methodology

2.2.1. WRF-Hydro

In this study, we used WRF-Hydro version 5.2 in conjunction with Noah-Multiparameterization (Noah-MP) (Niu et al. [21]) LSM model. The employed WRF-Hydro model (Figure 3) integrates atmospheric and hydrological processes, comprising LSM, hydrological, and separate-aggregate modules. Among various LSM models available within the current WRF-Hydro framework, Noah-MP LSM has an advantage over Noah LSM in replicating surface flux, surface temperature during dry periods, snow characteristics (snow water equivalent and depths), and runoff (Niu et al. [21]). The Noah-MP LSM operates as a spatially distributed 1D model based on four soil layers, addressing surface and subsurface flow paths vertically with respect to meteorological forcing. The ranges of soil depth configuration are 0–0.1 m, 0.1–0.4 m, 0.4–1 m, and 1–2 m. Both LSM resolution and hydrological routing resolution were consistently set at 100 m, 250 m, and 500 m. For instance, if the LSM resolution is 100 m, hydrological routing (i.e., overland flow and channel routing) also operates at 100 m resolution. The spatial resolution characteristics in the WRF-Hydro simulation experimental setup can be summarized in terms of the following: number of the grid (Table 2), channel grid (Figure 4), and streamflow order. Figure 5 is the distribution of streamflow order according to spatial resolution. The streamflow order in WRF-Hydro comprises parameters such as roughness coefficients and width, corresponding to orders from 1st to 10th. However, no channels higher than the 5th order have been generated in the medium-sized Geumho River catchment.

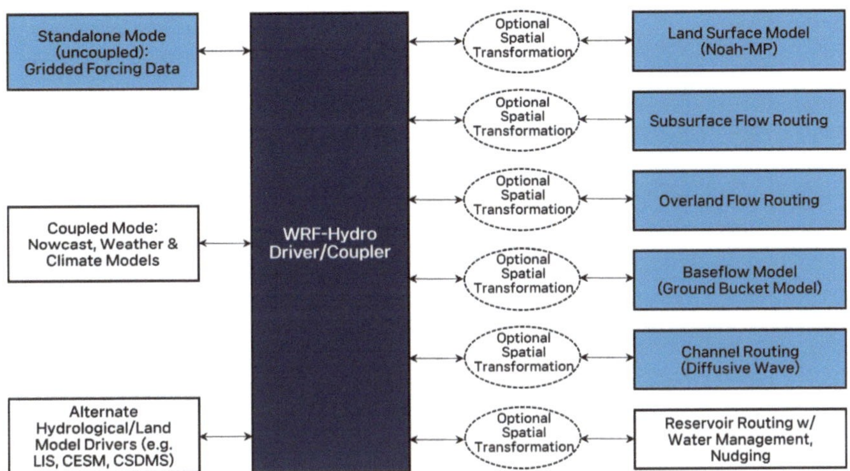

Figure 3. Schematic diagram of the WRF-Hydro modular modeling structure (adapted from [22]). Submodules and mode (Standalone Mode) opted for modeling are colored in blue.

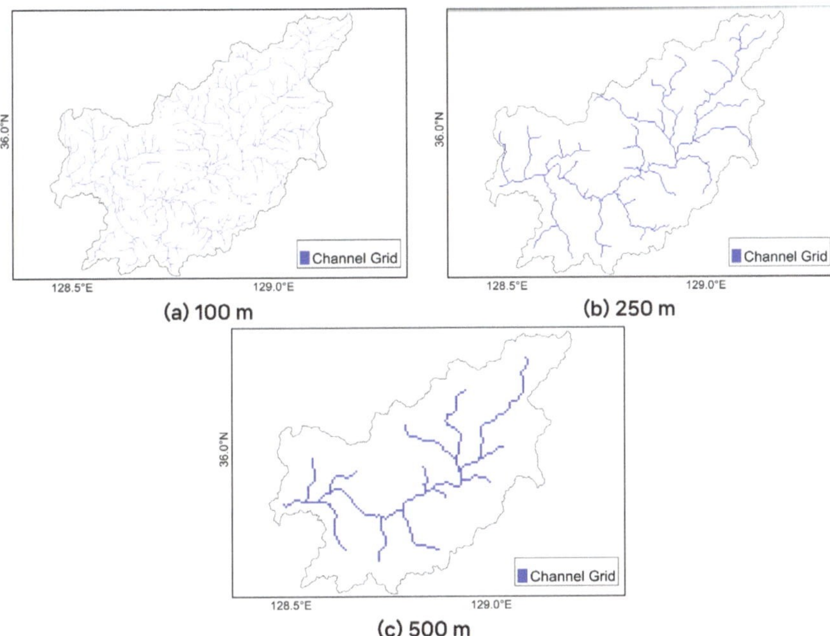

Figure 4. Distribution of channel grids according to resolution.

Hydrological modules enhance the descriptions of the infiltration excess process of Noah-MP and the lateral movement in saturated subsurface processes. Overland routing, subsurface flow, baseflow, and channel routing are incorporated. The methods for channel routing include vector-based routing such as Muskingum and Muskingum–Cunge, as well as grid-based routing based on diffusive wave approximation. Of these methods, the explicit, one-dimensional, variable time-stepping diffusive wave (Downer et al. [23]) was employed for the gridded channel network in this study. The equations for the variable time-stepping diffusive wave are as follows:

$$\frac{\partial A}{\partial t} + \frac{\partial Q}{\partial x} = q_1 \tag{1}$$

$$-\frac{\partial h}{\partial x} + S_o = S_f = \left(\frac{nQ}{AR^{2/3}}\right)^2 \tag{2}$$

where A denotes the wetted channel cross-sectional area, Q denotes the flow rate, q_1 denotes the lateral inflow, h denotes the water depth, S_o denotes the channel bed slope, n denotes the Manning's roughness coefficient for the channel bed, and R denotes the hydraulic radius of the channel cross-section.

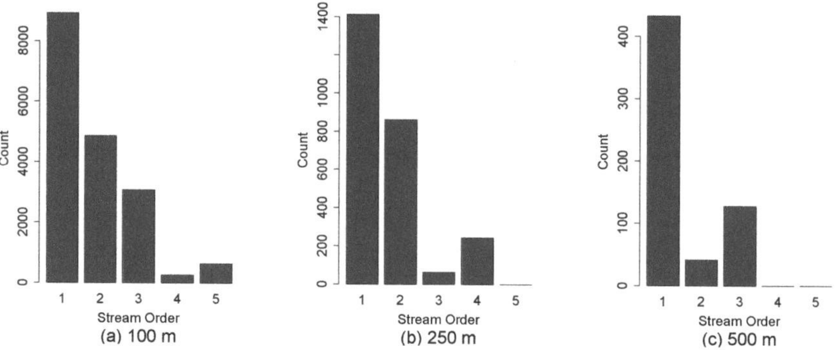

Figure 5. Distribution of streamflow order according to resolution.

To maintain computational stability and prevent numerical dispersion, a 6 s time interval was chosen for the overland and channel routing, satisfying the Courant condition criteria for diffusive wave routing at resolutions of 100 m, 250 m, and 500 m. Both one-way (standalone) and fully coupled integration of WRF and WRF-Hydro are supported within the current WRF-Hydro modeling system. In this study, a standalone version of WRF-Hydro was configured actively considering overland flow, saturated subsurface flow, gridded channel routing, and conceptual baseflow while without the lake and reservoir modules.

We adopt the following approach to focus on integrating PEST and WRF-Hydro in analyzing the impact of spatial resolution on model calibration. The aim is to identify the most influential parameters and adjust them with multiplicative scaling factors across the entire catchment to maintain spatial variation and model relationships, as suggested by Gupta et al. [24]. As a result of the literature review, decay coefficient 'k' is identified as the most influential parameter in Equation (3) for infiltration capacity (Kim et al. [15]; Lee et al. [11]; Zhang et al. [14]; Tolson et al. [6]; Chen et al. [4]). Infiltration capacity 'I_c' in this equation is modeled as follows (Schaake et al. [25]):

$$I_c = D_x(1 - e^{-kt}) \tag{3}$$

where D_x represents the maximum water-holding capacity of the soil column, k indicates the decay coefficient, and t denotes the elapsed time. Decay coefficient k is defined by the following equation:

$$k = \left(REFKDT\frac{DKSAT}{REFDK}\right) \cdot \left(\frac{DT}{REFDK}\right) \tag{4}$$

where $DKSAT$ represents the saturated hydraulic conductivity, while $REFKDT$ and $REFDK$ stand for parameters related to surface streamflow (Gochis et al. [26]). DT denotes the time step in seconds. Both $REFKDT$ and $REFDK$ are adjustable parameters. As the effect of adjusting *REFDK* is equivalent to the effect of adjusting *REFKDT* for parameter '*k*', it is not necessity to calibrate both (Kim et al. [15]). As such, we calibrate only *REFKDT* in this work. Regarding Equation (4), if *REFKDT* increases or decreases, *k* increases or decreases. *k* influences infiltration capacity when the maximum water holding capacity, D_x, is given. Accordingly, *REFDK* may be considered as controlling the streamflow. For PEST calibration of *REFKDT*, the default values are set at 3, with minimum and maximum values of 1×10^{-3} and 1×10^{2}, respectively. All other parameters in the LSM are set to default values in WRF-Hydro (Gochis et al. [26]).

REFKDT is calibrated based on simulations spanning 8 days for Rainfall Event 1 (from 5th to 13th, August 2020) and Rainfall Event 2 (from 1st to 9th, September 2022). Prior to PEST calibration, we manually adjusted groundwater bucket model parameters by comparing them with observed streamflow at the gauge of Gangchang during the warm-up periods for each rainfall event to calibrate the initial conditions. Table 3 outlines the parameters and their ranges that were tested in an effort to match model conditions.

Table 3. Manual calibration parameters and ranges. The groundwater bucket model equation is as follows: $Qexp = C(exp(E(Z/Z_{\max})) - 1)$. As E increases, C increases, and as $Z_{\max}$ decreases, discharge increases.

Calibrated Parameter	Range
$Z_{\max}$	50–200
C	0.5–1.5
E	1–4

2.2.2. PEST

We adopted an automated calibration procedure based on PEST software version 17.5 (Doherty et al. [27]). This procedure minimizes the objective function, which is the sum of the mean squared differences between the modeled and observed streamflow, employing the Gauss–Marquardt–Levenberg nonlinear least squares method. We calibrated a single parameter using 8-day observation data and one observation gauge along with prior information items. For each prior information item, we assigned a value equal to the default value provided by WRF-Hydro v5.2 (or the logarithm of that default value) to the adjustable parameter, assuming that the default parameter set is preferred.

2.2.3. Assessment Index

In this study, simulated streamflow is assessed using two statistical assessment criteria: Root Mean Square Error (*RMSE*) and the Nash–Sutcliffe Efficiency (*NSE*; Nash et al. [28], Moriasi et al. [29]). *RMSE* is a measure of the difference between the model-predicted and actual observed values in which the square root of the average of squared differences between predicted and observed values is calculated in evaluating how close the model is to the observed data. The equation defining the variables for *RMSE* calculation is

$$RMSE = \sqrt{\frac{1}{n}\sum_{t=0}^{n}(Y_t^{obs} - Y_t^{sim})^2} \tag{5}$$

where n represents the total number of observed data points, Y^{obs} denotes the observed discharge values, and Y^{sim} indicates the WRF-Hydro simulated results.

NSE quantifies the accuracy of modeled discharge by comparison to the mean of observed data. The equation defining the variables for *NSE* calculation is

$$NSE = 1 - \frac{n\sum_{t=0}^{n}(Y_t^{obs} - Y_t^{sim})^2}{n\sum_{t=0}^{n}(Y_t^{obs} - \overline{Y}_{mean}^{obs})^2} \tag{6}$$

where $\overline{Y}_{mean}^{obs}$ is the mean observed discharge. The *NSE* values range from $-\infty$ to 1. The closer the *NSE* value is to 1, the higher the accuracy of the evaluated model, and the higher the degree of agreement between the model prediction and the observed data. An *NSE* value below 0 means that the model's performance is worse than the prediction by an average of the observed data.

3. Results

3.1. Comparative Analysis of Streamflow Predictions with Varying Spatial Resolution

In the analysis of streamflow predictions using WRF-Hydro, the model spatial input data were generated at resolutions of 100 m, 250 m, and 500 m, and simulations were performed for two rainfall events in 2020 and 2022, utilizing default parameters in all cases. This section analyzed the results of no-calibration streamflow predictions, with emphasis on the outcomes based on model spatial resolution. Figure 6 shows a comparison of the simulated streamflow for Rainfall Event 1, from 5th to 13th, August 2020, at three different model spatial resolutions in comparison with the observation at two gauges: Gangchang (in the lowermost region of the Geumho River) and Ansim (in the mid-lower region of the same river). The black solid, green dashed, light blue dashed lines illustrate the observation and the simulations at 100 m and 250 m resolutions, respectively. In Figure 6, the hydrograph at 100 m resolution has a similar pattern to the observed hydrograph with a slightly overestimated peak flow. At resolutions of 250 m and 500 m, the simulated hydrographs are found to be underestimated in a magnitude of approximately half of the observed data. For Event 1, the observed peak flow at Gangchang (lowermost location) was 2305 m^3/s. Simulated peak flows at resolutions of 100 m, 250 m, and 500 m were 2775.3 m^3/s, 1042.2 m^3/s, and 1214.2 m^3/s, respectively. While the peak flow at 100 m resolution was higher than the observed value, it showed the smallest difference. This trend is also evident in terms of *NSE* for no calibration cases in Event 1 shown in Table 4, with the highest *NSE* value of 0.868 at 100 m resolution. The Root Mean Square Error (*RMSE*) and *NSE* values of the no-calibration cases were lower for Rainfall Event 2 than for Event 1. It was presumably because Rainfall Event 2 was triggered by a sudden heavy rain event due to Typhoon Hinnamnor in the absence of preceding rainfall implicating different hydrological initial conditions from Event 1 which could be characterized as a typical torrential rainfall during a monsoon season. Nonetheless, among the three resolutions, 100 m resolution demonstrated the smallest associated errors in no-calibration cases for Event 2.

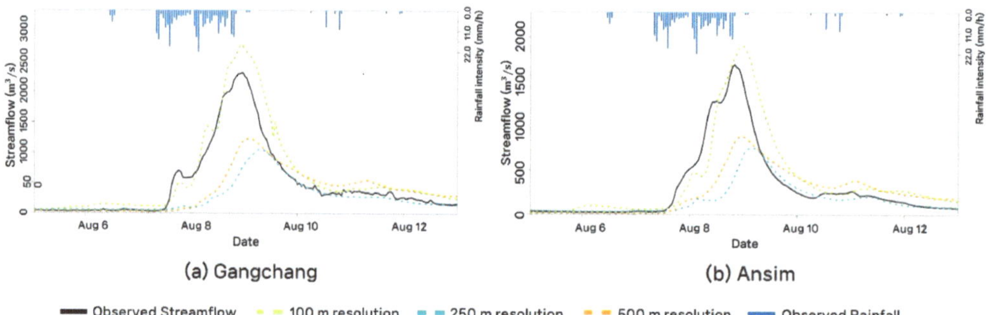

Figure 6. Streamflow prediction results based on model spatial resolution. The simulation of streamflow prediction results is performed for Rainfall Event 1 and compared with observed streamflow data at the gauges of Gangchang and Ansim.

Table 4. Evaluation metrics before (no calibration) and after calibration (calibrated) by PEST across different grid resolutions.

Event	Forecast Gauges	Resolution	RMSE (m³/s)		NSE	
			No Calibration	Calibrated	No Calibration	Calibrated
Event 1	Gangchang	100 m	147.9	135.5	0.868	0.889
		250 m	245.1	168.5	0.266	0.843
		500 m	250.4	134.7	0.476	0.872
	Ansim	100 m	124.6	109.9	0.791	0.841
		250 m	160.6	111.9	0.255	0.794
		500 m	169.1	89.5	0.524	0.906
	Geumchang	100 m	85.5	85.4	0.832	0.848
		250 m	118.5	106.0	0.277	0.644
		500 m	108.8	83.2	0.427	0.808
Event 2	Gangchang	100 m	113.6	43.7	0.058	0.938
		250 m	127.0	37.5	−0.187	0.972
		500 m	153.4	33.5	−0.322	0.971
	Ansim	100 m	100.3	60.2	−0.053	0.742
		250 m	111.1	44.7	−0.145	0.934
		500 m	129.0	38.6	−0.244	0.928
	Geumchang	100 m	75.8	38.4	−0.058	0.684
		250 m	74.9	29.8	−0.087	0.903
		500 m	82.2	26.3	−0.142	0.830

3.2. Impact of Scale-Specific Parameter Estimation on Streamflow Simulation

To analyze the impact of spatial resolution and model calibration, event-specific calibration was implemented for Rainfall Events 1 and 2 using streamflow data from the Gangchang station (downstream observatory) while those from the Ansim and Geumchang gauges were not included in the calibration process.

Table 4 presents the results from evaluation of streamflow predictions before and after model calibration, utilizing performance metrics such as *RMSE* and *NSE* for each grid resolution. For Rainfall Event 1, the resolution that yielded the smallest *RMSE* (m³/s) error was 500 m for Gangchang and Geumchang and 250 m for Ansim. For Gangchang, the error was significantly reduced from 250.4 to 134.7 at 500 m resolution. In terms of *NSE*, the best results were 100 m for Gangchang and Geumchang and 500 m for Ansim, with the greatest improvement of *NSE* from 0.266 to 0.843 for Gangchang at a resolution of 250 m. In particular, the calibration for 250 m resolution enhanced the performance in terms of the *NSE* with values higher than 0.9, improved by about twofold compared to no-calibration cases in all three locations (Gangchang, Ansim, and Geumchang). In the case of 500 m resolution, the *NSE* increased by 80 % compared to no-calibration cases. At 100 m spatial resolution, the *NSE* varied by less than 5 % after calibration due to superior streamflow simulations without calibration for Event 1.

In Rainfall Event 2, the spatial resolution with the smallest *RMSE* error for each point was 500 m for Gangchang, Ansim, and Geumchang. The error was significantly reduced from 153.4 to 33.5 for Gangchang. In terms of *NSE*, the best results were 250 m for Gangchang, Ansim, and Geumchang, with the greatest improvement of *NSE* from −0.322 to 0.971 for Gangchang at a resolution of 500 m.

Upon comparing the pre- and post-calibration results of the two rainfall events, Rainfall Event 2 exhibited significant improvements based on *NSE* and *RMSE*, as shown in Table 4. When calibration was implemented, significant enhancements were observed in the results for grid resolutions of 250 m and 500 m in comparison to the 100 m grid resolution for Event 2.

Figure 7 presents streamflow simulations with and without calibration across varying grid resolutions. Figure 7a,c,e correspond to Rainfall Event 1 while Figure 7b,d,f correspond to Rainfall Event 2. The black solid, red dashed, and purple dashed lines indicate the observations, no-calibration, and calibrated simulations, respectively. No-calibration cases were simulated using a default *REFKDT* parameter value of three. For both Rainfall Events 1 and 2, calibrated streamflow simulations exhibit significant improvement compared to no-calibration ones.

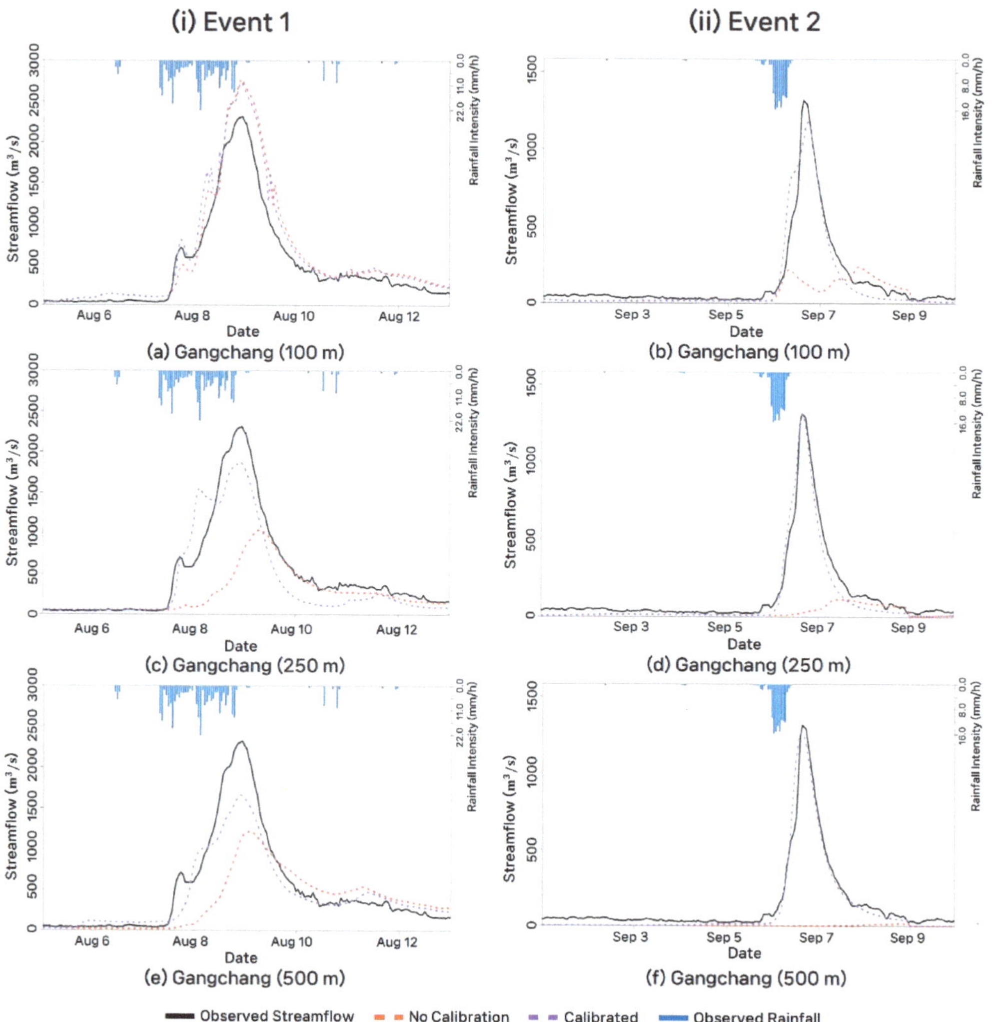

Figure 7. A comparison of resolution-specific simulations before (red) and after (purple) calibration. The inverted y-axis at the top of the graph represents rainfall.

In Figure 8, the analysis is focused on the results of streamflow prediction before and after calibration at Ansim and Geumchang during Rainfall Event 1. Figure 8a,c,e display WRF-Hydro streamflow simulations at various grid resolutions in Ansim, both before and after calibration by PEST. Similarly, Figure 8b,d,f illustrates WRF-Hydro simulations in Geumchang at multiple resolutions. Comparable patterns to those observed in Gangchang were identified at both the Ansim and Geumchang locations for resolutions of 100 m, 250 m, and 500 m.

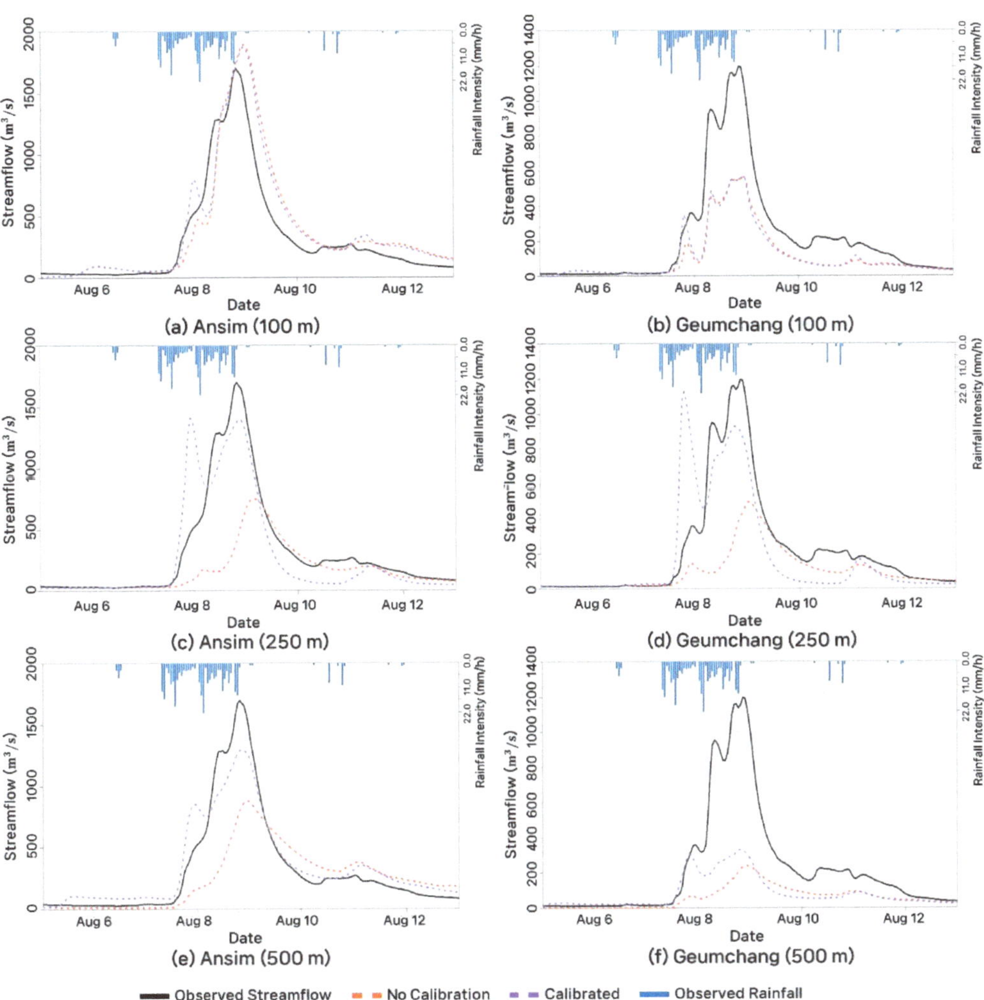

Figure 8. Model calibration results by resolution. Results of the calibration of the WRF-Hydro model were obtained using streamflow from Rainfall Event 1 at Gangchang Bridge and compared for Gangchang, Ansim, and Geumchang.

As depicted in Figure 8a,b, 100 m resolution showed the most improved predictions in terms of peak flow after calibration. The calibrated results at 250 m resolution, as illustrated in Figure 8c,d, also demonstrated significant improvements in both the maximum peak flow and its timing. In contrast, a higher peak flow for the previous rainfall was observed in the calibrated results. In Figure 8e,f, representing 500 m resolution, enhancements in both peak and post-peak flows were observed, despite the high peak for the prediction of previous rainfall streamflow, such as at the Ansim gauge.

Table 5 presents the calibrated parameters for the WRF-Hydro model, differentiated by spatial resolution, for Rainfall Events 1 and 2, along with calibration runtime. The runtime for PEST parameter calibration exhibited significant variation across different resolutions and rainfall events. Specifically, during Rainfall Event 2, which had considerably less preceding rainfall compared to Rainfall Event 1, calibration runtime for 100 m and 500 m resolutions more than doubled. There was no clear pattern between the resolution and the calibrated parameters. In Rainfall Event 2, the calibrated parameters of 100 m and 500 m differed by more than twofold compared to Rainfall Event 1. For 250 m resolution, there was no significant difference in the calibrated parameters between Rainfall Events 1 and 2 and only a small variation in calibration time. This implies that with default parameters set for 250 m resolution, both the computational time and the required adjustments to parameter values are minimal. This suggests there is an advantage in using a 250 m resolution for constructing new models.

Table 5. Optimized PEST parameters and calibration runtimes on different resolutions. With the same parameters, the single model runtimes were approximately 263 ± 10 s for 100 m, 84 ± 10 s for 250 m, and 68 ± 10 s for 500 m, respectively.

ID		Resolution		
		100 m	**250 m**	**500 m**
Event 1	*REFKDT*	1.283	0.203	0.555
	PEST Runtime (min)	216	71	54
Event 2	*REFKDT*	0.079	0.158	0.070
	PEST Runtime (min)	695	90	125

Our findings show differences and similarities compared to previous spatial resolution studies. Kim et al. [15] reported that in the absence of parameter calibration, 250 m spatial resolution was a good choice in terms of performance and calculation requirements for both LSM and routing models. Meanwhile, in our case, without calibration, the streamflow predictions were the most accurate at 100 m resolution. However, in terms of calibration performance and calibration calculation requirements, the strong advantage of a 250 m spatial resolution was found, in line with the suggestions of Kim et al. [15]. In other studies, such as Abbazadeh et al. [2], streamflow prediction accuracy was reported to increase with finer resolution from 36 to 9 to 1 km. However, simulation experiments in this study revealed that the accuracy did not increase linearly when the resolution became finer than 1 km such as 100 m, 250 m, and 500 m. At 250 m resolution, both calibration runtime and changes in the calibrated parameter values were minimized.

4. Conclusions

This study investigated the effects of spatial resolution on streamflow predictions by employing distributed hydrological modeling in conjunction with the parameter estimation tool, PEST. Simulation experiments for the Geumho River catchment in South Korea were performed using WRF-Hydro with spatial resolutions of 100 m, 250 m, and 500 m for land surface and routing components, focusing on flood periods. To assess the impact of model spatial resolution, no-calibration simulations were examined with default parameter sets at different model spatial resolutions. We then evaluated event-specific calibration results in terms of calibrated parameters and calibration runtimes. The key findings are summarized as follows:

1. In the simulations without calibration using the default parameter set, 100 m resolution exhibited superior performance in terms of *NSE*, although calibration was deemed necessary for Rainfall Event 2 (Rainfall Event 1 *NSE*: 0.868; Rainfall Event 2 *NSE*: 0.058).
2. For Rainfall Event 2, the *NSE* and *RMSE* results of calibrated simulations indicated significant improvement compared to those for Rainfall Event 1. In particular, at 250 m resolution, the *NSE* was 0.9 or higher at all gauges, with the evaluation index value more than doubled relative to no-calibration cases, thereby indicating more effective calibration compared to other resolutions.

3. Calibration runtime for calibrating PEST parameters varied significantly across resolutions and rainfall events. In particular, for Event 2 with a drier hydrological initial condition, the calibration runtimes at 100 m and 500 m resolutions nearly doubled compared to those for Event 1. For 250 m resolution, there was no significant difference in the calibrated parameters between Rainfall Events 1 and 2 (calibrated parameter of Rainfall Event 1: 0.203; Rainfall Event 2: 0.158).

We evaluated the effect of spatial resolution on the parameters and streamflow simulations, postulating that there might be a pattern in the variation of calibrated parameter values as spatial resolution changes. However, no scale-dependent patterns were found in calibrated parameters at least for the two selected rainfall events. This phenomenon is partly due to the changes in configuration and interaction among hydrological components at a finer spatial resolution. Therefore, we propose that resolution-aware parameter regionalization schemes be developed as a potential future research area. Such schemes would enable effective calibration in high-resolution models by integrating insights from lower-resolution calibrations and taking into account resolution-specific discrepancies. In the era of digital twins and hyperconnectivity, distributed modeling using new information with increasing volume and finer spatial resolution is expected to provide an improved understanding of hydrological processes and their interactions.

Author Contributions: Conceptualization, S.J.N. and B.K.; methodology, S.J.N. and B.K.; software, S.J.N., B.K., G.L., S.K. and Y.L.; validation, B.K., G.L. and Y.L.; formal analysis, B.K., G.L. and Y.L.; investigation, B.K., G.L., S.K. and Y.L.; resources, S.J.N.; writing—original draft preparation, B.K.; writing—review and editing, S.J.N., B.K., G.L., S.K. and Y.L.; visualization, B.K. and Y.L.; supervision, S.J.N.; project administration, S.J.N. and B.K.; funding acquisition, S.J.N. All authors have read and agreed to the published version of the manuscript.

Funding: This work was supported by the Korea Environment Industry and Technology Institute (KEITI) grant, funded by the Korea Ministry of Environment (MOE) (RS-2023-00218973) and the National Research Foundation of Korea (NRF) grant, funded by the Korea government (MSIT) (No. 2022R1A4A5028840, RS-2023-00246532). This support is gratefully acknowledged.

Data Availability Statement: The data used in this study are available upon request.

Acknowledgments: We express our gratitude to Euisang Jeong of the HydroCore Ltd. and Eun-Jeong Lee of National Institute of Environmental Research for providing valuable data and comments. We also appreciate two anonymous reviewers and an academic editor for their comments to improve the quality of this research.

Conflicts of Interest: The authors declare no conflicts of interest.

References

1. Maxwell, R.M.; Condon, L.E.; Kollet, S.J. A high-resolution simulation of groundwater and surface water over most of the continental US with the integrated hydrologic model ParFlow v3. *Geosci. Model Dev.* **2015**, *8*, 923–937. [CrossRef]
2. Abbaszadeh, P.; Gavahi, K.; Moradkhani, H. Multivariate remotely sensed and in-situ data assimilation for enhancing community WRF-Hydro model forecasting. *Adv. Water Resour.* **2020**, *145*, 103721. [CrossRef]
3. Moges, E.; Demissie, Y.; Li, H. Uncertainty propagation in coupled hydrological models using winding stairs and null-space Monte Carlo methods. *J. Hydrol.* **2020**, *589*, 125341. [CrossRef]
4. Chen, Y.; Li, J.; Xu, H. Improving flood forecasting capability of physically based distributed hydrological models by parameter optimization. *Hydrol. Earth Syst. Sci.* **2016**, *20*, 375–392. [CrossRef]
5. Verri, G.; Pinardi, N.; Gochis, D.; Tribbia, J.; Navarra, A.; Coppini, G.; Vukicevic, T. A meteo-hydrological modelling system for the reconstruction of river runoff: the case of the Ofanto river catchment. *Nat. Hazards Earth Syst. Sci.* **2017**, *17*, 1741–1761. [CrossRef]
6. Tolson, B.A.; Shoemaker, C.A. Cannonsville Reservoir Watershed SWAT2000 model development, calibration and validation. *J. Hydrol.* **2007**, *337*, 68–86. [CrossRef]
7. Setegn, S.G.; Srinivasan, R.; Melesse, A.M.; Dargahi, B. SWAT model application and prediction uncertainty analysis in the Lake Tana Basin, Ethiopia. *Hydrol. Process. Int. J.* **2010**, *24*, 357–367. [CrossRef]
8. Senatore, A.; Mendicino, G.; Gochis, D.J.; Yu, W.; Yates, D.N.; Kunstmann, H. Fully coupled atmosphere-hydrology simulations for the central Mediterranean: Impact of enhanced hydrological parameterization for short and long time scales. *J. Adv. Model. Earth Syst.* **2015**, *7*, 1693–1715. [CrossRef]

9. Naabil, E.; Kouadio, K.; Lamptey, B.; Annor, T.; Chukwudi Achugbu, I. Tono basin climate modeling, the potential advantage of fully coupled WRF/WRF-Hydro modeling System. *Model. Earth Syst. Environ.* **2023**, *9*, 1669–1679. [CrossRef]

10. Wang, W.; Liu, J.; Li, C.; Liu, Y.; Yu, F.; Yu, E. An Evaluation Study of the Fully Coupled WRF/WRF-Hydro Modeling System for Simulation of Storm Events with Different Rainfall Evenness in Space and Time. *Water* **2020**, *12*, 1209. [CrossRef]

11. Lee, J.; Kim, Y.; Wang, D. Assessing the characteristics of recent drought events in South Korea using WRF-Hydro. *J. Hydrol.* **2022**, *607*, 127459. [CrossRef]

12. Cho, K.; Kim, Y. Improving streamflow prediction in the WRF-Hydro model with LSTM networks. *J. Hydrol.* **2022**, *605*, 127297. [CrossRef]

13. Liu, S.; Wang, J.; Wang, H.; Wu, Y. Post-processing of hydrological model simulations using the convolutional neural network and support vector regression. *Hydrol. Res.* **2022**, *53*, 605–621. [CrossRef]

14. Zhang, J.; Lin, P.; Gao, S.; Fang, Z. Understanding the re-infiltration process to simulating streamflow in North Central Texas using the WRF-hydro modeling system. *J. Hydrol.* **2020**, *587*, 124902. [CrossRef]

15. Kim, S.; Shen, H.; Noh, S.; Seo, D.J.; Welles, E.; Pelgrim, E.; Weerts, A.; Lyons, E.; Philips, B. High-resolution modeling and prediction of urban floods using WRF-Hydro and data assimilation. *J. Hydrol.* **2021**, *598*, 126236. [CrossRef]

16. Abbas, S.A.; Bailey, R.T.; White, J.T.; Arnold, J.G.; White, M.J.; Čerkasova, N.; Gao, J. A framework for parameter estimation, sensitivity analysis, and uncertainty analysis for holistic hydrologic modeling using SWAT+. *Hydrol. Earth Syst. Sci.* **2024**, *28*, 21–48. [CrossRef]

17. De Wit, J.A.; Van Huijgevoort, M.H.; Van Dam, J.C.; Van Den Eertwegh, G.A.; Van Deijl, D.; Ritsema, C.J.; Bartholomeus, R.P. Hydrological consequences of controlled drainage with subirrigation. *J. Hydrol.* **2024**, *628*, 130432. [CrossRef]

18. Fersch, B.; Senatore, A.; Adler, B.; Arnault, J.; Mauder, M.; Schneider, K.; Völksch, I.; Kunstmann, H. High-resolution fully coupled atmospheric–hydrological modeling: a cross-compartment regional water and energy cycle evaluation. *Hydrol. Earth Syst. Sci.* **2020**, *24*, 2457–2481. [CrossRef]

19. Sofokleous, I.; Bruggeman, A.; Camera, C.; Eliades, M. Grid-based calibration of the WRF-Hydro with Noah-MP model with improved groundwater and transpiration process equations. *J. Hydrol.* **2023**, *617*, 128991. [CrossRef]

20. Wang, J.; Wang, C.; Rao, V.; Orr, A.; Yan, E.; Kotamarthi, R. A parallel workflow implementation for PEST version 13.6 in high-performance computing for WRF-Hydro version 5.0: a case study over the midwestern United States. *Geosci. Model Dev.* **2019**, *12*, 3523–3539. [CrossRef]

21. Niu, G.Y.; Yang, Z.L.; Mitchell, K.E.; Chen, F.; Ek, M.B.; Barlage, M.; Kumar, A.; Manning, K.; Niyogi, D.; Rosero, E.; et al. The community Noah land surface model with multiparameterization options (Noah-MP): 1. Model description and evaluation with local-scale measurements. *J. Geophys. Res. Atmos.* **2011**, *116*, D12109. [CrossRef]

22. Gochis, D.J.; Yu, W.; Yates, D.N. *The WRF-Hydro Model Technical Description and User's Guide*; Version 1.0, NCAR Technical Document; NCAR: Boulder, CO, USA, 2013; 120p.

23. Downer, C.W.; Ogden, F.L.; Martin, W.D.; Harmon, R.S. Theory, development, and applicability of the surface water hydrologic model CASC2D. *Hydrol. Process.* **2002**, *16*, 255–275. [CrossRef]

24. Gupta, H.; Sorooshian, S.; Hogue, T.; Boyle, D. Advances in automatic calibration of watershed models. *Calibration Watershed Model.* **2003**, *6*, 9–28. [CrossRef]

25. Schaake, J.C.; Koren, V.I.; Duan, Q.Y.; Mitchell, K.; Chen, F. Simple water balance model for estimating runoff at different spatial and temporal scales. *J. Geophys. Res. Atmos.* **1996**, *101*, 7461–7475. [CrossRef]

26. Gochis, D.; Barlage, M.; Dugger, A.; FitzGerald, K.; Karsten, L.; McAllister, M.; McCreight, J.; Mills, J.; Pan, L.; Rafieeinasab, A.; et al. *WRF-Hydro Model Source Code Version 5*; Medium: Fortran90; UCAR/NCAR: Boulder, CO, USA, 2018. [CrossRef]

27. Doherty, J. *Manual for PEST*, 5th ed.; Watermark Numerical Computing: Brisbane, Australia, 2002.

28. Nash, J.E.; Sutcliffe, J.V. River flow forecasting through conceptual models part I—A discussion of principles. *J. Hydrol.* **1970**, *10*, 282–290. [CrossRef]

29. Moriasi, D.N.; Arnold, J.G.; Van Liew, M.W.; Bingner, R.L.; Harmel, R.D.; Veith, T.L. Model Evaluation Guidelines for Systematic Quantification of Accuracy in Watershed Simulations. *Trans. ASABE* **2007**, *50*, 885–900. [CrossRef]

Article

Spatial–Temporal Water Balance Evaluation in the Nile Valley Upstream of the New Assiut Barrage, Egypt, Using WetSpass-M

Zhanchao Li [1,2,*], Ahmed S. Eladly [1,3], Ehab Mohammad Amen [4,5,6], Ali Salem [7,8,*], Mahmoud M. Hassanien [3,9], Khailah Ebrahim Yahya [1,10] and Jiaming Liang [1]

[1] College of Water Resources Science and Engineering, Yangzhou University, Yangzhou 225009, China; adly_hoo.2014@yahoo.com (A.S.E.); ebrahimkhailah@gmail.com (K.E.Y.); fnliangjiaming@163.com (J.L.)

[2] Intelligent Water Conservancy Research Institute, Nanjing Jurise Engineering Technology, Nanjing 211899, China

[3] Egyptian Ministry of Water Resources and Irrigation, Imbaba 12666, Egypt; mahmoud_kyr1@yahoo.com

[4] Department of Applied Geology, College of Science, Tikrit University, Tikrit 34001, Iraq; ehab.m86@tu.edu.iq

[5] Doctoral School of Science, Technology, and Engineering, University of Granada, 18071 Granada, Spain

[6] Natural Resources Research Center (NRRC), Tikrit University, Tikrit 34001, Iraq

[7] Civil Engineering Department, Faculty of Engineering, Minia University, Minia 61111, Egypt

[8] Structural Diagnostics and Analysis Research Group, Faculty of Engineering and Information Technology, University of Pécs, 7624 Pécs, Hungary

[9] Institute of Water and Hydropower Resources, Beijing 100038, China

[10] Civil Engineering Department, College of Engineering, Thamar University, Dhamar 504408, Yemen

* Correspondence: 006520@yzu.edu.cn (Z.L.); alisalem@gamma.ttk.pte.hu (A.S.)

Citation: Li, Z.; Eladly, A.S.; Amen, E.M.; Salem, A.; Hassanien, M.M.; Yahya, K.E.; Liang, J. Spatial–Temporal Water Balance Evaluation in the Nile Valley Upstream of the New Assiut Barrage, Egypt, Using WetSpass-M. *Water* **2024**, *16*, 543. https://doi.org/10.3390/w16040543

Academic Editors: Agnieszka Rutkowska and Katarzyna Baran-Gurgul

Received: 27 December 2023
Revised: 25 January 2024
Accepted: 31 January 2024
Published: 9 February 2024

Abstract: The components of water balance (WBC) that involve precipitation, evapotranspiration, runoff, irrigation, and groundwater recharge are critical for understanding the hydrological cycle and water management of resources in semi-arid and arid areas. This paper assesses temporal and spatial distributions of surface runoff, actual evapotranspiration, and groundwater recharge upstream of the New Assiut Barrage (NAB) in the Nile Valley, Upper Egypt, using the WetSpass-M model for the period 2012–2020. Moreover, this study evaluates the effect of land cover/land use (LULC) alterations in the study period on the WBC of the NAB. The data provided as input for the WetSpass-M model in the structure of raster maps using the Arc-GIS tool. Monthly meteorological factors (e.g., temperature, rainfall, and wind speed), a digital elevation model (DEM), slope, land cover, irrigation cover, a soil map, and depth to groundwater are included. The long-term temporal and spatial mean monthly irrigation and precipitation (127 mm) is distributed as 49% (62 mm) actual evapotranspiration, 15% (19 mm) groundwater recharge, and 36% (46 mm) surface runoff. The replacement of cropland by built-up areas was recognized as the primary factor responsible for the major decrease in groundwater, an increase in evapotranspiration and an increase in surface runoff between LCLU in 2012 and 2020. The integration of the WetSpass model with GIS has shown its effectiveness as a powerful approach for assessing WBC. Results were more accurate and reliable when hydrological modeling and spatial analysis were combined. The results of this research can help make well-informed decisions about land use planning and sustainable management of water resources in the upstream area of the NAB.

Keywords: WetSpass-M; LULC; actual evapotranspiration; groundwater recharge; arid areas; Egypt

1. Introduction

Water shortage has been the most obvious consequence of climate change in the Middle East and North Africa (MENA) region; experts have frequently referred to this region as "the world's most water-stressed". The MENA region is expected to be among the first in the world to "effectively run out of water", which is alarming due to water resources being depleted faster than precipitation can replenish them [1]. Egypt is an African country in the northeast. The majority of its area is desert, which the Nile River cuts across from south to north. Since the Nile River is the nation's primary source of freshwater, the majority of people reside in the Nile Valley. Egypt has a hot desert climate, with nearly zero annual average precipitation in the driest regions of Upper Egypt—the south and western deserts [2]. In order to fully utilize Egypt's most valuable natural resource, the water from the Nile, several regulating structures, such as dams and barrages, have been constructed along the river since the early 20th century. One of these is the Assiut barrage, which is currently in need of significant repairs. The rehabilitation design for the barrage includes the creation of a head-pond with water levels that are 0.6 m higher. This development has resulted in an increased flow of seepage water towards agricultural lands and villages, which may potentially lead to a rise in groundwater levels in affected areas, including floodplains located 60 km upstream of the NAB [3,4]. Groundwater is considered the second primary source of drinkable and irrigation water in the Assiut province [5]. Groundwater recharge is a vital aspect in assessing groundwater resources; nevertheless, it is challenging to compute [6]. Due to increased human demand and climate change, as well as the need to compensate for declining surface water supplies during dry months, groundwater usage is anticipated to increase in the future [7]. In addition, evapotranspiration is a critical component of the hydrological cycle that has a direct link with temperature and is one of the most essential results in the water balance equation for any natural region or water body [8]. Recharge estimation is challenging in arid and semi-arid areas where potential evapotranspiration surpasses average precipitation [9,10].

Consequently, different supplies of water must be researched and managed, and the assessment of aspects of water balance is vital for the proper oversight of water management and land, for instance, estimating water availability, quantifying the sustainable rate of groundwater depletion, and preventing land degradation and desertification [11]. The Grand Ethiopian Renaissance Dam (GERD) has significantly increased the vulnerability and sensitivity to water supplies [12]. Several methods have traditionally been used for assessing groundwater recharge, such as experimental techniques, empirical methods, statistical approaches such as water table fluctuation (WTF), the Rorabaugh method, the hydrological budget (HB), and numerical methods such as the simulation of water balance [13,14]. There are various hydrological models accessible today for predicting groundwater recharge, including the soil and water assessment tool (SWAT), a simple daily soil–water balance (SWB), the Système Hydrologique (MIKE SHE), a GPU-accelerated (GPU: Graphics Processing Unit) and LTS-based (LTS: local time step) finite volume shallow water model, Topmodel (topographic hydrologic model), and other physically distributed models that work well in assessing runoff regions in mountainous terrain [15–19]. Recently, water and energy transfer between plants, the atmosphere, and the soil (WetSpass model) [20] has been seen in a quasi-steady state in several studies, and many authors have used it in different zones as it gives good results. It has been constantly used to estimate WBC [21] and has been updated to a WetSpass-M model by reducing the temporal resolution to a monthly scale [22]. A calibrated WetSpass model for the Nile Delta was created by changing the parameters for Nile Valley conditions [23], Palestine, the GAZA strip [24], the Drava basin, Hungary [25,26], the Moulouya basin, Morocco [21], and in Khadir Canal Sub-Division, Pakistan [27]. The WetSpass model has appeared as a very good method for assessing the water balance budget under many parameters as an input dataset: LULC, slope, groundwater depth, and soil texture are taken into account, which are not involved in other methods. Table A1 in Appendix A summarizes the purpose/scope, key features, advantages, and disadvantages of different hydrological/hydrodynamic models.

The spatial and temporal distribution of WBC in the upstream region of the NAB has not yet been studied. Better awareness of the spatial and temporal changes of WBC, particularly surface runoff, actual evapotranspiration, and recharge, is vital for the long-term sustainability and efficient management of water resources upstream of the NAB. The main aims of this work are (1) the evaluation of the temporal and spatial distribution of groundwater recharge, actual evapotranspiration, and surface runoff using a WetSpass-M model under the Geographic Information System (GIS) framework, (2) the evaluation of the relationship between WBC with various land-use classes and years, and (3) the assessment of the effect of land use/land cover changes on the total water budget of the NAB between the LULC in 2012 and 2020. This study is also the first to assess the geographic variability of long-term annual, seasonal, and monthly WBC upstream of the NAB. The data, along with boundary conditions and aquifer geometry, will be applied to develop a groundwater model.

2. Methods and Materials

2.1. Area of Study

The investigated area covers the southern part of Assiut City, located in Upper Egypt, approximately 1080 km^2 along the Nile River (Figure 1). The Nile River splits the region into two portions—the Eastern bank expands between the Tema district in the south and the Abnoub district in the north, while the western part expands between the Tema district in the south and the Assiut District in the north [3]. Its geographical coordinates are 27°20′ N, 31°30′ E. The NAB region extends 60 km towards the south (upstream) and 20 km towards the north (downstream). The upstream area of the NAB includes 8 districts [4], serving an irrigation area of 795 km^2 according to the change in cropland as the main area in 2018.

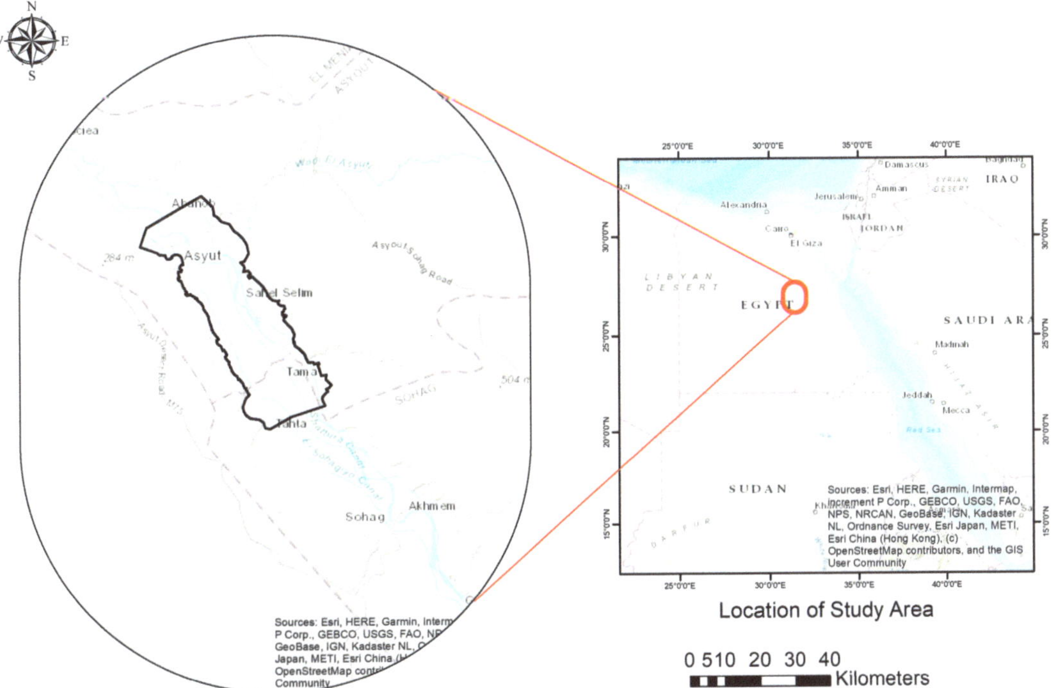

Figure 1. Location of the study area in Egypt.

2.2. Hydrological Simulation (WetSpass-M)

The WetSpass model is a physically based technique that has been developed to estimate the long-term mean of WBC [20,28]. For quasi-steady conditions of spatially distributed water balance, a WetSpass-M model is used in this study to estimate WBC in annual, seasonal, and monthly periods. The total WBC of the vegetated, open-water, bare soil, and impervious fractions per raster cell are determined using the subsequent equations [20]:

$$S_{raster} = a_s S_s + a_o S_o + a_i S_i + a_v S_v \tag{1}$$

$$ET_{raster} = a_sE_s + a_oE_o + a_iE_i + a_vET_v \tag{2}$$

$$R_{raster} = a_sR_s + a_oR_o + a_iR_i + a_vR_v \tag{3}$$

where each following symbol represents S_{raster} (surface runoff), ET_{raster} (total evapotranspiration), and R_{raster} (recharge). Each of them has open water (o), bare soil (s), impervious area (i), and vegetation (v), while a_s, a_o, a_i, and a_v are the fractions of each LULC in a grid cell. Figure 2 depicts the WetSpass-M model's scheme. The WetSpass-M model equations used to compute monthly WBC are available in [29]. The calibrated WetSpass model of Armanuos et al. [23] in the Nile Delta aquifer is used in this study.

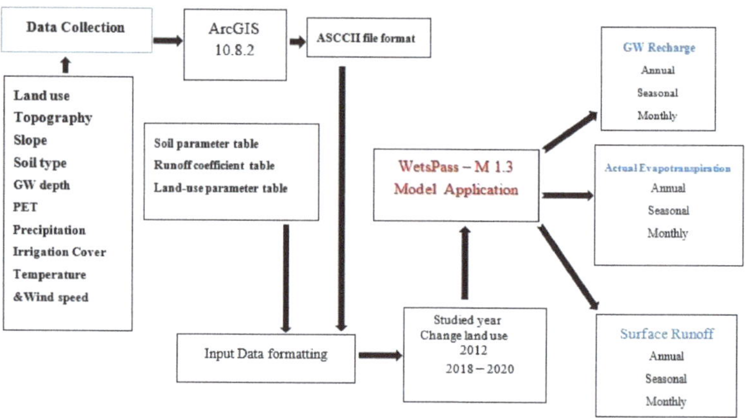

Figure 2. Schematic representation for the modeling process of the WetSpass–M model.

2.3. Input Parameter

The WetSpass-M model's input data are divided into two categories: GIS grid maps and parameter tables [30]. The first category is ASCI maps, including meteorological data, potential evapotranspiration (PET), wind speed, precipitation and average temperature, topography, slope, LULC, irrigation cover, soil type, groundwater depth, and leaves area index (LAI). Secondly, the parameter tables of soil type and LULC are attached to the model via soil and LULC attribute tables. Furthermore, the attribute tables enable researchers to alter parameter values associated with the defined soil or LULC types in the future [28]. All input data were set as a raster map derived from the DEM with 100 m × 100 m cells with a total of 495,614 raster cells in ESRI ASCII grid format accumulated between the years 2012 and 2020. Table 1 shows the WetSpass-M model's input parameters.

Table 1. Input data and sources of the WetsPass-M model.

Input Parameter	Periods	Source of Data	Cell Size
Topography DEM and slope	constant	https://earthexplorer.usgs.gov/ (accessed on 2 January 2023)	100 m × 100 m
land use land cover	2012	https://earthexplorer.usgs.gov/ (accessed on 12 March 2023)	100 m × 100 m
	2018–2020	https://livingatlas.arcgis.com/landcoverexplorer// (accessed on 10 March 2023)	
soil	Constant	FAO-UNESO 1988	100 m × 100 m
Groundwater depth	2012–2020	72 borehole piezometer monitoring monthly during the NAB project establishing, MWRI	100 m × 100 m
Precipitation mm/month	2012–2020	https://crudata.uea.ac.uk/cru/data/hrg// (accessed on 12 January 2023)	100 m × 100 m
Wind speed	2012–2020	https://crudata.uea.ac.uk/cru/data/hrg/// (accessed on 12 January 2023)	100 m × 100 m
Temperature	2012–2020	https://crudata.uea.ac.uk/cru/data/hrg/// (accessed on 12 January 2023)	100 m × 100 m

Table 1. *Cont.*

Input Parameter	Periods	Source of Data	Cell Size
Irrigation cover	2012–2020	GAD M.I., ElGamal M. m. 2020—MWRI	100 m × 100 m
Potential evapotranspiration	2012–2020	Calculated from the Thornthwaite formula	100 m × 100 m
Lookup table land use land cover	-	WetSpass model processing	-
Lookup table runoff coefficient	-	WetSpass model processing	-
A lookup table of soil parameter	-	WetSpass model processing	-

2.3.1. Topographic Features and Slope

Based on the majority of the investigations, geomorphology is the most essential component of groundwater [31]. The DEM upstream of the NAB region is derived from the Shuttle Radar Topography Mission (SRTM) (Figure 3a). The investigated area's highest point is 178 m in the southeast portion of the Assiut mountains; however, the lowest point is 42 m in the middle section of the valley, and the mean elevation of the study area is found to be 57 m. The slope map is generated directly using the DEM and the slope analysis tool under the GIS environment. The slope ranges from 0% to 49%, with an average of 0.97% (Figure 3b). The grade of the slope directly impacts surface water infiltration. Steep slopes have restricted groundwater recharge due to excessive surface runoff [32]. On a low slope, on the other hand, the gradient inhibits the flow of water and hence increases the rate of infiltration [33].

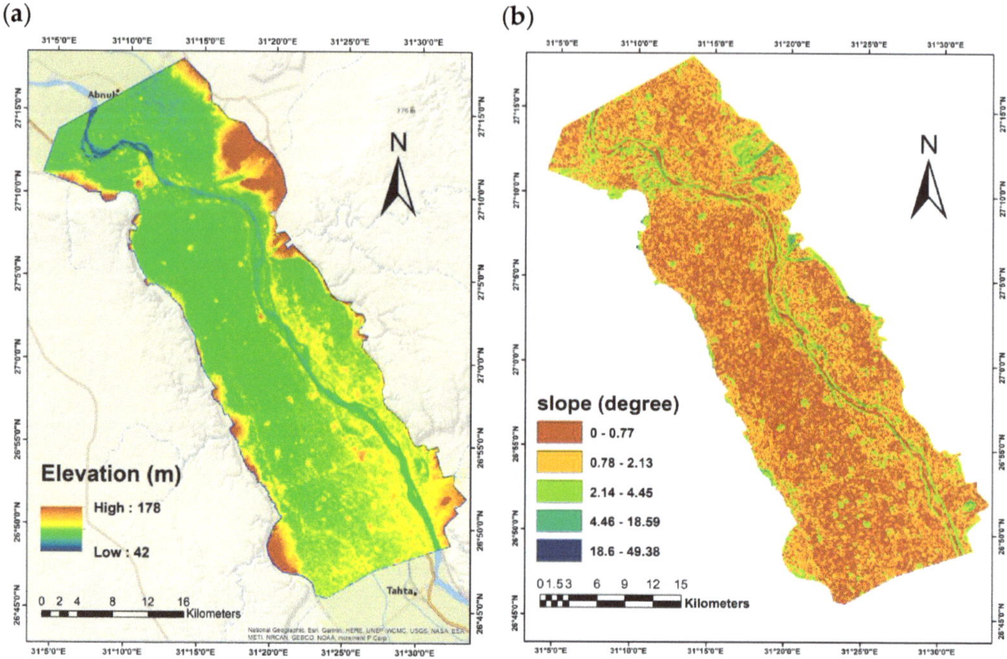

Figure 3. (**a**) Elevation; (**b**) slope.

2.3.2. Land Use/Land Cover (LULC) Data

Upstream of the NAB is surrounded by agricultural land, and the amount of usable land is rapidly decreasing [34]. Moreover, among the most critical controlling factors in valley hydrology is LULC [35]. The LULC also has the ability to determine the values of vegetative parameters like LAI and evaporative zone depth. The parameter of LAI drives both surface evaporation and transpiration [36]. The data were derived from multi-temporal satellite images, as shown in Table 1, for the years 2012 and 2014, downloaded from the USGS Earth Explorer, while for the years 2018–2020, data were sourced from ESRI/Sentinal-2 land cover 10 m resolution. The study area is distinguished by 7 land cover types, as shown in Figure 4a. The region is predominantly characterized by agricultural land (73%), built area (20%), water bodies (4%), bare ground (2.2%), and a total area of range land, trees, and flooded vegetation (1.8%). LULC type is re-coded into 12 classes according to the standard code of WetSpass.

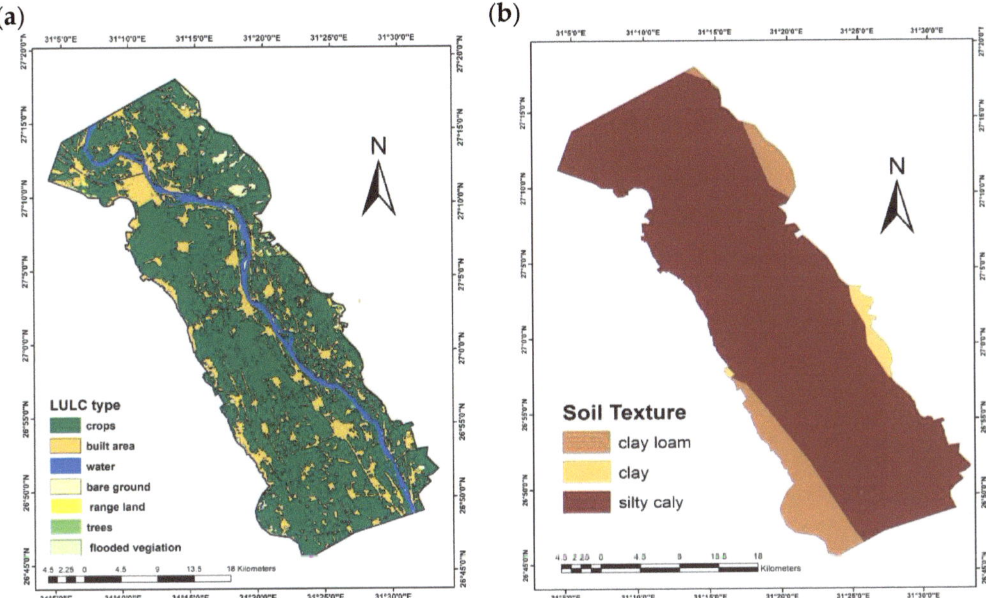

Figure 4. (**a**) Land use/land cover, (**b**) soil texture of study area.

2.3.3. Soil Data

Runoff and recharge are controlled by soil properties. Soil infiltration capacity is determined by soil texture and permeability, which determines storage capacity and controls the speed at which water penetrates deep layers. Sandy soil has the most rapid rates of infiltration, but loamy and heavy clay soil have lower rates of infiltration and more surface runoff [37]. The soil map is derived from the Harmonized World Soil Database (HWSD) (Figure 4b) [38]. The prevailing soil type of the region is silty clay [39], which covers 88% (949.50 km^2) of the total study region, while clay loam and clay soil represent 10.17% (109.55 km^2) and 1.64% (17.74 km^2), respectively.

2.3.4. Meteorological Data

The monthly dataset of meteorological parameters, e.g., precipitation, wind speed, and temperature for the period 2012–2020, was obtained from CRU TS (Climatic Research Unit gridded Time Series). CRU TS is a widely utilized climate dataset that encompasses all land areas of the world on a 0.5° longitude by 0.5° latitude grid [40]. The research area experiences an average annual precipitation ranging from 2.11 mm year^{-1} to 6.91 mm year^{-1}, with a mean rate of 4.74 mm year^{-1} (Figure 5a). The precipitation amount is significantly restricted, being relatively minor in comparison to the water used for irrigation and the extensive irrigation canal system throughout the year. Approximately 60% of the total amount of precipitation falls during the winter and autumn seasons, with the remaining 40% occurring in the summer and spring. The studied region experiences an average maximum temperature of 31 °C in July and an average lowest temperature of 12 °C in January. Additionally, the average yearly wind speed is recorded at 5 m/s.

The calculation potential evapotranspiration (PET) is determined using the Thornthwaite formula depending on latitude and temperature [41], taking into consideration the mean monthly temperature and the thermal index:

$$\mathrm{PET} = 1.6k \left(\frac{10T}{I} \right)^a \tag{4}$$

where

PET: monthly potential evapotranspiration in cm
T: average monthly temperature in Celsius
k: daylight and days in the month related to the altitude of the place

$$a = 0.000000675\ I3 - 0.0000771\ I2 + 0.01792\ I + 0.49239\ I$$

$$I = \sum_{m=1}^{12} im \quad im = \left(\frac{tm}{5}\right)^{1.5}$$

where
I: annual thermal index
im: monthly thermal index
tm: main temperature for the month
m: take value from January to December

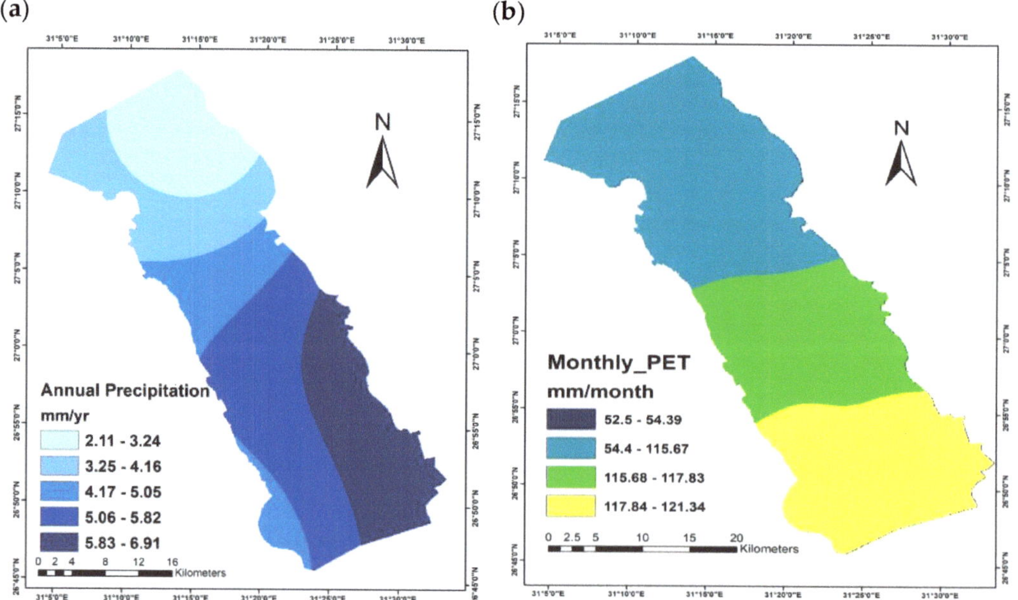

Figure 5. (**a**) Annual precipitation of study area from 2012–2020, (**b**) average potential ET.

The average monthly PET varies from 52 mm/month to 121 mm/month, with an average of 116 mm/month (Figure 5b). The highest PET is in July, with a total of 246 mm, while January has the lowest at 15 mm. The annual PET of the NAB is 1392 mm. About 88% of the PET is observed in dry seasons (summer and spring), while the remaining 12% is in wet seasons (winter and autumn). The Aridity Index (AI) of the upstream of the NAB region was determined using the following equation [42]:

$$AI = P/PET \tag{5}$$

where P represents annual rainfall, and PET represents annual potential evapotranspiration. AI is a climatic measure that can be applied to quantify the extent of availability of precipitation relative to the atmospheric water demand. Based on AI classification limits in Table A2 in Appendix A, the study area was classified as a hyper-arid climatic zone.

2.3.5. Groundwater Depth and Irrigation Cover

The records of groundwater level data for 72 observation wells were obtained from the Reservoir and Grand Barrage sector of Egypt (RGBS) from 2012 to 2019 (Figure 6a). Kriging interpolation was employed to generate a spatial distribution map of the average monthly groundwater depth, as depicted in Figure 6b. The groundwater depth ranged from 7.11 m to 7.56 m, with a mean average of 7.40 m. Moreover, the majority of the irrigated land in the investigated area relies heavily on water sourced from the Nile River via a sophisticated network of irrigation canals, serving the area of agricultural land (795 km^2) an average of 127 mm/month [4]. The monthly irrigation map prepared depends on the amount of irrigation water under land use/cover in the study area. Then WetSpass-M model is able to add the irrigation water cover to rainfall, as shown in Figure 6c.

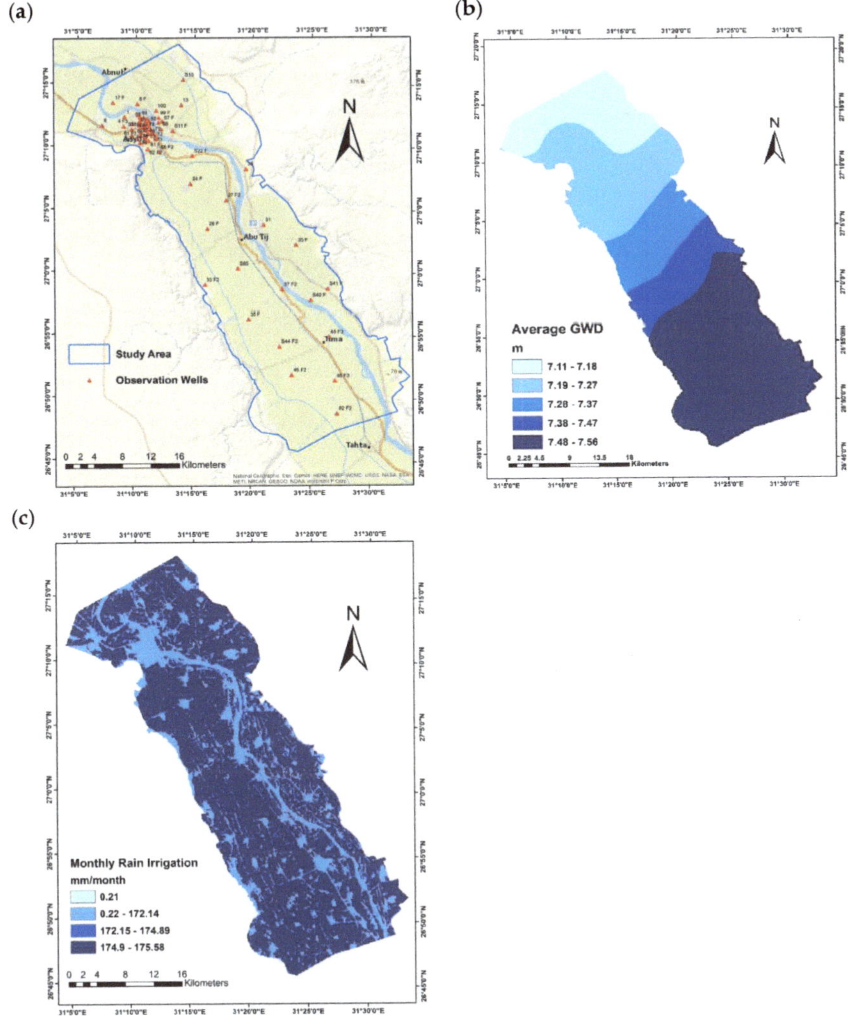

Figure 6. (**a**) Location of observations wells; (**b**) monthly groundwater depth; (**c**) monthly rainfall and irrigation water.

3. Results

3.1. Water Balance Component

The WetSpass-M model simulation produces digital maps that display the spatial distribution in addition to the numerical values of WBC. The digital maps are composed of raster maps, with each pixel representing the magnitude of the corresponding WBC (in mm/month) for the period 2012–2020. The WetSpass-M model has been used to compute actual evapotranspiration, which encompasses the combined values of vegetation water evaporation, vegetative cover transpiration, and bare soil evaporation occurring between the plants. The calculation of surface runoff relies on a reasoning technique that incorporates soil moisture coefficients and real surface runoff. Groundwater recharge, on the other hand, is determined as the residual elements derived from subtracting the combined values of actual evapotranspiration and surface runoff from the total amount of precipitation and irrigation water [43]. This study is the initial investigation into WBC upstream of the NAB region. An evaluation of the many components that contribute to the annual water balance is necessary in order to analyze the water budget upstream of the NAB region. Furthermore, it is crucial to evaluate these components periodically, both on a monthly and seasonal basis, to ascertain the precise water demands for agricultural pursuits. The results obtained from WetSpass describing the different components of the water balance will be used as boundary conditions and inputs for incorporating groundwater modeling for the upstream portion of the NAB [44]. The spatial representation of annual, seasonal, and monthly actual evapotranspiration, as simulated by the WetSpass model, is provided in Table 2.

Table 2. Long-term monthly, seasonal, and annual WetSpass simulated components of the upstream of the NAB area during 2012–2020.

Period	Value	Prec. and Water Irrig. (mm)	Groundwater Recharge (mm)	Surface Runoff (mm)	Evapotranspiration (mm)
Monthly	Range	0–175	0–32	0–99	0–242
	avg.	127	19	46	62
	st.dev.	77	11	35	45
Annual	Range	2.11–2106	0–385	0–1189	0.6–2910
	avg.	1533	228	566	739
	st.dev.	930	139	426	542
Winter	Range	1.36–464	0–169	0–347	0.44–124
	avg.	338	97	192	49
	st.dev.	204	59	119	28
Spring	Range	0.71–530	0–91	0–265	0–722
	avg.	386	53	126	207
	st.dev.	234	33	93	141
Summer	Range	0.3–587	0–67	0–257	0–1497
	avg.	428	39	104	285
	st.dev.	258	25	97	266
Autumn	Range	0–524	0–70	0–318	0–566
	avg.	381	39	144	198
	st.dev.	230	23	116	122

The WetSpass-M model estimated the monthly actual evapotranspiration in the upstream area of the NAB to vary between 0 mm/month and 242 mm/month, with an average of 62 mm/month and a standard deviation of 45 mm/month (Figure 7b). The annual actual evapotranspiration is calculated by summing up monthly data for the whole year. The study period yielded annual actual evapotranspiration values ranging from a low of 0.6 mm to a maximum of 2910 mm, with a mean value of 739 mm (Figure 7e). The average annual actual evapotranspiration contributes 48% of the combined average annual precipitation and irrigation water. The average long-term actual evapotranspiration values throughout the wet seasons (autumn and winter) and dry seasons (spring and summer) are 247 mm and 492 mm, respectively. The spring and summer seasons account for approximately 66% of the total evapotranspiration, with the remaining 34% occurring in other seasons (Table 2 and Figure 8). The variation in water demand between the two seasons accounts for this inequality. Furthermore, numerous farmers utilize their cultivated land for irrigation, particularly during the summer season when there is a significant demand for water in upper Egypt [45]. The southern section of the Nile River experienced the highest actual evapotranspiration 520 mm, while the northern half has a slightly lower value of 460 mm (Figure 7).

The spatial distribution of the annual mean interception is presented in Figure 7d. The annual interception varies from 0 mm/year to 300 mm/year, with an average rate of 100 mm/year. Approximately 61% of the simulated interception takes place during the dry seasons, specifically summer and spring. The remaining 39% occurs in the wet seasons (winter and autumn). The WetSpass-M model determines the monthly surface runoff in millimeters per month through a logical approach that takes into account both the current surface runoff and the coefficient of soil moisture [28]. The annual observed surface runoff has significant regional variability, ranging from 0 mm to 1189 mm (Figure 7g). The monthly surface runoff ranges from 0 mm month^{-1} to a maximum of 99 mm month^{-1}, with an average of 46 mm month^{-1} and a standard deviation of 35 mm month^{-1}. The annual mean and standard deviation of this distribution are 566 mm year^{-1} and 426 mm year^{-1}, respectively (Figure 7c and Table 2).

The estimation of annual surface runoff involves the accumulation of monthly simulated data over the whole time. The average surface runoff in the study area accounts for around 36% of the annual average precipitation and irrigation water. The average surface runoff during the summer and spring seasons is 230 mm, while the runoff during the winter and autumn seasons is roughly 338 mm. The middle of the upstream area of the NAB along the Nile River exhibits the greatest average annual and seasonal surface runoff values due to its gradual incline and the prevalence of silty clay, clay loam, and clay soils with limited permeability.

(a) **(b)**

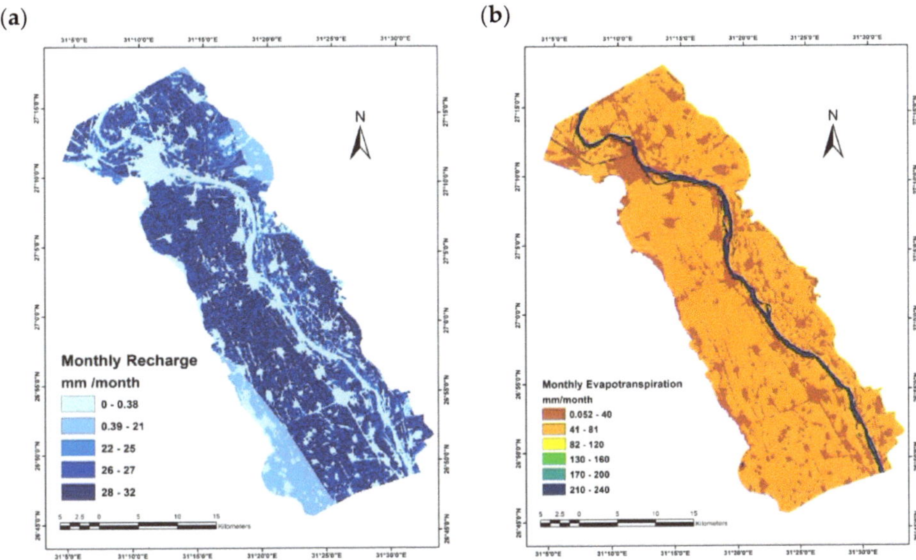

Figure 7. *Cont.*

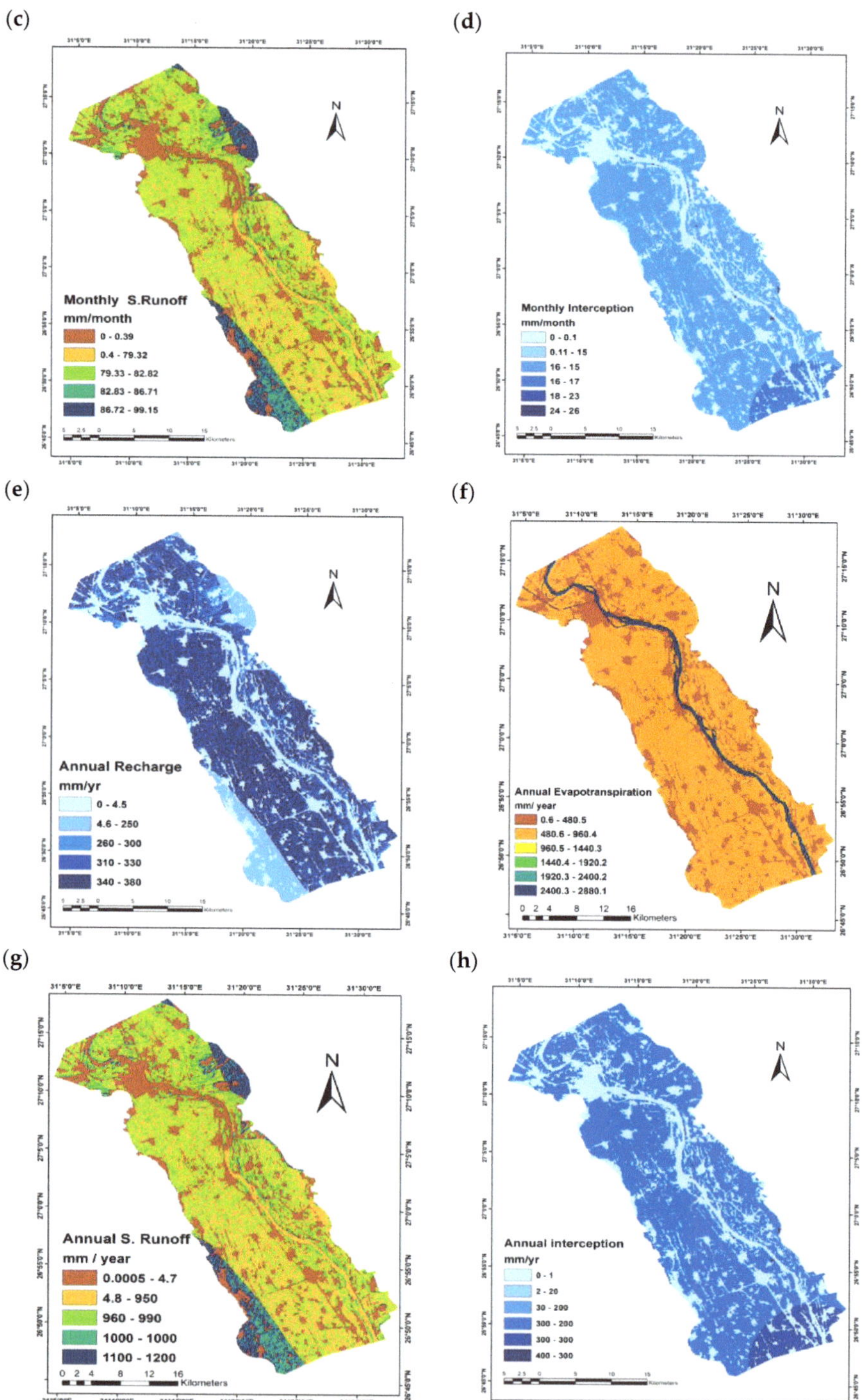

Figure 7. (**a**) Monthly GW recharge; (**b**) monthly evapotranspiration; (**c**) monthly s.runof (**d**) monthly interception; (**e**) annual GW recharge; (**f**) annual evapotranspiration; (**g**) annual s.runoff; (**h**) annual interception.

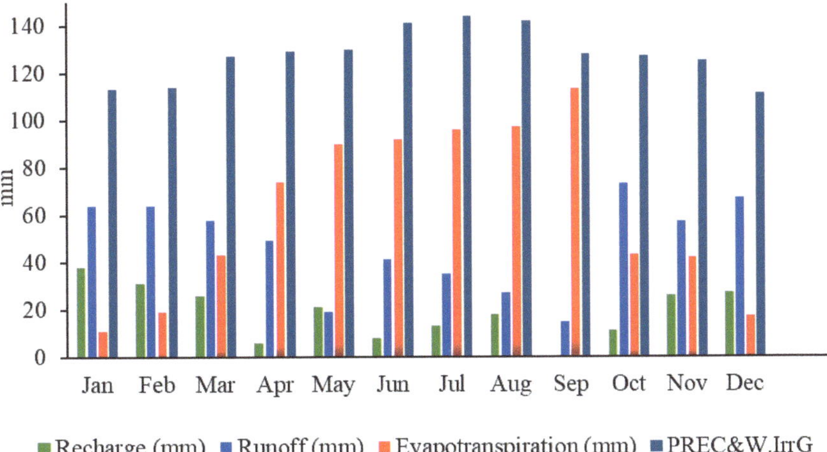

Figure 8. Average monthly WBC upstream of the NAB between 2012 and 2020.

Groundwater recharge is a critical determinant in the evaluation of groundwater resources; nevertheless, its assessment presents inherent challenges [6,46]. As a residual parameter of WBC, the WetSpass simulates groundwater recharge for the upstream area of the NAB by deducting evapotranspiration and discharge from the monthly precipitation and irrigation water, climate conditions, slope, topography, LULC, soil type, and groundwater depth, which all influence the spatial variation of groundwater recharge. The spatial distribution of seasonal groundwater recharge in the investigated area is influenced by the valley's topography and other distinctive features (Figure 7a). The estimation of the annual mean of groundwater recharge is performed using simulated monthly data. The mean value of groundwater recharge is 228 mm year^{-1}, with a standard deviation of 139 mm year^{-1}. The annual mean of groundwater recharge ranges from 0 mm year^{-1} to 384 mm year^{-1} (Figure 7e and Table 2). The monthly groundwater recharge of the upstream of the NAB region, as simulated, varies between 0 mm and 32 mm month^{-1}. The mean and standard deviations are 19 and 11 mm month^{-1}, respectively (Table 2). Fifteen percent of the average annual precipitation and irrigation water represents the amount of average groundwater recharge. The simulated monthly groundwater recharge in the investigated area is presented in Figure 9. Approximately 61% of the annual recharge of groundwater happens during the wet seasons. The remaining 39% occurs in the dry seasons, as shown in Figure 9c,d. The mean long-term groundwater recharge during the wet seasons and dry seasons is 134 mm and 92 mm, respectively (Table 2 and Figure 9). The highest value of groundwater recharge is observed in agricultural regions in the east and west parts of the Nile Valley.

The northeast and southwest account for less groundwater recharge, which is related to the existence of hot and barren regions with less-permeable clay loam soils. Additionally, the urban area experiences the lowest recharge due to limited water use for irrigation and low rainfall.

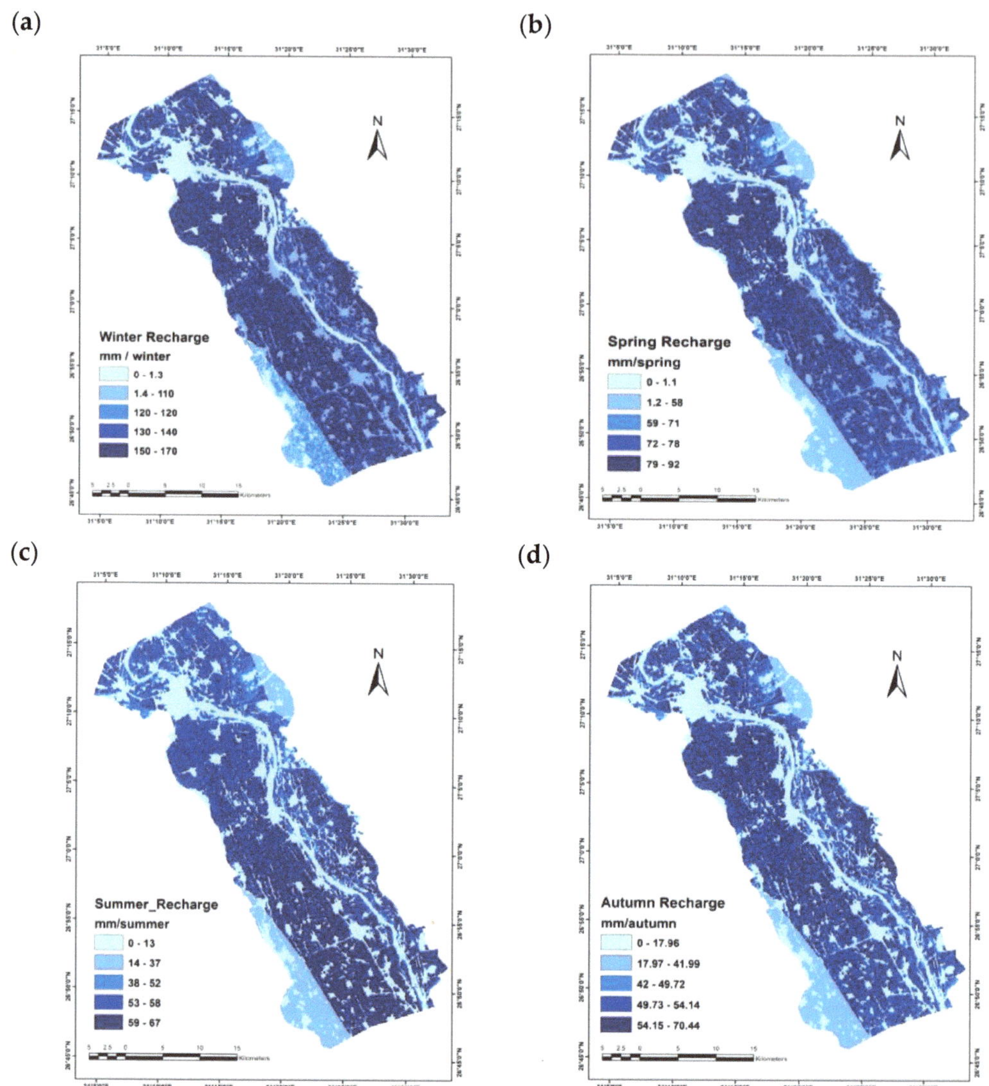

Figure 9. Long-term seasonal groundwater recharge during the period 2012–2020. (**a**) Winter; (**b**) spring; (**c**) summer; (**d**) autumn.

3.2. Water Balance Components Values under Different LULC Types and Soil Textures

The WBC may exhibit variability across different LULC types and soil textures [47]. The amount of groundwater recharge, actual evapotranspiration, and surface runoff are influenced by LULC, as shown in Figure 10. Around 73% of the upstream NAB region is covered by agricultural land, which is distributed throughout the surveyed region. Agricultural land exhibits a significant amount of groundwater recharge, averaging 858 mm year^{-1}, as well as has the highest surface runoff, averaging 966 mm year^{-1}. Built-up areas are defined by a surface that does not allow water to pass through easily, resulting in a limited ability to replenish groundwater and release water through evapotranspiration, with an average of 3 mm year^{-1}. In the WetSpass model, open water (i.e., lakes and rivers) is given a zero-groundwater recharge value because open water surfaces are presumed to be groundwater discharge destinations, while it has a significant amount of evapotranspiration of 2788 mm year^{-1}. Increased runoff in the region is a result of the conventional flood irrigation system that is being followed [48]. In addition, the study area characterized by silt clay soil and clay loam has increased surface runoff. Therefore, studying the temporal and spatial distribution of surface runoff might help us understand the main elements that affect the variability of runoff in the Nile Valley.

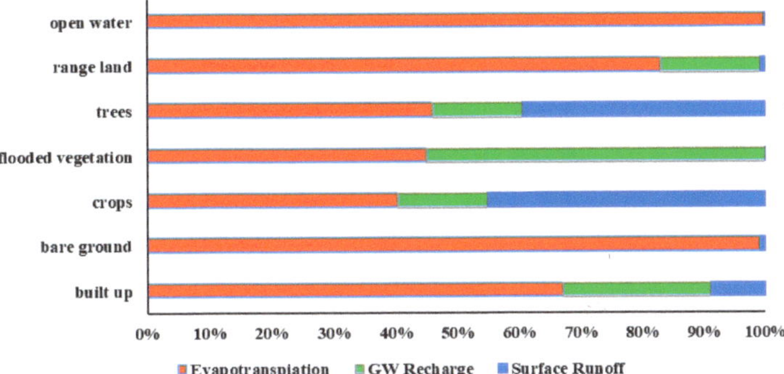

Figure 10. Water balance components under different types of LULC.

The WBC are significantly influenced by soil textures. The varying spatial distribution of soil textures significantly influences the hydraulic properties at the local and regional scales [46]. Heavy soils (silty clay and clay loam) upstream of the NAB region exhibit high surface runoff due to their low hydraulic conductivity. Clay soils have approximately two-thirds of the groundwater recharge of loamy soils. The WetSpass-M model simulated the annual actual evapotranspiration of silty clay soils as 667 mm year^{-1}, whereas the surface runoff was 257 mm year^{-1} (Figure 11). Additionally, the model estimated that the annual groundwater recharge of silty clay soil is 89 mm year^{-1}. The increasing variability in the amount of WBC across different types of soil indicates that the WBC rate in the investigated area is more dependent on soil texture.

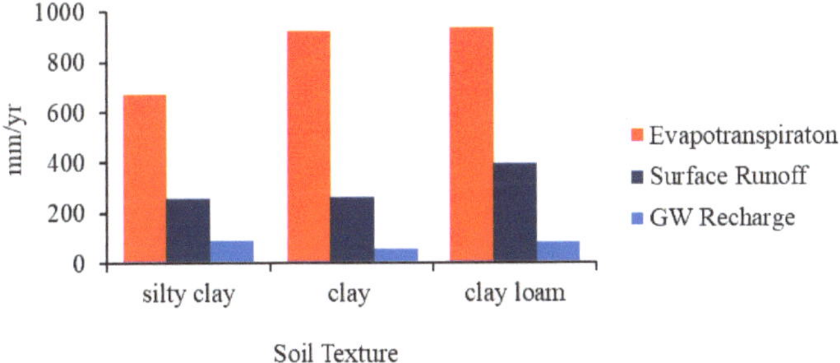

Figure 11. Water balance components under different types of soil.

3.3. Effects of LULC Changes on WBC

Approximately 75% of the total area consists of agricultural land, which is primarily located along the banks of the Nile River on both the eastern and western banks of the Nile. Figure 12 displays the proportion of land use categories from 2012 to 2020. According to Figure 13, agriculture is the most prevalent and influential land use in the NAB, followed by built-up areas and open water. Rangeland, bare ground, flooded vegetation, and trees have a scattered distribution with low percentages. From 2012 to 2020, there was a noticeable decrease in the agricultural area (cropland), with a fall from 82% to 72%. This decrease can be attributed to the considerable expansion in urban areas, as shown in Figure 12.

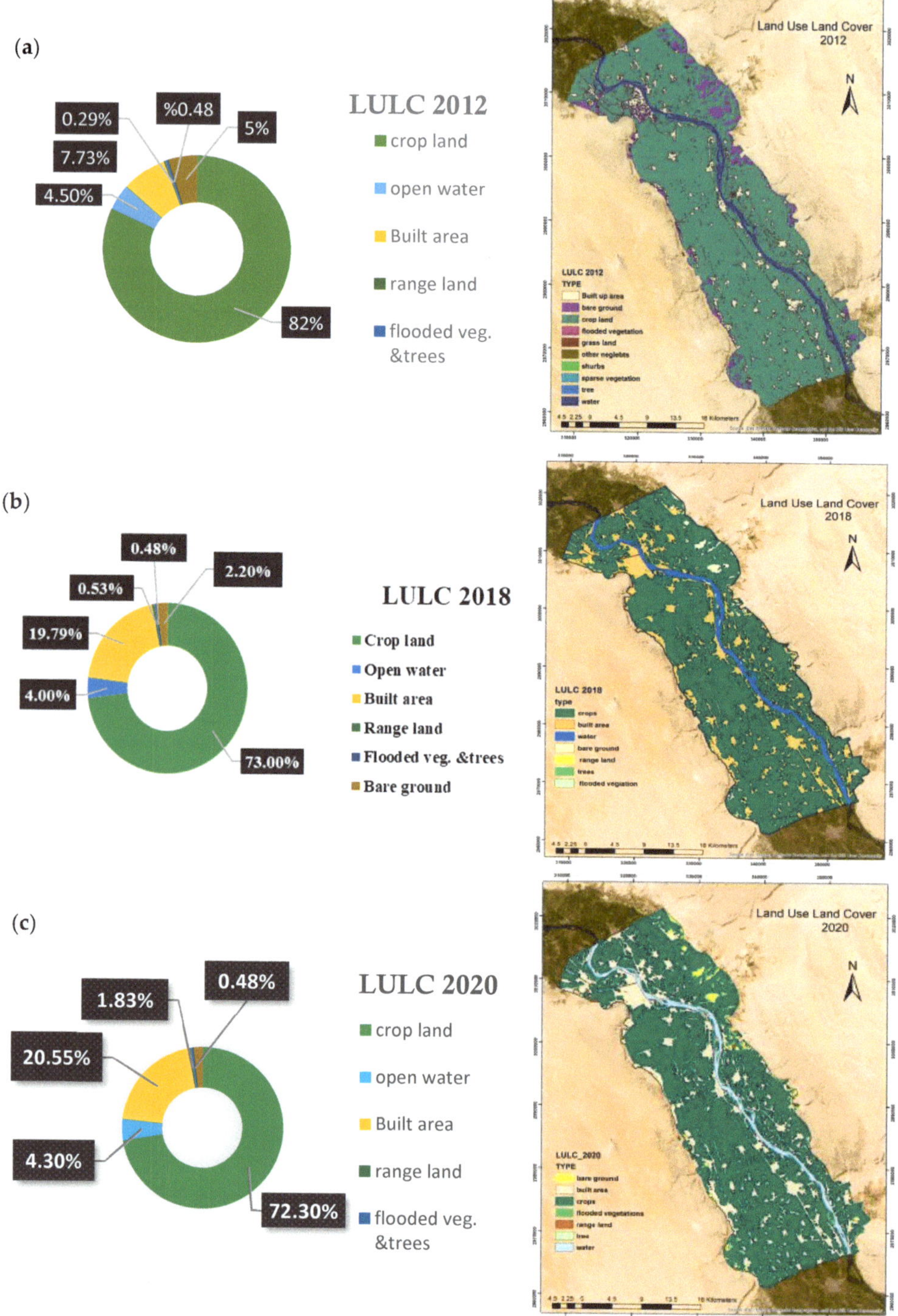

Figure 12. Area percentage of LULC classes in (**a**) 2012, (**b**) 2018 and (**c**) 2020.

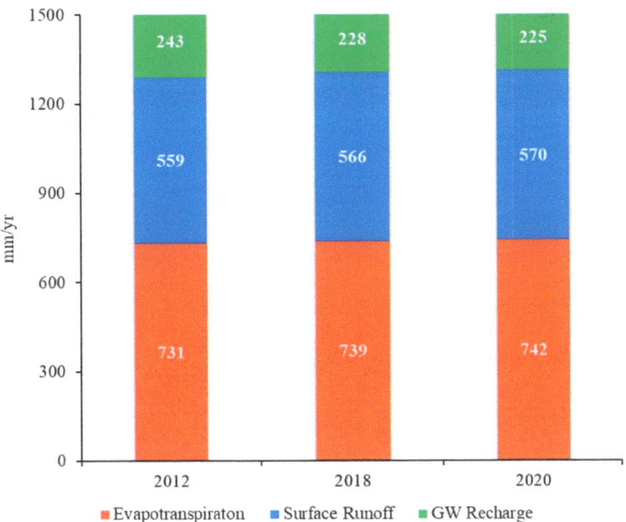

Figure 13. Average annual WBC upstream of the NAB region from 2012 to 2020 (value in mm year^{-1}).

To investigate the impact of LULC changes on WBC, four separate runs were initiated, each matching a certain LULC from the years 2012, 2018, and 2020, respectively. The other input variables, including meteorological data, soil types, topography (slope and digital elevation model), and distributed groundwater depth, were maintained at a constant level for all four trials. The simulation utilized meteorological parameters from the period 2012–2020 as climate input, together with the Digital Elevation Model (DEM), soil data, and slope of the research area for all four iterations. Each run replicated the long-term average WBC over 9 years, specifically from 2012 to 2020, which consisted of 108 time steps. Between 2012 and 2020, there was a significant rise in built-up areas, from 7.73% to 21.55%. This expansion was mostly achieved by transforming cropland into urban areas. Therefore, an evaluation was conducted to assess the effects of urbanization on the variations in WBC in built-up areas during this period.

The simulated average annual WBCs upstream of the NAB for each LULC of the years 2012, 2018, and 2020 are presented in Figure 13. The replacement of cropland by built-up areas was recognized as the primary factor responsible for the major decrease in groundwater recharge by 15 and 18 mm year^{-1}, an increase in evapotranspiration by 8 and 11 mm year^{-1}, and an increase in surface runoff by 7 and 11 mm year^{-1} for LCLU between 2012–2018 and 2018–2020, respectively (Figures 12 and 13). From 2012 to 2020, there was an increase in surface runoff and evapotranspiration, while groundwater recharge experienced a decrease. The increase in surface runoff between 2012 and 2020 corresponds to the increase in built-up areas and range land (Figures 12 and 13). The analysis of alterations in LULC maps and surface runoff reveals that the rise in average yearly runoff may be ascribed to the expansion of built-up areas and range land from 2012 to 2020. These surfaces are largely or completely impermeable, and they were deemed to have a detrimental effect on the upstream of the NAB. The main driver of the shift in surface runoff from 2012 to 2020 was the observed growth in built-up areas, which was deemed to have a detrimental effect on the region upstream of the NAB.

The changes in built-up areas and cropland had the most significant impact on the changes in groundwater recharge and actual evapotranspiration. The biggest factor contributing to the significant decrease in groundwater recharge by 18 mm per year and the rise in evapotranspiration by 11 mm per year in the period 2012–2020 has been identified. The decrease in groundwater recharge is linked to the increase in urban areas and the decline of cropland. Furthermore, urbanization leads to a reduction in the recharge of groundwater [47]. Average groundwater recharge experienced a significant drop of 6% from 2012 to 2018. The findings indicate that LULC alterations significantly impact the comprehensive water balance upstream of the NAB. The technique used in this study enables the calculation of hydrological components, spatial and temporal, taking into account the alterations in LULC. This provides decision-makers and stakeholders with precise quantitative data to facilitate the implementation of effective and sustainable water resource management in the Nile Valley upstream of the NAB.

4. Conclusions

The upstream region of the NAB experiences substantial human impacts, resulting in water scarcity and increased vulnerability to drought events. In order to create a groundwater model for the valley, an accurate assessment of groundwater recharge and actual evapotranspiration is essential as boundary conditions. The upstream region of the NAB was analyzed using the WetSpass-M model to determine the annual, monthly, and seasonal rates of surface runoff, groundwater recharge, and actual evapotranspiration from 2012 to 2020. The WBC for each grid pixel was determined for the vegetated land, impervious fractions, bare soil, and open water. The primary input variables for the model consisted of climate data (such as air temperature, precipitation, wind speed, irrigation cover, and potential evapotranspiration), LAI, groundwater level, soil types, slope, DEM, and LULC. The input data were generated as raster maps using the ArcGIS framework. The upstream region of the NAB was mostly characterized by agricultural areas and silty clay soils, as observed in the LULC and soil texture analysis. The WBCs were assessed for different LULC and soil texture conditions.

The annual simulated evapotranspiration ranges from $1\ \mathrm{mm\ year^{-1}}$ to $2880\ \mathrm{mm\ year^{-1}}$, with an average of $739\ \mathrm{mm\ year^{-1}}$, which is 49% of the yearly average rainfall and water irrigation cover. Approximately 15% ($228\ \mathrm{mm\ year^{-1}}$) of the total annual precipitation and irrigation water is attributed to the recharge of groundwater. The lowest recorded recharge is $0\ \mathrm{mm\ year^{-1}}$, while the highest recorded recharge is $385\ \mathrm{mm\ year^{-1}}$. The annual surface runoff of the investigated area ranged from 0 mm to 1189 mm in the period between 2012 and 2020. The surface runoff accounts for 36% of the average annual precipitation and water irrigation cover at $566\ \mathrm{mm\ year^{-1}}$. The simulation outputs confirm the accurate utilization of the WetSpass-M model for estimating the various components of the water budget upstream of the NAB. This study can be employed to create a comprehensive groundwater model and assess potential locations for regulated artificial recharge by collection of runoff discharge to enhance groundwater storage. The findings suggest that LULC changes have a significant impact on groundwater level and water balance. Specifically, the primary factor responsible for the $11\ \mathrm{mm\ year^{-1}}$ rise in surface runoff in the NAB between 2012 and 2020 was the growth of built-up areas. Moreover, there is a downward trend in groundwater recharge in response to these changes, which primarily stems from human activities related to land use, particularly the reduction in agricultural land. Consequently, the agricultural conditions, which are the primary means of sustenance for the indigenous community and ecological services, became crucial. The proposed approach is a significant tool for assessing and managing the rehabilitation upstream of the NAB in an effective and sustainable manner. Given the rapid decline of water resources in the region due to human activity, it is imperative to take immediate and effective measures to mitigate the decrease in groundwater recharge and increase in surface runoff. Policymakers should take into account the impact of LULC change during the restoration of the region. They should also identify and implement measures to mitigate the negative impact of LULC alterations. Furthermore, the surveyed region necessitates effective water management and a modification in the irrigation system to enable the agricultural land to absorb an appropriate quantity of water.

Author Contributions: Conceptualization, A.S.E., E.M.A., A.S. and Z.L.; methodology, A.S.E. and E.M.A.; investigation A.S.E., E.M.A., A.S., M.M.H., K.E.Y., J.L. and Z.L.; software: A.S.E., E.M.A. and M.M.H.; validation, A.S.E. and E.M.A.; formal analysis, A.S.E. and E.M.A.; writing the paper, A.S.E.; review and editing the paper; J.L., A.S. and Z.L.; visualization, A.S.E., E.M.A., A.S., M.M.H., K.E.Y. and J.L.; supervision, Z.L. All authors have read and agreed to the published version of the manuscript.

Funding: This research was funded by the National Natural Science Foundation of China (Grant No. 51779215), the Natural Science Foundation of Jiangsu Province (Grant No. BK20171288), and the Major Scientific and Technological Project of Shandong Gangshiyuan Construction Engineering Group Co., Ltd. (Grant No. 2020RS-1058).

Data Availability Statement: The datasets used and/or analyzed during the current study are available from the corresponding author upon reasonable request.

Acknowledgments: The present scientific contribution is dedicated to the 120th anniversary of the foundation of the University of Yangzhou, CHINA. The first author would like to thank the Egyptian Ministry of Water Resources and Irrigation (MWRI) and Reservoir AND Grand Barrage Sector (RGBS) for providing him. Furthermore, the authors would like to acknowledge the support from the National Natural Science Foundation of China (Grant No. 51779215), the Natural Science Foundation of Jiangsu Province (Grant No. BK20171288), and the Major Scientific and Technological Project of Shandong Gangshiyuan Construction Engineering Group Co., Ltd. (Grant No. 2020RS-1058).

Conflicts of Interest: The authors declare that this study received funding from the Major Scientific and Technological Project of Shandong Gangshiyuan Construction Engineering Group Co., Ltd. The funder was not involved in the study design, collection, analysis, interpretation of data, the writing of this article or the decision to submit it for publication.

Appendix A

Table A1. Purpose/scope, key features, advantages, and disadvantages of different hydrological/hydrodynamic models.

Model Name	Purpose/Scope	Key Features	Advantages	Disadvantages
WETSPASS Model	The WETSPASS model integrates water and energy fluxes within the soil–vegetation–atmosphere system. It simulates various hydrological processes, including evapotranspiration, interception, runoff, and groundwater recharge.	The WETSPASS model adopts a simplified simulation approach that balances accuracy and computational efficiency. It provides a comprehensive representation of the hydrological cycle and its interactions with the energy balance. It can be applied at different spatial scales and is designed to be user-friendly.	The WETSPASS model is known for its simplicity, ease of use, and flexibility in parameterization. It can be customized to different ecosystems and climatic conditions, and it integrates well with GIS platforms.	It may not capture all the complexities of hydrodynamic processes or finer-scale spatial variations.
GPU-accelerated and LTS-based 2D Hydrodynamic Model	The GPU-accelerated and LTS-based 2D hydrodynamic model specifically focuses on simulating two-dimensional hydrodynamic processes, such as river flow, flood inundation, and stormwater runoff. It leverages the computational power of GPUs (Graphics Processing Units) to enhance simulation speed and efficiency.	This model utilizes parallel computing on GPUs to accelerate the simulation of complex hydrodynamic equations. It may employ adaptive time-stepping algorithms, such as the Local Time Stepping (LTS) method, to enhance numerical stability and efficiency.	The use of GPUs allows for faster simulations compared to traditional CPU-based models, enabling real-time or near-real-time simulations. The LTS method can improve computational efficiency by dynamically adjusting time steps based on local conditions.	GPU-accelerated models may require specialized hardware and software setups and expertise in GPU programming. Additionally, the applicability and performance of the model may depend on the availability and quality of high-resolution topographic and bathymetric data.

Table A1. *Cont.*

Model Name	Purpose/Scope	Key Features	Advantages	Disadvantages
TOPMODEL (Topographic Index-Based Hydrological Model)	TOPMODEL is a hydrological model used for simulating the spatial distribution of water flow and soil moisture within a watershed. It focuses on the influence of topography on hydrological processes.	TOPMODEL utilizes a topographic index, which represents the relative wetness of a location based on its position in the landscape. It considers the variable source area concept, where only a portion of the watershed contributes to the runoff generation.	TOPMODEL accounts for the spatial variability of soil moisture and flow connectivity based on topographic characteristics. It can capture the effects of landscape heterogeneity and preferential flow paths.	TOPMODEL may require accurate and high-resolution digital elevation models (DEMs) to capture the topographic variability. Calibration of the model can be challenging due to the sensitivity of the topographic index parameter.
SWAT (Soil and Water Assessment Tool)	SWAT is a widely used hydrological model for simulating water flow, sediment transport, and nutrient cycling in watersheds. It assesses the impacts of land management practices on water resources and quality.	SWAT integrates various components, including weather, land use, soil, and vegetation, to simulate hydrological processes at different spatial and temporal scales. It considers both surface runoff and groundwater flow.	SWAT provides a comprehensive representation of the hydrological cycle and can handle a wide range of land use and management scenarios. It allows for the evaluation of different conservation practices and their impacts on water resources.	SWAT requires extensive input data, including detailed soil, land use, and weather data. Calibration and parameterization can be time-consuming and challenging.
HEC-HMS (Hydrologic Engineering Center's Hydrologic Modeling System)	HEC-HMS is a widely used hydrological model for simulating rainfall-runoff processes in watersheds. It is primarily used for engineering and water resources planning purposes.	HEC-HMS employs a modular approach that allows users to build custom hydrological models by selecting and integrating various components. It can simulate different runoff generation mechanisms and has options for different routing methods.	It offers flexibility in model configuration and allows for a detailed representation of watershed characteristics. It is widely recognized and supported in the engineering community.	It requires substantial input data, including precipitation, soil properties, and land use. It may require expertise in hydrological modeling and engineering concepts.
MIKE SHE	It is a comprehensive, integrated hydrological model that simulates the entire hydrological cycle, including surface water and groundwater interactions.	It combines surface water flow, groundwater flow, and unsaturated zone flow in a coupled manner. It can simulate complex hydrological processes, such as overland flow, infiltration, evapotranspiration, and stream-aquifer interactions.	It provides a detailed representation of the hydrological system and can handle complex hydrological scenarios. It allows for the assessment of water resources, flooding, and groundwater management.	It requires extensive input data, including hydraulic properties, climatic data, and topographic information. Model setup and parameterization can be complex and require expertise in hydrological modeling.

Table A2. The UNEP classification limitations for the Aridity Index [42].

Climatic Zone	P/PET (Thornthwaite Method)
Hyper-arid	<0.05
Arid	0.05–0.2
Semi-arid	0.2–0.65
Sub-humid	0.5–0.65
Humid	>0.65

References

1. Wehrey, F.; Fawal, N. *Cascading Climate Effects in the Middle East and North Africa: Adapting through Inclusive Governance*; Carnegie Endowment for International Peace: Washington, DC, USA, 2022.
2. Nashwan, M.S.; Shahid, S. Spatial distribution of unidirectional trends in climate and weather extremes in Nile river basin. *Theor. Appl. Climatol.* **2019**, *137*, 1181–1199. [CrossRef]
3. Dawoud, M.A.; El Arabi, N.E.; Khater, A.R.; van Wonderen, J. Impact of rehabilitation of Assiut barrage, Nile River, on groundwater rise in urban areas. *J. Afr. Earth Sci.* **2006**, *45*, 395–407. [CrossRef]
4. MI, G.; Elgamal, M.; Khalaf, S.; Nassar, W. Effect of the Construction of New Assuit Barrage on Groundwater Regime, (Assuit, Egypt). *Al-Azhar Univ. Civ. Eng. Res. Mag.* **2020**, *42*, 158–174.
5. World Health Organization. *Guidelines for Drinking-Water Quality*; World Health Organization: Geneva, Switzerland, 2004; Volume 1.
6. Alley, W.M.; Healy, R.W.; LaBaugh, J.W.; Reilly, T.E. Flow and storage in groundwater systems. *Science* **2002**, *296*, 1985–1990. [CrossRef] [PubMed]
7. Harmsen, E.W.; Miller, N.L.; Schlegel, N.J.; Gonzalez, J.E. Seasonal climate change impacts on evapotranspiration, precipitation deficit and crop yield in Puerto Rico. *Agric. Water Manag.* **2009**, *96*, 1085–1095. [CrossRef]
8. Sepaskhah, A.R.; Razzaghi, F. Evaluation of the adjusted Thornthwaite and Hargreaves-Samani methods for estimation of daily evapotranspiration in a semi-arid region of Iran. *Arch. Agron. Soil Sci.* **2009**, *55*, 51–66. [CrossRef]
9. Kinzelbach, W.; Aeschbach, W. *A Survey of Methods for Analysing Groundwater Recharge in Arid and Semi-Arid Regions*; Division of Early Warning and Assessment, United Nations Environment Programme: Geneva, Switzerland, 2002; Volume 2.
10. Sekhar, M.; Rasmi, S.; Sivapullaiah, P.; Ruiz, L. Groundwater flow modeling of Gundal sub-basin in Kabini river basin, India. *Asian J. Water Environ. Pollut.* **2004**, *1*, 65–77.
11. Salem, A.; Dezső, J.; El-Rawy, M.; Lóczy, D. Water management and retention opportunities along the Hungarian section of the Drava River. In *Recent Advances in Environmental Science from the Euro-Mediterranean and Surrounding Regions, Proceedings of the 2nd Euro-Mediterranean Conference for Environmental Integration (EMCEI-2), Sousse, Tunisia, 10–13 October 2019*, 2nd ed.; Springer: Cham, Switzerland, 2021.
12. Heggy, E.; Sharkawy, Z.; Abotalib, A.Z. Egypt's water budget deficit and suggested mitigation policies for the Grand Ethiopian Renaissance Dam filling scenarios. *Environ. Res. Lett.* **2021**, *16*, 074022. [CrossRef]
13. El-Rawy, M.; Batelaan, O.; Buis, K.; Anibas, C.; Mohammed, G.; Zijl, W.; Salem, A. Analytical and numerical groundwater flow solutions for the FEMME-modeling environment. *Hydrology* **2020**, *7*, 27. [CrossRef]
14. Anuraga, T.; Ruiz, L.; Kumar, M.M.; Sekhar, M.; Leijnse, A. Estimating groundwater recharge using land use and soil data: A case study in South India. *Agric. Water Manag.* **2006**, *84*, 65–76. [CrossRef]
15. Singh, G.; Saraswat, D. Development and evaluation of targeted marginal land mapping approach in SWAT model for simulating water quality impacts of selected second generation biofeedstock. *Environ. Model. Softw.* **2016**, *81*, 26–39. [CrossRef]
16. Carrera-Hernández, J.J.; Gaskin, S.J. Spatio-temporal analysis of potential aquifer recharge: Application to the Basin of Mexico. *J. Hydrol.* **2008**, *353*, 228–246. [CrossRef]
17. Wu, J.; Hu, P.; Zhao, Z.; Lin, Y.T.; He, Z. A gpu-accelerated and lts-based 2d hydrodynamic model for the simulation of rainfall-runoff processes. *J. Hydrol.* **2023**, *623*, 129735. [CrossRef]
18. Zhang, D.; Madsen, H.; Ridler, M.E.; Refsgaard, J.C.; Jensen, K.H. Impact of uncertainty description on assimilating hydraulic head in the MIKE SHE distributed hydrological model. *Adv. Water Resour.* **2015**, *86*, 400–413. [CrossRef]
19. Gumindoga, W.; Rientjes, T.H.M.; Haile, A.T.; Dube, T. Predicting streamfow for land cover changes in the Upper Gilgel Abay River Basin, Ethiopia: A TOPMODEL based approach. *Phys. Chem. Earth A/B/C* **2014**, *76–78*, 3–15. [CrossRef]
20. Batelaan, O.; De Smedt, F.W. A flexible, GIS based, distributed recharge methodology for regional groundwater. In Proceedings of the Impact of Human Activity on Groundwater Dynamics: Proceedings of an International Symposium (Symposium S3) Held During the Sixth Scientific Assembly of the International Association of Hydrological Sciences (IAHS), Maastricht, The Netherlands, 18–27 July 2001.
21. Amiri, M.; Salem, A.; Ghzal, M. Spatial-Temporal Water Balance Components Estimation Using Integrated GIS-Based Wetspass-M Model in Moulouya Basin, Morocco. *ISPRS Int. J. Geo-Inf.* **2022**, *11*, 139. [CrossRef]
22. Abu-Saleem, A.; Al-Zu'bi, Y.; Rimawi, O.; Al-Zu'bi, J.; Alouran, N. Estimation of water balance components in the Hasa basin with GIS based WetSpass model. *J. Agron.* **2010**, *9*, 119–125. [CrossRef]
23. Armanuos, A.M.; Negm, A.; Yoshimura, C.; Valeriano, O.C.S. Application of WetSpass model to estimate groundwater recharge variability in the Nile Delta aquifer. *Arab. J. Geosci.* **2016**, *9*, 553. [CrossRef]
24. Aish, A.M.; Batelaan, O.; De Smedt, F. Distributed recharge estimation for groundwater modeling using WetSpass model, case study—Gaza strip, Palestine. *Arab. J. Sci. Eng.* **2010**, *35*, 155.
25. Salem, A.; Dezső, J.; El-Rawy, M. Assessment of groundwater recharge, evaporation, and runoff in the Drava Basin in Hungary with the WetSpass Model. *Hydrology* **2019**, *6*, 23. [CrossRef]
26. Salem, A.; Dezső, J.; El-Rawy, M.; Loczy, D.; Halmai, Á. Estimation of groundwater recharge distribution using Gis based WetSpass model in the Cun-Szaporca oxbow, Hungary. In Proceedings of the 19th International Multidisciplinary Scientific GeoConference SGEM 2019, Albena, Bulgaria, 28 June–7 July 2019; Volume 19, pp. 169–176.
27. Aslam, M.; Salem, A.; Singh, V.P.; Arshad, M. Estimation of spatial and temporal groundwater balance components in Khadir Canal Sub-Division, Chaj Doab, Pakistan. *Hydrology* **2021**, *8*, 178. [CrossRef]

28. Batelaan, O.; De Smedt, F. GIS-based recharge estimation by coupling surface–subsurface water balances. *J. Hydrol.* **2007**, *337*, 337–355. [CrossRef]
29. Abdollahi, K.; Bashir, I.; Verbeiren, B.; Harouna, M.R.; Van Griensven, A.; Huysmans, M.; Batelaan, O. A distributed monthly water balance model: Formulation and application on Black Volta Basin. *Environ. Earth Sci.* **2017**, *76*, 198. [CrossRef]
30. Batelaan, O.; De Smedt, F. *WetSpass: A Flexible, GIS Based, Distributed Recharge Methodology for Regional Groundwater Modelling*; IAHS Publication: Wallingford, UK, 2001; pp. 11–18.
31. Magesh, N.S.; Chandrasekar, N.; Soundranayagam, J.P. Delineation of groundwater potential zones in Theni district, Tamil Nadu, using remote sensing, GIS and MIF techniques. *Geosci. Front.* **2012**, *3*, 189–196. [CrossRef]
32. Kanagaraj, G.; Suganthi, S.; Elango, L.; Magesh, N. Assessment of groundwater potential zones in Vellore district, Tamil Nadu, India using geospatial techniques. *Earth Sci. Inform.* **2019**, *12*, 211–223. [CrossRef]
33. Gupta, M.; Srivastava, P.K. Integrating GIS and remote sensing for identification of groundwater potential zones in the hilly terrain of Pavagarh, Gujarat, India. *Water Int.* **2010**, *35*, 233–245. [CrossRef]
34. Mahmoud, H.; Divigalpitiya, P. Modeling future land use and land-cover change in the asyut region using markov chains and cellular automata. In *Smart and Sustainable Planning for Cities and Regions: Results of SSPCR, 2015*; Springer: Cham, Switzerland, 2017; pp. 99–112.
35. Fetter, C.W. *Applied Hydrogeology, Prentice Hall Upper Saddle River*; Academic Science: Cambridge, MA, USA, 2001.
36. Ala-Aho, P.; Rossi, P.; Kløve, B. Estimation of temporal and spatial variations in groundwater recharge in unconfined sand aquifers using Scots pine inventories. *Hydrol. Earth Syst. Sci.* **2015**, *19*, 1961–1976. [CrossRef]
37. SMEC. *Hydrological Study of the Tana-Beles Sub-Basins: Surface Water Investigation*; Ethiopian Ministry of Water and Energy: Addis Ababa, Ethiopia, 2007.
38. Fischer, G.; Nachtergaele, F.; Prieler, S.; Van Velthuizen, H.; Verelst, L.; Wiberg, D. *Global Agro-Ecological Zones Assessment for Agriculture (GAEZ 2008)*; IIASA: Laxenburg, Austria; FAO: Rome, Italy, 2008; Volume 10.
39. Saleem, H.A.; El-Tahlawi, M.R.; Kassem, A.-E.; Boghdady, G.Y. Effect of the Nile aquifer lithological characteristics on groundwater chemistry in Assiut Governorate, Egypt. *J. Ecol. Eng.* **2019**, *20*, 73–83. [CrossRef]
40. Harris, I.; Osborn, T.J.; Jones, P.; Lister, D. Version 4 of the CRU TS monthly high-resolution gridded multivariate climate dataset. *Sci. Data* **2020**, *7*, 109. [CrossRef]
41. Thornthwaite, C.W. An approach toward a rational classification of climate. *Geogr. Rev.* **1948**, *38*, 55–94. [CrossRef]
42. Franklin, W.; Cardy, G. The United Nations data bases on desertification. In *Rangeland Desertification*; Springer: Berlin/Heidelberg, Germany, 2000; pp. 131–141.
43. Batelaan, O.; De Smedt, F.; Triest, L. Regional groundwater discharge: Phreatophyte mapping, groundwater modelling and impact analysis of land-use change. *J. Hydrol.* **2003**, *275*, 86–108. [CrossRef]
44. Salem, A.; Dezső, J.; El-Rawy, M.; Lóczy, D. Hydrological Modeling to Assess the Efficiency of Groundwater Replenishment through Natural Reservoirs in the Hungarian Drava River Floodplain. *Water* **2020**, *12*, 250. [CrossRef]
45. Gabr, M.E.; Soussa, H.; Fattouh, E. Groundwater quality evaluation for drinking and irrigation uses in Dayrout city Upper Egypt. *Ain Shams Eng. J.* **2021**, *12*, 327–340. [CrossRef]
46. Dezső, J.; Salem, A.; Lóczy, D.; Slowik, M.; Pall, D.G. Randomly layered fluvial sediments influenced groundwater-surface water interaction. In Proceedings of the International Multidisciplinary Scientific GeoConference: SGEM, Albena, Bulgaria, 29 June–5 July 2017; Volume 17, pp. 331–337.
47. Salem, A.; Abduljaleel, Y.; Dezső, J.; Lóczy, D. Integrated assessment of the impact of land use changes on groundwater recharge and groundwater level in the Drava floodplain, Hungary. *Sci. Rep.* **2023**, *13*, 5061. [CrossRef]
48. Allam, M.; El Gamal, F.; Hesham, M. Irrigation systems performance in Egypt. *Irrig. Syst. Perform. Options Méditerr. Sér. B Etudes Rech.* **2005**, *52*, 85–98.

Article

Assessing the Effects of Miedzyodrze Area Revitalization on Estuarine Flows in the Odra River

Robert Mańko 

Faculty of Civil and Environmental Engineering, West Pomeranian University of Technology in Szczecin, 70-310 Szczecin, Poland; robert.manko@zut.edu.pl

Abstract: The estuarine section of the Odra River network is influenced by various phenomena that shape its hydrological regime. The Lower Odra region includes "Miedzyodrze," an area between the main branches of the Odra River that was previously used for agriculture. However, due to a lack of maintenance in the 20th century, Miedzyodrze's infrastructure suffered significant damage, resulting in blockages and channel shallowing. Previous models of the lower Odra River network overlooked Miedzyodrze's hydrodynamic impact on flow. To address this, a study aimed to assess Miedzyodrze's influence on flows within the network. Three computational scenarios were developed: one treating Miedzyodrze as an uncontrolled floodplain, another excluding it from the flow like past models, and a third incorporating the hydraulic capacity of selected Miedzyodrze channels with hypothetical restoration. The construction of the models involved extensive field research, including bathymetric surveys and an inventory of channels and structures. Challenges arose from legal and technical constraints during the research. The hydraulic network model was developed using Hec-Ras software and underwent calibration and verification processes for accuracy and reliability. The study focused on analyzing changes in water distribution, flow reduction along the East Odra, flow ratios at specific points, and downstream flow alterations based on different scenarios and the aperture extent of the Widuchowa weir. The conducted analyses and deductions validate the thesis proposed in this study that the potential process of channel dredging and renovation of the hydraulic infrastructure in Miedzyodrze will significantly influence the flow distribution within the lower Odra River network. The significant impact of the Międzyodrze area on water distribution in the lower course of the Odra River has been successfully demonstrated. Under specific hydrological scenarios, a potential increase in flow through the Międzyodrze area from approximately 10–100 m^3/s to a range of 60–420 m^3/s has been identified. This dynamic alteration of river flow exerts a pronounced influence on further water distribution within the entire river network. For the purpose of addressing the matter at hand within this study, the following procedures were undertaken: → analysis of characteristic flow regimes and states, → bathymetric measurements, → flow measurements at selected cross-sections, → construction of a numerical model of the river network, → model calibration, → formulation of a set of boundary conditions, → modeling, → results analysis.

Keywords: Miedzyodrze; Odra River; Hec-Ras

Citation: Mańko, R. Assessing the Effects of Miedzyodrze Area Revitalization on Estuarine Flows in the Odra River. *Water* **2023**, *15*, 2926. https://doi.org/10.3390/w15162926

Academic Editor: Anargiros I. Delis

Received: 14 July 2023
Revised: 9 August 2023
Accepted: 11 August 2023
Published: 14 August 2023

1. Introduction

The issue of river floods and storm surges, as well as flood protection in river estuaries, has been a subject of interest for hydrologists for many years. The processes shaping water level profiles in river estuaries are highly complex problems that do not have unique mathematical solutions. Water level profiles in these areas are determined by multiple factors, such as flow values, the influence of the sea, and wind conditions, which do not have strict relationships. Currently, the most effective tool for describing and forecasting flow in open channels is mathematical modeling. For the model to function accurately, it requires verification and calibration based on field studies. The model will represent the hydrodynamics of water flow more accurately when more measurement data is available for the verification process, obtained under different hydrological conditions (during low and high flows and sea level profiles).

River estuaries have a significant impact on human settlements, serving as a source of water supply and wastewater discharge as well as supporting economic activities such as maritime ports and fisheries. Despite the enormous economic and social benefits associated with locating settlements in estuarine regions, it is important to acknowledge the significant flood risk associated with these areas. In the case of river estuaries, floods can occur in two ways: floods resulting from increased flow within the river channel and floods caused by storm surges during storm events. One approach to passive flood protection is the construction of flood embankments, retention reservoirs, and polder areas. These measures aim to mitigate flood risks and provide a level of flood protection to the surrounding communities and infrastructure.

A particular example is the Lower Odra area between Gozdowice and Roztoka Odrzanska. This area is highly complex from a hydrographic perspective. The dense river network with a ring-like structure, the presence of the Miedzyodrze area, and the existence of Lake Dabie make it an exceptionally interesting research subject for hydrologists [1]. Currently, due to the devastation of hydrotechnical infrastructure, the Miedzyodrze area serves primarily as an embanked floodplain, significantly influencing the hydraulic conditions of the lower Odra River network. A characteristic feature of the lower Odra River channels is their significant depth and very small water surface gradients.

An important factor influencing the water flows in the lower Odra River network is the weir in Widuchowa, which determines the discharge values into the East and West Odra branches. Hydrotechnical structures located within the river channels require separate consideration during the construction of numerical models [2,3], for which a separate calibration process should also be conducted.

The first comprehensive measurements of water flow in the channels of the Lower Odra network were conducted in 2009–2010 by Kurnatowski [1]. Despite numerous studies in the Lower Odra network, no measurements of water flow in the Miedzyodrze channels have been conducted. Research in the Miedzyodrze area from the Widuchowa node to the Skosnica channel has been limited to qualitative water quality studies, while hydraulic calculations focused on determining the distribution of flows into the East and West Odra branches and the contribution of Skosnica to the flow balance. As part of this work, field research was conducted, including point measurements of the depths of the Miedzyodrze channels, to create an up-to-date bathymetry model of this area for simulating various hydrological scenarios.

The main branches of the lower Odra River network are traversed by international waterways [4], and therefore, carrying out revitalization works on the Miedzyodrze can significantly impact the hydrodynamic conditions along these routes.

The primary objective of this research is to undertake a comprehensive numerical analysis of hydraulic phenomena in the lower Odra River network and assess the influence of channel clearance and the renovation of hydraulic infrastructure on flow characteristics within the network. To accomplish this objective, a one-dimensional water flow model has been constructed, employing three computational variants:

- Current situation—the Miedzyodrze area is treated as an uncontrolled floodplain.
- Situation after the renovation of hydraulic infrastructure—all hydraulic structures are closed, preventing the involvement of the Miedzyodrze area in water flow.
- Situation after channel clearance—selected channels within the Miedzyodrze area are hydraulically passable, while the areas between the channels are considered small floodplain zones.

Before achieving the stated objectives, a fundamental requirement must be met, namely the positive verification of the model based on measurement data. Due to the absence of a correlation between water levels and flows in the lower Odra River network and the consequent lack of rating curves, even a well-established measurement network can only provide reliable water level hydrographs, while the essential flow hydrographs, necessary for the process of model verification, calibration, and subsequent validation, will still be unavailable.

Due to the nature of the study, the author did not undertake a detailed analysis of the anticipated repair and modernization works. The focus of the study is the outcome, which is the restoration of the functionality of the hydraulic structures. The author limited the scope to the inventory of hydrotechnical devices in the Miedzyodrze area, emphasizing that any potential repair work should be preceded by a thorough investigation. One should take into consideration the scope of pertinent preparations. In [5], it was indicated that the example of Indian water management policy highlights the consideration of numerous factors, primarily emphasizing the focus on conservation activities, which must form an integral part of climate change strategies. Furthermore, it has been underscored that water management systems must account for the requirements of clean water while maintaining suitable cultural and recreational objectives.

In the first half of 2022, a deepened waterway section with a depth of 12.5 m and simultaneous widening to 100 m was commissioned for use between Szczecin and Świnoujście, Poland. Currently, the maximum ship draft is approximately 9 m, which is insufficient for conducting international cargo exchanges [6]. The deepening significantly enhances the accessibility of inland ports in the Szczecin area [7,8]. The contract value amounted to around 450 million USD. Due to the high value of the contract, the impact of the deepened waterway on flow hydraulics must be carefully considered. As early as 2021 [9], concerns were raised about the potential effects of the deepened waterway on flow patterns in the lower Odra River network, leading to the potential destabilization of sediment transport dynamics. The present study serves as the initial step in further analyzing the hydraulic implications within the lower Odra River network.

2. General Characteristics of the Lower Odra River and the Miedzyodrze Area

In a typical river, three stages of its course can be observed: the upper, middle, and lower reaches. The division into these sections is based on the occurring channel-forming processes [1]. The lower reach of the river is characterized by a smaller channel gradient, reduced bed erosion, and increased sediment accumulation. The Odra River serves as a model example of this three-part division in the Polish context. According to this division, the Odra can be classified into three distinct sections:

- Upper Odra,
- Middle Odra,
- Lower Odra.

The Odra River originates in the Odra Mountains in the Czech Republic. The total length of the river from its source to the mouth is 854 km, with 742 km located in Poland. The catchment area of the Odra River covers nearly 120,000 km^2, of which 106,000 km^2 are located within Poland (Figure 1).

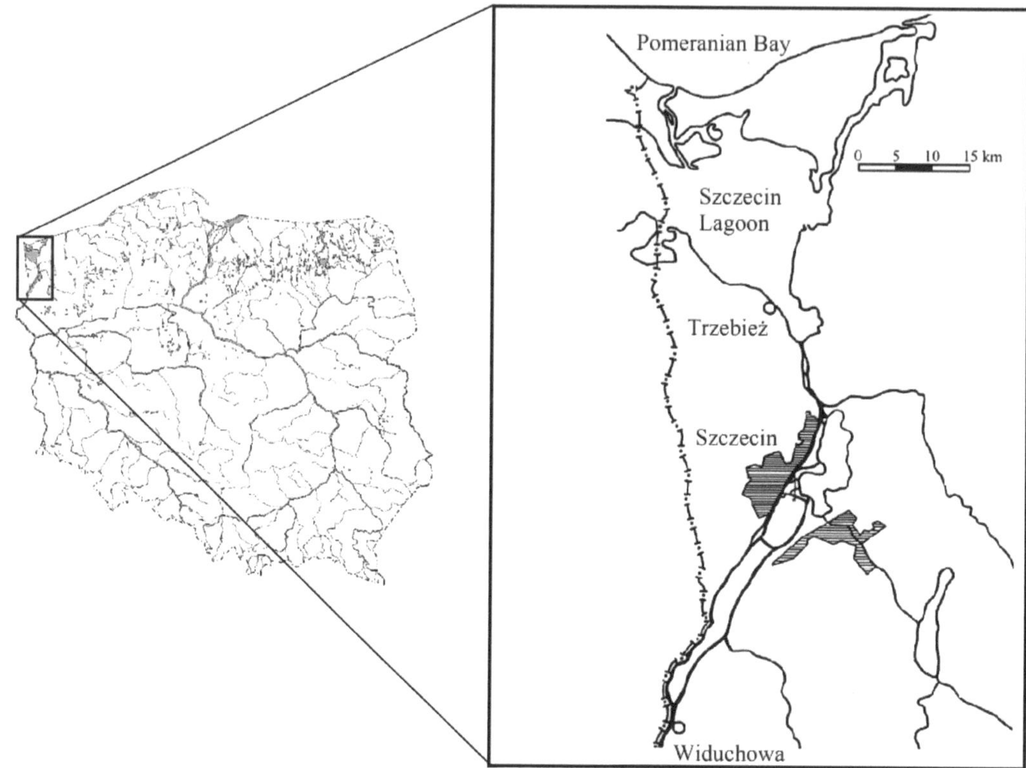

Figure 1. Lower Odra River location [10].

Depending on the adopted research criteria, the concept of the "Lower Odra" can be ambiguous. Over the years, four criteria have been developed to define the area of the lower Odra:

- Hydrological criterion: The Lower Odra is defined as the section of the river from Roztoka Odrzanska, near Trzebiez, to the water gauge section in Gozdowice, beyond the Baltic Sea influence [11].
- Geographic criterion: The Lower Odra begins at the mouth of the Warta River and ends at Lake Dabie.
- Hydrographic criterion: The term Lower Odra refers to the section of the Odra River from the Warta River to Roztoka Odrzanska.
- Navigational criterion: The Lower Odra is considered the section from Zaton Gorna (Odra-Havel Canal) to the Long Bridge in Szczecin and Lake Dabie.

Buchholz [11,12] and later Kowalewska-Kalkowska [13] introduced the term "Odra River Mouth Area", which encompasses: The river network consisting of the section of the Odra River from Gozdowice to Roztoka Odrzanska near Trzebiez, including Lake Dabie; The Szczecin Lagoon; and the maritime straits through which the Odra River flows into the Baltic Sea: Dziwna, Swina, and Piana.

According to [14,15], the Odra River Mouth Area can be classified into three types of estuaries based on the upper and lower boundaries:

- First-order estuary: Pomeranian Bay,
- Second-order estuary: Greifswald Bay and Szczecin Lagoon, including Achterwasser Bay and Lake Wrzosowskie,

- Third-order estuary: Lake Dabie, which is connected to the Szczecin Lagoon via Inski Nurt, West Odra, and Roztoka Odrzanska.

From Gozdowice (at km 645.3) to Widuchowa (at km 701.8), the Odra River flows in a wide curve with an initial northwest deviation of about 17 km, then northward for approximately 14 km, and finally turns northeast for about 27 km. The entire stretch of the Odra River in this section forms the border between the Republic of Poland and the Federal Republic of Germany. The average depths on this stretch range from 2.5 to 3.0 m, and the average water surface width is around 140–150 m [9,13]. Between the Widuchowa and Gozdowice water gauge cross-sections, the Slubia River (at km 652.5), Cedynski Canal (at km 673.4), and Rurzyca River (at km 695.2) are tributaries [16]. At km 704.1, the Odra River branches into the East Odra, which is the main navigation channel, and the West Odra, separated by a weir called the Widuchowa weir. After the bifurcation, both branches of the Odra generally run parallel to each other and are oriented northeast. The West Odra, from Widuchowa to Roztoka Odrzanska, has varying depths ranging from 6 to 10 m and water surface widths between 140 and 200 m. The East Odra flows into the large flow-through Lake Dabie and has average depths of around 7 m and water surface widths ranging from 150 to 160 m throughout its length. The area between the West and East Odra is called Miedzyodrze and was used for agricultural purposes before World War II. In the northern part of Miedzyodrze, at km 730.5, the East Odra connects to the West Odra (29.8 km of its course) through the Skosnica channel, which is approximately 2.5 km long. The average depth of Skosnica is about 3.5–4 m, and the water surface width is around 110 m [11,12]. Other important connections between the East and West Odra include the Odynca, Gryfino, Mielenski, and Parnica channels. From this point, the East Odra is referred to as Regalica, and from Parnica to its mouth, it is known as the Mienia River. The Marwice-Gartz Canal is a significant connection between the Odra branches. It is the only canal crossing Miedzyodrze that allows vessels with outboard combustion engines to navigate. A comprehensive depiction of the lower Odra valley's attributes can be examined within reference [16]. Due to the typical nature of a river estuary, the lower Odra is less susceptible to alterations in its morphology compared to its middle and upper reaches [17]. One characteristic feature of the Odra River Mouth is its kilometrage, which has been divided into three ways due to various factors such as navigation, the Widuchowa weir, and the administrative division between maritime inland waters and inland waters [18]:

According to the Institute of Meteorology and Water Management, the Odra River stretches from Gozdowice (645.3 km) through the East Odra and then Skosnica (starting at 730.5 km and ending at 733.2 km), and further via the West Odra to the Inski stream (753.1 km of the Odra River).

According to the Maritime Office in Szczecin, the waterway extends from the central breakwater (formerly known as the east breakwater) in Swinoujscie to the Long Bridge in Szczecin, known as the Swinoujscie-Szczecin waterway (66.5 km).

According to the Maritime Institute in Szczecin, the West Odra is a separate kilometer from the Widuchowa weir to the Long Bridge (36.6 km).

Due to various kilometrage systems, distances between cross-sections or even individual objects calculated along the river axis may often only be approximated [1], resulting in divergent values presented in the scientific literature.

Lake Dabie, where the East Odra flows, has an area of 56 km^2. The maximum length of Lake Dabie is approximately 15 km, the maximum width is about 7 km, and the maximum depth is around 4.2 m [6,7]. Lake Dabie is a relatively shallow body of water. Between 1962 and 1996, the average depth of Lake Dabie decreased from 2.84 to 2.61 [19], indicating a reduction in the lake's volume of approximately 13 million cubic meters [20].

In Figure 2 [1], the area of the lower Odra network (from the Widuchowa junction to Roztoka Odrzanska) is presented.

Figure 2. Lower Odra River network [1].

The area bounded by the East Odra and West Odra, from the Widuchowa junction to the Inski stream, is called Miedzyodrze [21]. In the southern part of the Miedzyodrze area (from the Widuchowa junction to the Odyniec Canal), the Lower Odra Valley Landscape Park is located, designated as an area of exceptional natural and cultural value since 1 April 1993. The Park is a unique collection of peat bogs and wetlands on a European scale [1].

As a consequence of World War II, numerous hydro-technical structures in the Miedzyodrze area were damaged. It was not until the 1960s that these structures were reconstructed and put back into operation, allowing for the renewed agricultural utilization of the Miedzyodrze region. However, due to the escalating operational costs of the hydro-technical facilities, agriculture became financially unviable. The lack of maintenance and oversight of the hydraulic devices led to the re-destruction of the majority of these structures. Subsequently, near-complete naturalization and ecological succession took place.

Due to the prevailing marshiness of most of the Miedzyodrze area, the region has become poorly accessible and, in certain locations, entirely inaccessible to humans. The numerous instances of canal and hydro-technical equipment devastation have rendered them currently devoid of any regulatory functions within Miedzyodrze. Despite the distinctive natural values presented by Miedzyodrze in comparison to Europe (and possibly the world), it is essential to bear in mind and frequently emphasize that the current state is attributable to human influence, specifically the political circumstances of the 1950s and 1960s, which facilitated the degradation and naturalization of this area.

One characteristic feature of the Miedzyodrze area, from the Widuchowa Junction to the Skosnica channel, is the presence of polders. These are flat areas designated for inundation during the autumn-winter period, slightly elevated above the average water level in the river. The elevations of the polders range from approximately 0.00 to 0.40 m above sea level. Following the network arrangement, Miedzyodrze has been artificially divided hydrographically into three independent polders:

- Widuchowa Polder, also known as the southern polder, encompasses the area from the Widuchowa Junction to the embankment of the Gryfino-Mescherin road.
- Gryfino Polder, known as the middle polder, covers the area from the embankment of the Gryfino-Mescherin road to the embankment of the A6 Motorway.
- Szczecin Polder, referred to as the northern polder, extends from the embankment of the motorway to the Skosnica.

The Miedzyodrze area also includes the region between Skosnica and the Odynca Canal, which is also known as the Regatta Track. [11]

Detailed information about the Miedzyodrze polders, including their surface area, watercourse area, and average elevations, is presented in Table 1 [18].

Table 1. Miedzyodrze's polders.

No	Polder Name	Polder Area [ha]	Watercourse Area [ha]	Average Elevation (m above Sea Level)
1	Widuchowski	2465.44	348.5	0.10
2	Gryfinski	2232.32	135.0	0.20–0.30
3	Szczecinski	758.64	137.3	0.20–0.30

Hydrotechnical structures in the Miedzyodrze area

During the construction works aimed at proper water management, 35 hydrotechnical structures were built, including:

1. 18 chamber-economic locks,
2. 2 chamber-navigation locks,
3. 6 weirs (embankment overflow structures),
4. 5 embankment culverts,
5. 4 pumping stations.

Chamber-economic locks:

The purpose of the chamber-economic locks was economic transportation, mainly for transporting hay from the Miedzyodrze area. The majority of these locks, specifically 10 of them, were located in the southern polder. Each of the middle and northern polders had four locks.

Chamber-navigation locks:

The main function of the chamber-navigation locks was to provide a navigable connection between the East Odra and the West Odra. Both locks facilitated the passage of barges with a displacement of up to 400 tons.

The locks are single-chamber locks without a drop gate. They were entirely made of concrete, with reinforced concrete main walls and concrete retaining walls. The upper parts of the chamber navigation locks were faced with clinker bricks. Each lock was equipped with double-leaf supporting gates operated manually from the shore using lifting rods and gear mechanisms. The gates had closing orifices controlled by valves, which allowed for changing the water level in the chambers during the passage of floating vessels [11,12,14]. Both locks are located on the Marwice-Gartz channel, which is the only channel in the Miedzyodrze area that allows the movement of motor-powered vessels.

Embankment overflow structures:

Six embankment weirs let floodwaters into the Miedzyodrze area during the summer period and inundated the polders during the autumn-winter season. These weirs are concrete structures with one or two spans equipped with flat gates (2 embankments) or sliding flaps (4 embankments). The closures were manually operated through a system of gear transmissions, lifting rods, and coupling devices [22]. Three weirs are located in the southern polder, two in the middle polder, and only one in the northern polder.

Embankment culverts:

The purpose of the embankment culverts was to assist in water level regulation in the Miedzyodrze area. The self-closing supporting gates installed in the culverts protected the Miedzyodrze area from excessive water inflow during high water levels in the East Odra. During low water levels, the culverts discharged excess water from the Miedzyodrze area, thus supporting the pumping stations. The embankment culverts are concrete structures with one or multiple rectangular-shaped channels. The southern polder has four embankment culverts, while the middle polder has one.

Stations pumping

The main purpose of the pumping stations was to maintain the optimal water table level in the Miedzyodrze area, ensuring optimal water conditions for the green areas located there. There are two pumping stations in the southern polder and one each in the middle and northern polders.

Flood embankments

The Miedzyodrze area is protected by embankments from the east, west, and northern sides. The flood embankments, with an average height of 1.5 m, protected the Miedzyodrze polders up to an Odra flow rate of 1600 m^3/s (according to the Hohensaaten water gauge) [22]. Buchholz [11], based on archival materials, determined the lengths of the flood embankments in the respective polders, which are presented in Table 2.

Table 2. Lengths of flood embankments in the Miedzyodrze area.

River	Lengths of Embankments in Polder [km]		
	Widuchowa	Gryfino	Szczecin
East Odra	14.08	9.77	2.55
West Odra	14.55	10.85	4.18
Skosnica	-	-	2.57
Total	28.63	20.62	9.30

Similar to the hydrotechnical structures in the Miedzyodrze area, the flood embankments have undergone extensive deterioration due to a prolonged lack of maintenance. The embankment slopes have experienced significant slippage, resulting in localized reductions in the height of the flood embankments. Animals, particularly beavers, have had a significant impact on the degradation of these embankments. An example of embankment slippage at the beginning of the East Odra is depicted in Figure 3.

Figures 4 and 5 depict selected deteriorated hydrotechnical structures in the Miedzyodrze area.

The exact causes behind the devastation of water infrastructure remain partially elusive. The destruction of locks can be attributed to a multitude of factors. Primarily, the aging of infrastructure emerges as a significant cause, particularly in the case of older locks that have endured inadequate maintenance or upgrading over prolonged periods. Moreover, deliberate acts of sabotage, acts of vandalism, or unforeseen technical malfunctions can precipitate damage to locks.

Miedzyodrze's channels

The absence of regular maintenance practices and unimpeded sedimentation and debris accumulation have been key factors contributing to the shallowing of the Miedzyodrze channels, ultimately leading to the cessation of unimpeded surface water flow within the area. The exchange of water within the channels now predominantly occurs during significant flood events, effectively submerging the entire Miedzyodrze area. The lack of continuous water exchange within the channels has resulted in the complete colonization of certain channel segments by vegetation (see Figures 6 and 7).

Figure 3. The damaged embankment on the East Odra at km 705.0.

Figure 4. Devastated Lock (East Odra at km 725.4).

Figure 5. Devastated Pumping Station Building (West Odra at km 28.9).

Figure 6. Overgrown channels of Miedzyodrze (bird's eye view) [23].

Figure 7. Overgrown channel in Miedzyodrze.

Due to the current condition of the channels, aerial photographs are insufficient for identifying all the channels in Miedzyodrze. Therefore, on-site inspections were necessary to accurately assess the actual state of the channels. It is estimated that there are over 200 km of channels in the Miedzyodrze area, which significantly extended the time dedicated to field surveys. The water quality across the entire Międzyodrze area until the year 2022 was of a significantly poor nature [24]; however, following the ecological catastrophe spanning nearly the entire length of the Odra River, the quality of these waters further deteriorated [25].

Widuchowa weir

The Widuchowa weir is located at the beginning of the West Odra (approximately 100 m from the bifurcation). Its main function is to control the flow distribution at the Widuchowa node. By diverting a significant portion of the flow through the East Odra, it increases the difference in water levels in the main branches of the Odra, allowing for the gravitational drainage of excess water from the Miedzyodrze area. Higher water levels in the East Odra improve the navigability of the river.

The weir has a width of 78 m and is divided into five spans by four piers. Each span, with a width of 15.6 m, is equipped with guides that divide it into eight sections, each 1.72 m wide [26]. The gates are made of wide-flange steel profiles measuring 260 mm, allowing for the operation of stoplogs to close individual sections. Each section is closed using three stoplogs measuring 1.87×1.50 m, operated by a gantry crane that moves along rails on the weir. The weir is equipped with a concrete threshold with an elevation at the crown of -3.00 m above sea level, which is a permanent part of the structure. The height and width of the threshold are 2.05 m and 70 m, respectively. The piers are made of concrete, while the casing of the pier heads, upper edges of the piers, and outer edges of the abutments are made of granite slabs [27]. Figures 8 and 9 depict the Widuchowa weir (Figure 8—viewed from the upstream side, Figure 9—viewed from the downstream side).

Figure 8. Widuchowa Weir—view from the upstream side.

Figure 9. Widuchowa Weir—view from the downstream side.

The occurrence of weirs in looped river networks can introduce additional challenges in the determination of coefficients. A novel approach to deducing the chasing coefficients in looped river networks with weirs is presented in [28].

3. Hydrology of the Lower Odra Area

Table 3 presents characteristic water levels and flows in the lower Odra network between Trzebiez and Gozdowice. There are six cross-sections in this section where observations have been continuously conducted by the Institute of Meteorology and Water Management for several decades. Characteristic water levels and flows for the long-term period are determined based on annual data. For each year, three flow values are analyzed: the minimum, average (mean), and maximum, along with their corresponding water level values. Subsequently, these data are arranged in a time series, from which the minimum, mean, and maximum characteristic values are derived for the entire long-term period.

Table 3. Characteristic flow and stages in the Lower Odra area [29].

Gauge	Statistical Analysis Period	Characteristic Values				
Characteristic flows		NNQ	SNQ	SSQ	SWQ	WWQ
Gozdowice	1959–2007	158	252	535	1251	3180
Characteristic stages		NNW	SNW	SSW	SWW	WWW
Gozdowice	1959–2007	144	208	322	492	659
Bielinek	1954–2007	147	209	333	532	754
Widuchowa	1949–2007	440	479	545	652	771
Gryfino	1952–2007	440	466	523	601	685
Podjuchy	1993–2007	442	474	524	598	628
Long Bridgw in Szczecin	1959–2000	433	459	512	587	622
Trzebiez	1949–2000	429	456	510	583	614

Notes: NNQ/NNW—Lowest flow/stage of the lowest from the multi-year period. SNQ/SNW—Mean flow/stage of the lowest from the multi-year period. SSQ/SSW—Mean flow/stage of the averages from the multi-year period. SWQ/SWW—Mean flow/stage of the highest from the multi-year period. WWQ/WWW—Highest flow/stage of the highest from the multi-year period.

An influential determinant of the states and flows in the lower Odra network is the Baltic Sea. In shaping the water surface elevations at the Odra estuary, it is the Baltic Sea that exerts the most significant influence [30–32]. Despite its closed nature, the Baltic Sea manifests pronounced fluctuations in water surface elevation, with a potential amplitude of up to 3 m between extreme values. The key factors contributing to the variability of the water level encompass the following factors [33].

Over the years, a gradual rise in the average annual sea level has been observed, amounting to approximately 0.7 ± 0.1 mm per year for the Swinoujscie region. This systematic increase in sea level in the Swinoujscie area over the coming decades may substantially alter the flow regime within the lower Odra network [19].

Due to the significant influence of the sea on the formation of flows, the lower Odra River area is not exposed to the risk of floods like the middle and upper sections of the Odra River [34]. The influence of the sea also translates into the development of hypothetical flood hydrographs [35].

4. Mathematical Description and Field Research

The study utilized the widely available Hec-Ras software, employing its one-dimensional module for calculating flows and water levels in open channels. The mathematical foundations of the equations used were extensively described in [36,37].

The fundamental equations used to solve flow problems in the Hec-Ras program are the Saint-Venant equations.

$$\frac{\partial Q}{\partial t} + \frac{\partial}{\partial x}\left(\frac{Q^2}{A}\right) + gA\frac{\partial z}{\partial x} + \frac{Q\,|Q|n^2 g}{R_H^{\frac{4}{3}}\,A} = 0 \tag{1}$$

$$\frac{\partial A}{\partial t} + \frac{\partial Q}{\partial x} = 0 \tag{2}$$

The aforementioned equations are known as the Saint-Venant system of equations, often referred to in the literature as the full dynamic model. These equations are a widely used mathematical model for describing one-dimensional flow in open channels. In the literature, various forms and extensions of these equations can be encountered. In specific scenarios, lateral inflows, channel bifurcation, and the storage capacity of floodplain cross-sections can be considered.

Individual terms in the equation represent energy losses caused by variations in flow intensity over time, flow intensity changes along the channel length, water surface slope, and bed shear stresses (friction forces). In the continuity equation, the individual terms account for the changes in the cross-sectional area over time and the change in flow intensity along the channel length.

When the Saint-Venant system of equations is expanded to include:

- The storage capacity of retention areas—$\frac{\partial S}{\partial t}$
- Lateral momentum inflow at the node—$\varsigma \frac{\partial Q_B{}^2}{\partial x}$
- The coefficient of non-uniformity in velocity distribution (Boussinesq coefficient)—β

We obtain the complete governing equations in the HEC-RAS model.

The Manning coefficient is one of the most significant parameters in hydraulic calculations of open channels. The accuracy of the estimation of the roughness coefficient has a significant impact on the accuracy of the calculations. The magnitude of the coefficient is influenced by various factors, including bed roughness, vegetation, channel regularity, type of transported and entrained sediment and its flow intensity, channel shape and depth, seasonality, and any obstacles (such as tree trunks, bridge piers, etc.) affecting flow hydrodynamics [38].

Calculations involving looped river networks can be reviewed in, among others, references [10,39], where [39] applied methods including A Prediction-Correction Solver for Real-Time Simulation of Free-Surface Flows.

If it is not possible to estimate the value of the Manning coefficient, it should be determined based on field surveys and mathematical models where the roughness coefficient is treated as an identifiable parameter involving the minimization process of a specific function dependent on the sought coefficient [29]. Due to the non-uniformity of the bed material in the cross-sectional and longitudinal channel profiles, only the concept of equivalent roughness (Manning roughness) can be applied, which integrates the values of individual channel parts [40–42]. Adopting an equivalent roughness as a global value is acceptable due to the small variations in channel morphology and bed and bank coverings [43].

The identification of Manning coefficients for the lower Odra River channels has been extensively analyzed by researchers from the Department of Hydraulic Engineering at the Szczecin University of Technology. Orlewicz [40] reported that for average flow conditions, the Manning coefficient for the lower Odra River channels is 0.030 m$^{-1/3}$s. Kurnatowski in 2004 [13,43,44] determined the influence of changes in the height reference system and geoid position on the values of roughness coefficients. It was established that for different reference systems, the global roughness ranges from 0.0175 to 0.0418 for the "Amsterdam zero", and from 0.0163 to 0.0306 for the "zero Kronsztad". Roszak [45] investigated the impact of sediment grain size on the roughness of the lower Odra River channels. The forms of the channel bed play a significant role in hydraulic roughness. Arcement and Schneider [46] identified changes in the roughness coefficient for various types of channel bed forms, obtaining the highest values for dunes in the final phase of development and the lowest values for flat beds.

Insight into solving the problem of identifying the roughness coefficient and its variations as flow parameters can be gained from references such as [38,47]. Interestingly, the modification of the Manning coefficient can also be employed to expedite the modeling of tsunamis [48].

The values of roughness coefficients for river channel sections should be systematically verified and calibrated based on new data obtained from field surveys.

Field studies of the lower Odra River network involve measurements of hydrological and meteorological parameters, as well as bathymetric surveys. The measurement of water levels and flows has been continuously conducted for several decades by the Institute of Meteorology and Water Management. Kurnatowski [1] performed an analysis of hydrological data, complemented by water level measurements at the Widuchowa weir, resulting in 34 independent steady flow situations, including water levels in Trzebiez (W_T), water levels in Widuchowa (W_W), flows in the Odra River at the Gozdowice section (Q_G), and the difference in water levels between the upper and lower sections at the Widuchowa weir. Results are presented in Table 4.

Table 4. Measurements of steady flow in the lower Odra River network—Part 1.

Lp.	W_T [cm]	W_W [cm]	Q_G [m^{3}s^{-1}]	ΔW [cm]
1	490.6	520.0	454.3	8
2	495.4	523.4	387.3	10
3	504.1	536.2	445.0	9
4	502.0	537.9	479.0	16
5	508.4	611.1	1264.0	10
6	531.2	561.4	578.0	4
7	496.1	584.2	864.1	47
8	496.9	599.5	1031.3	50
9	505.0	532.7	417.3	18
10	522.2	544.9	423.6	11
11	491.9	567.9	771.5	33
12	505.8	542.7	509.3	16
13	499.2	531.8	458.3	19
14	502.1	533.6	467.2	19
15	506.1	617.0	1300.0	6
16	503.0	631.6	1383.3	11
17	492.8	559.6	696.0	41
18	490.6	518.5	453.0	17
19	502.5	535.5	465.2	20
20	504.9	541.0	520.3	20
21	491.6	508.0	291.0	5.5
22	496.2	514.2	362.3	9
23	514.4	532.3	269.7	7
24	487.6	531.3	565.8	29
25	495.8	516.3	292.0	7
26	487.2	509.4	274.0	7
27	500.4	518.6	235.5	3
28	501.4	519.4	373.2	10
29	520.7	560.8	703.5	4
30	504.9	525.0	321.0	9
31	502.9	532.0	424.0	10
32	496.7	526.6	399.3	10
33	477.8	570.4	928.7	51
34	506.0	583.8	924.4	40

Recent studies of the lower Odra River network were conducted using a state-of-the-art ultrasonic velocity meter called the ADCP (Acoustic Doppler Current Profiler). The research focused on the main node of the lower Odra River network, which includes the junction of the West Odra (26), the Odra—Pucka part (17), Skosnica (23), Regalica (24), and the East Odra (27) (numerical labels according to Figure 2). The schematic diagram of the river node where the measurements were conducted is presented in Figure 10.

The measurements conducted allowed for the identification of a group of measurements characterized by flow steadiness, indicating small errors in flow balance closure within the network nodes. The adjusted measurement values for the selected section of the river network are presented in Table 5.

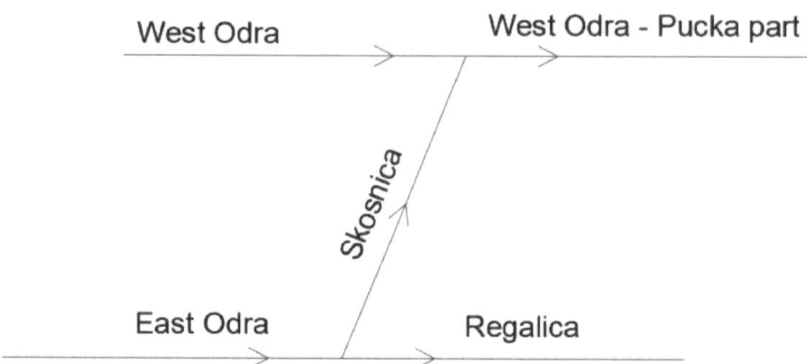

Figure 10. Selected node from the lower Odra river network.

Table 5. Measurements of steady flow in the lower Odra River network—Part 2.

Lp.	Flow [m^3s^{-1}]					Elevation of the Water at Long Bridge [Meters above Sea Level]
	East Odra	West Odra	Skosnica	Regalica	Odra—Pucka Part	
1	524.5	350.5	64.5	460.0	415.0	0.07
2	373.0	221.0	57.4	315.6	278.4	0.04
3	268.0	176.0	36.3	231.7	212.3	0.01
4	553.0	363.1	63.8	489.2	426.9	0.05
5	553.0	346.0	63.3	489.7	409.3	0.04
6	747.0	497.0	84.2	662.8	581.2	0.01

Due to an absolute prohibition on the movement of combustion engine-powered vessels in the Miedzyodrze channels, it was not possible to conduct comprehensive measurements using specialized measuring equipment. Therefore, the study was limited to conducting 25 series of individual sounding surveys using an electric-powered vessel, which resulted in a limited time for each series. From these surveys, average depths were obtained at selected points within the Miedzyodrze channel network. Ultimately, over 500 depth measurements were performed in the Miedzyodrze area. The measurements of average channel depths in Miedzyodrze were used by the author to estimate the geometric parameters of the channels that could be achieved after dredging operations.

The river network in the Hec-RAS model can be treated as a system of individual river segments (channels) connected at nodes. Within this study, three variants of the lower Odra model have been prepared for which calculations and comparative analyses will be conducted.

The first analyzed variant assumes the possibility of utilizing the retention area of Miedzyodrze. The exchange of water between the polders and the East and West Odra occurs through completely open hydraulic structures located on both branches of the Odra River and Skosnica. Due to the "slenderness" (length-to-width ratio) of the Widuchowa polder, it has been divided into two separate cooperating polders: Polder 1a—the southern part of the Widuchowa polder, and Polder 1b—the northern part of the Widuchowa polder. Polders 2 and 3 marked in Figure 11 represent the Gryfino and Szczecin polders.

Based on Variant 0, it is relatively easy to construct Variant 1 by disconnecting the Miedzyodrze area from the surrounding river channels. Variant 1 includes the section of the lower Odra River network from Widuchowa to Regalica and the "Odra—Pucka part". The schematic diagram of the Variant 1 river network is presented in Figure 12.

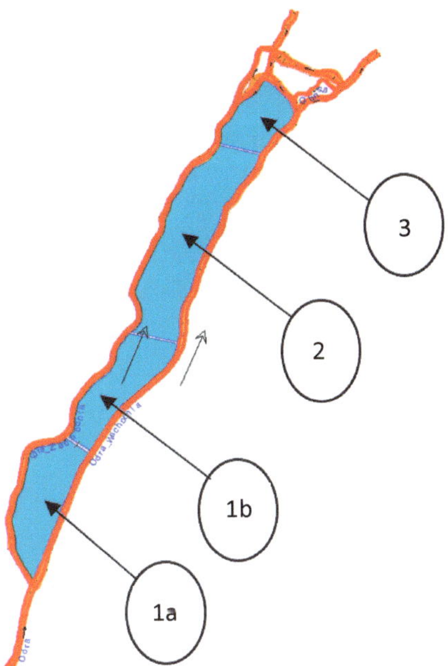

Figure 11. River structure—Variant 0.

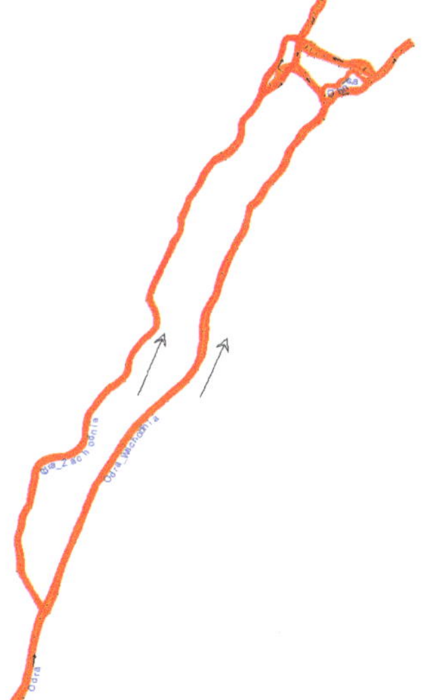

Figure 12. River structure—Variant 1.

Currently, due to the devastation of hydrotechnical devices, the situation where the Miedzyodrze area remains "dry" is not possible. Therefore, this variant should be treated as an idealized scenario, corresponding to a situation where the technical condition of all devices (locks, culverts, etc.) in Miedzyodrze has been restored to its original state, and during flood events, all devices are closed, preventing water from flowing into the Miedzyodrze area.

Variant 1 is composed of 14 river and canal sections connected at 8 nodes. The remaining cross-sections were obtained through interpolation. The average distance between measured cross-sections is approximately 430 m. The total length of all channels in this variant exceeds 81 km.

Variant 2 is the final variant of the prepared model for the lower Odra River network. As the first model developed to date, it considers the possibility of free flow through the Miedzyodrze channels. Using aerial photographs of the Miedzyodrze area, a Preliminary Variant 2 was prepared, which is presented in Figure 13. Due to the large number of channel sections in Miedzyodrze, a selection process was initially conducted to determine which channels would ultimately be included in Variant 2.

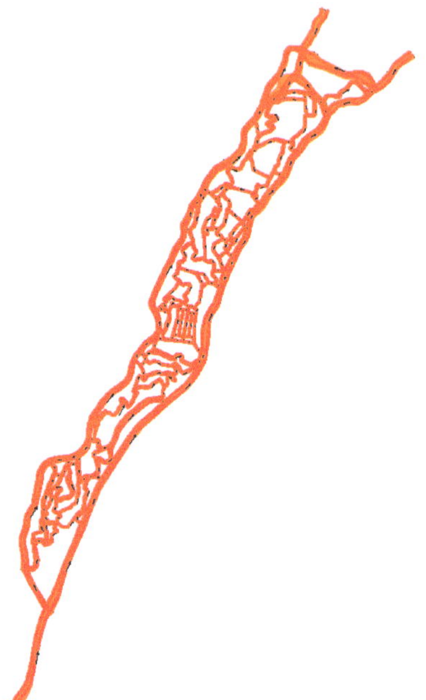

Figure 13. River structure—Preliminary Variant 2.

Preliminary Variant 2 comprises 215 interconnected canal segments at 134 nodes, spanning a total length of over 200 km. The variant encompasses more than 5500 cross-sections, which were determined based on field measurements to estimate the average depths at specific locations within the channel network.

The process of selecting the channel network structure involved the exclusion of canal segments from the initial "preliminary Variant 2" that exhibited average flow velocities below 0.025 m/s. This determination was made considering a fixed hydrological scenario and specific model parameters:

- The upper boundary condition was set as Q = 500 m^3/s.
- The lower boundary conditions were defined as WOP = WR = 0.00 m above mean sea level.
- The global roughness coefficient was established as n = 0.030 m$^{-1/3}$s.

During the model optimization process, numerous canal segments within "preliminary Variant 2" were identified where the average velocities did not surpass the predefined threshold. Consequently, these segments were excluded from the final version of Variant 2, as illustrated in Figure 14.

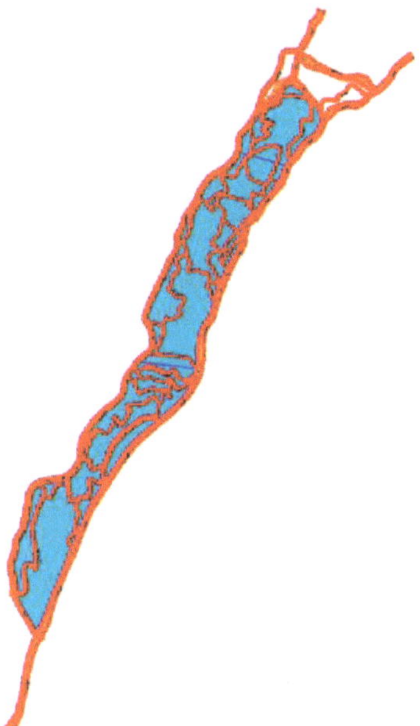

Figure 14. River structure—Variant 2.

Variant 2 consists of 116 segments interconnected at 76 nodes. The river network in Variant 2 comprises over 4000 cross-sections. During the optimization process, the canal network was reduced, leading to the redefinition of the boundaries of retention areas. The resulting areas were included in the model following a similar approach as in Variant 0, thus creating Variant 2, which includes 39 polder areas.

5. Boundary Conditions

The boundary conditions in the model consist of flow hydrographs at the upstream section of the Odra River in Widuchowa, as well as water level hydrographs at the outlets of Regalica and the Odra—Pucka part. Due to the lack of correlation between water levels and flows at arbitrary cross-sections along the lower Odra River from Widuchowa to the outlets, there is significant flexibility in choosing these boundary conditions.

The basis for establishing the boundary conditions is characteristic water levels. Considering the absence of systematic flow measurements in the Widuchowa profile, the lack of correlation between water levels and flows, and the small increase in the Odra River catchment area downstream of Gozdowice, it is assumed that the flow in Widuchowa is equal to the flow in Gozdowice. The author has selected 10 upper boundary conditions based on established characteristic water levels. The minimum flow is represented by the lowest recorded flow, and subsequent flows are increased by 40%, resulting in the final flow approximation of the highest recorded flow. As a result of the analysis, 10 different flows were obtained, ranging from 158 m^3/s to over 3200 m^3/s.

The lower boundary conditions (Table 6), in the form of water level hydrographs at the mouths of the Odra—Pucka part and Regalica, are also based on characteristic values. Due to the significant influence of the sea on water levels in Dabie Lake and the channels in the estuary section of the Odra, the author of the study assumed identical boundary conditions for the profiles at the mouths of the Odra—Pucka part and Regalica. The simplification made by the author is also a result of the fact that no comparative statistical analysis of water levels in the Odra—Pucka part and Regalica has been conducted so far, which could be crucial in developing the boundary conditions of the model.

Table 6. Lower boundary conditions (as water stages at outlets) [29].

No.	Stages [m a.s.l]	No.	Stages [m a.s.l]	No.	Stages [m a.s.l]
W_1	−0.70	W_2	−0.50	W_3	−0.30
W_4	−0.10	W_5	0.00	W_6	0.10
W_7	0.30	W_8	0.50	W_9	0.70
W_{10}	0.90				

The lack of correlation between water levels and flows in the downstream section of the Odra necessitates the analysis of all possible combinations of lower and upper boundary conditions, resulting in a total of 100 hydrological scenarios.

Despite using the term "steady flow", the author refers only to the steady boundary conditions, while the entire computational process for the steady flow regime is based on unsteady flow equations. This is directly due to limitations in the Hec-Ras system, which cannot accurately distribute flows in ring-shaped river networks during steady-state calculations.

The manipulation of weir gates has a significant impact on the flow distribution in the lower Odra network, resulting in the division of flow between the East and West Odra. It can be inferred that increasing the number of open gate spans diverts a larger portion of the flow to the West Odra. For this study, an internal boundary condition was assumed, represented by the maximum gate opening.

6. Model Calibration

A model calibration was conducted for Variant 0, which currently provides the best representation of the actual situation. The process of model calibration involves adjusting the roughness coefficient to minimize the discrepancies between measured and calculated values. In the study, a manual calibration approach was employed, similar to that in [49], despite the availability of numerous tools for calibration automation within the Hec-Ras software package [50]. Table 7 presents the measured and calculated flow values in the indicated segment of the lower Odra network, where the calculated values represent the global roughness coefficient from this variant that yielded the smallest model fit error. Furthermore, the table includes information about the difference between measured and calculated values and provides the error value calculated using the formula:

$$\varepsilon = \left| \frac{Q_m - Q_c}{Q_c} \right| \times 100\%, \tag{3}$$

where Q_m is the flow measured and Q_c is the flow calculated.

The conducted analysis revealed the agreement of results at verification points for the majority of cases. The best agreement was obtained for cases 2, 4, 6, 8, 9, 11, and 12, with an average error of approximately 1.9% and an average difference of about 2.9 m^3/s. For the remaining 2 cases, the corresponding averages were 6.8% and 11.4 m^3/s. The average percentage error for all measurements was 4.1%. The best fit was achieved for measurement number 8, with an average flow mismatch of around 0.2 m^3/s and an average percentage error of 0.17%. The poorest fit was observed for case 13, where the difference between calculated and measured values reached 47 m^3/s for the West Odra section, with a percentage error of 21.4% at Skosnica. In case number 3, an even greater percentage error is observed for Skośnica, reaching as high as 34.8%. This significant discrepancy indicates substantial differences between the actual values and the analysis results for this specific scenario. The main reason for such a large percentage of errors in Skośnica is primarily the presence of low flows. Compared to other analyzed streams, the flows in Skośnica are several times smaller, leading to larger deviations between the actual flow values and the estimated results. Additionally, in some instances, the differences between flow values are similar, further contributing to the increased percentage of errors. It is crucial to consider these factors during the analysis and interpretation of results to ensure more accurate flow estimations for the Skośnica area in future studies. In general, it can be concluded that the model has been properly calibrated and can be applied in further stages of the study.

Table 7. Comparison of measured and calculated steady flow rates.

Case		Manning Coefficient	Odra	East Odra	West Odra	Skosnica	Regalica	Odra—Pucka Part
1	Q_m	0.021	875.0	524.5	350.5	64.5	460.0	415.0
	Q_c			537.9	334.1	67.0	470.0	402.0
	$|\Delta Q|$			13.4	16.4	2.5	10.0	13.0
	ε			2.5	4.9	3.7	2.1	3.2
2	Q_m	0.030	378	233.1	145.7	31.7	201.4	177.4
	Q_c			232.4	146.4	30.7	201.4	177.4
	$|\Delta Q|$			0.7	0.7	1.0	0.0	0.0
	ε			0.3	0.5	3.2	0.0	0.0
3	Q_m	0.033	256.7	171.9	84.8	28.9	143.0	113.7
	Q_c			189.1	98.4	21.4	135.3	120.4
	$|\Delta Q|$			17.2	13.6	7.5	7.7	6.7
	ε			9.1	13.8	34.8	5.7	5.6
4	Q_m	0.027	404.9	250.4	154.6	33.4	216.9	188.0
	Q_c			238.2	147.1	33.0	203.5	179.7
	$|\Delta Q|$			12.1	7.5	0.4	13.4	8.3
	ε			5.1	5.1	1.2	6.6	4.6
5	Q_m	0.027	594.0	373.0	221.0	57.4	315.6	278.4
	Q_c			362.2	230.6	46.9	316.2	276.7
	$|\Delta Q|$			10.8	9.6	10.5	0.6	1.7
	ε			3.0	4.2	22.4	0.2	0.6
6	Q_m	0.033	384.3	234.3	150.0	35.0	199.3	185.0
	Q_c			233.3	147.8	31.6	201.6	179.5
	$|\Delta Q|$			1.0	2.2	3.4	2.3	5.5
	ε			0.4	1.5	10.8	1.2	3.1
7	Q_m	0.030	363.4	217.8	145.7	31.3	186.4	177.0
	Q_c			210.9	131.8	30.1	179.7	160.8
	$|\Delta Q|$			6.8	13.9	1.2	6.7	16.2
	ε			3.2	10.5	4.0	3.7	10.0
8	Q_m	0.030	339.2	208.3	130.9	28.1	180.2	159.0
	Q_c			208.5	131.0	28.1	180.0	159.5
	$|\Delta Q|$			0.2	0.1	0.0	0.2	0.5
	ε			0.1	0.1	0.1	0.1	0.3
9	Q_m	0.033	444.0	268.0	176.0	36.3	231.7	212.3
	Q_c			268.5	170.8	36.3	232.4	206.9
	$|\Delta Q|$			0.5	5.2	0.0	0.7	5.4
	ε			0.2	3.0	0.0	0.3	2.6
10	Q_m	0.021	916.1	553.0	363.1	63.8	489.2	426.9
	Q_c			563.7	348.4	70.5	491.6	420.5
	$|\Delta Q|$			10.7	14.7	6.7	2.4	6.4
	ε			1.9	4.2	9.5	0.5	1.5
11	Q_m	0.021	899.0	553.0	346.0	63.3	489.7	409.3
	Q_c			553.0	342.5	69.0	482.6	412.8
	$|\Delta Q|$			0.0	3.5	5.7	7.1	3.5
	ε			0.0	1.0	8.3	1.5	0.8
12	Q_m	0.024	659.7	407.5	252.2	56.8	350.7	309.0
	Q_c			406.4	250.5	56.0	347.8	306.8
	$|\Delta Q|$			1.1	1.7	0.8	2.9	2.2
	ε			0.3	0.7	1.4	0.8	0.7
13	Q_m	0.027	1244.0	747.0	497.0	84.2	662.8	581.2
	Q_c			767.2	450.4	107.2	648.4	568.8
	$|\Delta Q|$			20.2	46.6	23.0	14.4	12.4
	ε			2.6	10.3	21.4	2.2	2.2

The result of calibration will be a functional relationship between the input flow to the model and the channel roughness coefficient. After excluding the results from cases 6 and 13, the dependency curve can be presented in Figure 15.

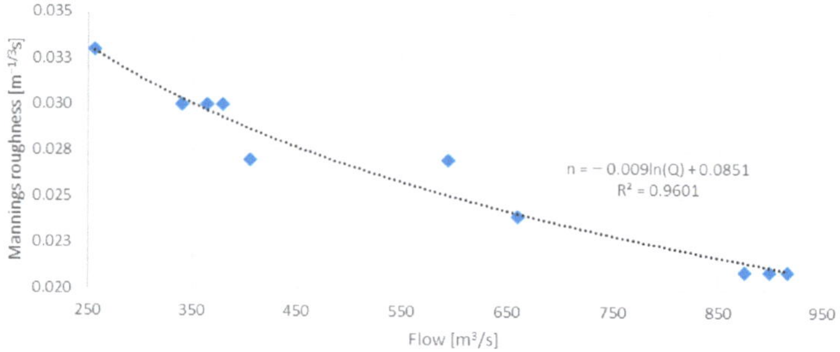

Figure 15. Flow—Mannings coefficient correlation curve.

7. Calculations

The main objective of this study is to comprehensively investigate the impact of channel clearance and hydraulic structure repairs in the Miedzyodrze area on the flow patterns in the lower Odra River network. The lower Odra River network covers a significant area; therefore, the author of the study focused on analyzing selected cross-sections that were considered crucial for understanding and monitoring the influence of these flows. These cross-sections were carefully selected and presented in Figure 16, which provides a graphical representation of their locations.

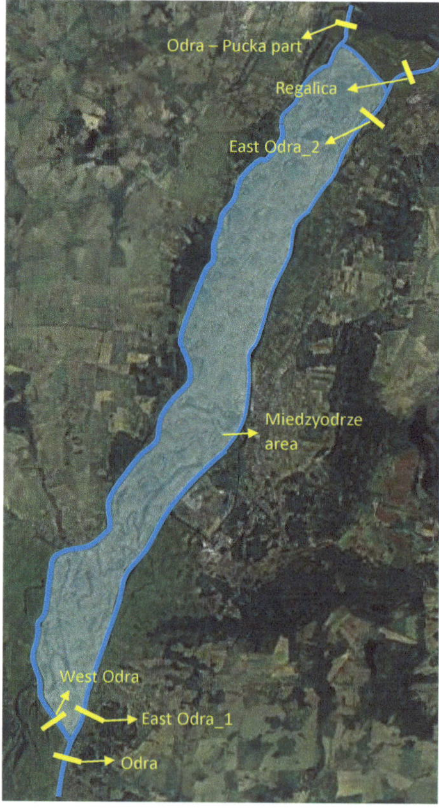

Figure 16. Cross-section locations [51].

Based on the obtained simulation results, the impact of channel clearance in the Miedzyodrze area and the influence of hydraulic structure repairs were assessed on the flows in selected cross-sections of the lower Odra River network. The calculations conducted allowed for the following analyses of steady-state flow in the network:

- Distribution of water from the Odra to West Odra_1 and East Odra_1.

$$dQ_1 = \frac{Q_{EO,1}}{Q_{WO,1}} \tag{4}$$

- Changes in flow values in the cross-sections of East Odra_1 and East Odra_2.

$$dQ_2 = Q_{EO,1} - Q_{EO,2} \tag{5}$$

- The ratio of flows from Regalica to Odra Pucka

$$dQ_3 = \frac{Q_{OP}}{Q_R} \tag{6}$$

where:

$Q_{EO,1}$—Flow in East Odra_1 cross-section [m^3/s]
$Q_{EO,2}$—Flow in East Odra_2 cross-section [m^3/s]
$Q_{WO,1}$—Flow in West Odra_1 cross-section [m^3/s]
$Q_{WO,2}$—Flow in West Odra_2 cross-section [m^3/s]
Q_R—Flow in Regalica cross-section [m^3/s]
Q_{OP}—Flow in Odra—Pucka part cross-section [m^3/s]

In Figures 17 and 18, the changes in the ratio of flows between East Odra and West Odra are presented as a function of variations in the input flow to the model and as a function of changes in water level in the outlet cross-sections of the model ($Z_{R,OP}$).

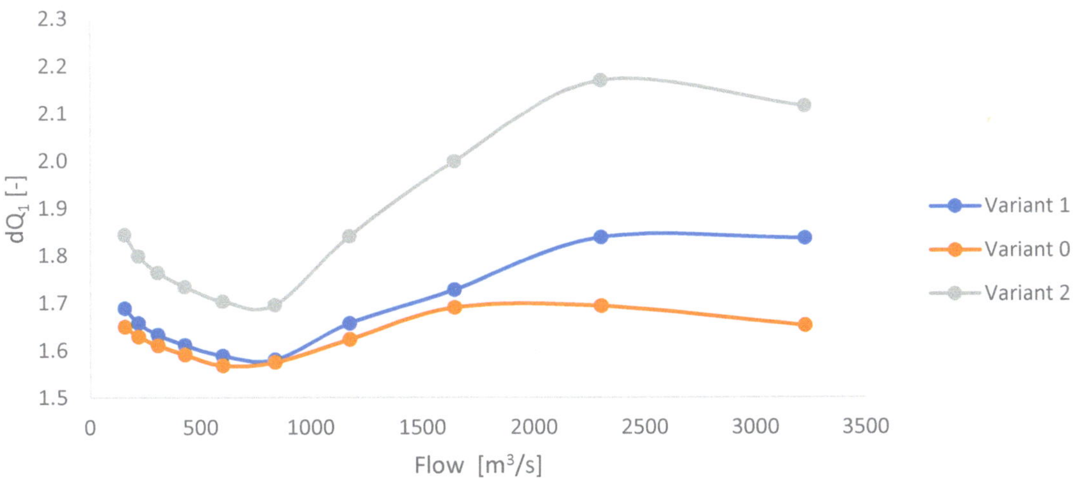

Figure 17. Changes in the value of dQ1 at a steady water level in the outlet cross-section $Z_{R,OP}$ = 0.00 [m a.s.l.].

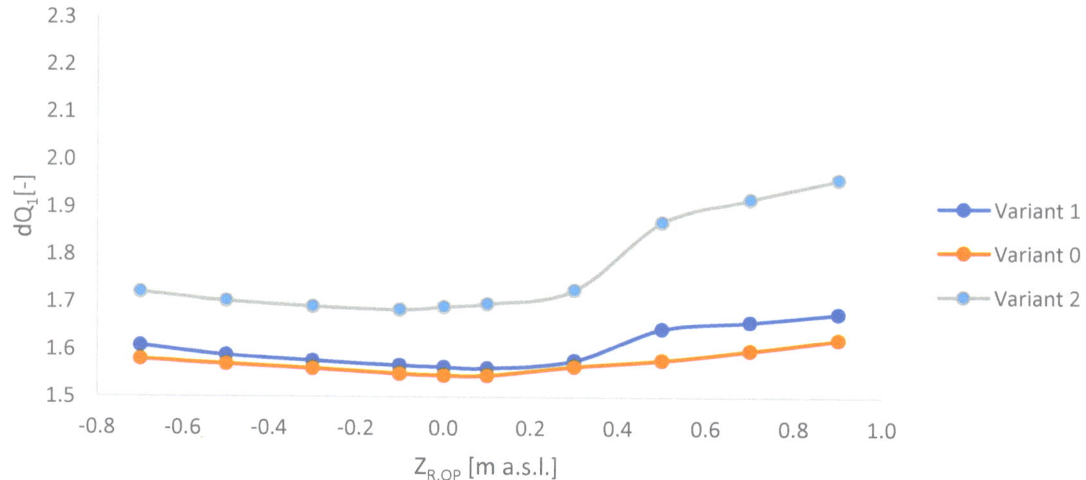

Figure 18. Changes in the value of dQ_1 at a steady water flow in the input cross-section $Q_O = 599$ [m^3/s].

The analysis of the graphs indicates that the highest flow distribution values are observed in Variant 2, which assumes the complete dredging of all channels in the Miedzy-odrze area. Furthermore, it is notable that the changes in flow values do not show a monotonous relationship on the graphs, and for each variant, there exist two local extrema.

Regarding the analysis of the graphs presented in Figure 18, there is a noticeable lack of significant changes in the dQ1 values (flow change) relative to the water level elevations of approximately 0.3 m above sea level. Only exceeding this threshold leads to a noticeable increase in the dQ_1 values.

The following figures (Figures 19 and 20) depict variations in flow reduction along the longitudinal profile of the East Odra. Significant dependencies are observed only for variants 0 and 2, as variant 1 completely excludes the hydraulic connection within the Miedzyodrze area.

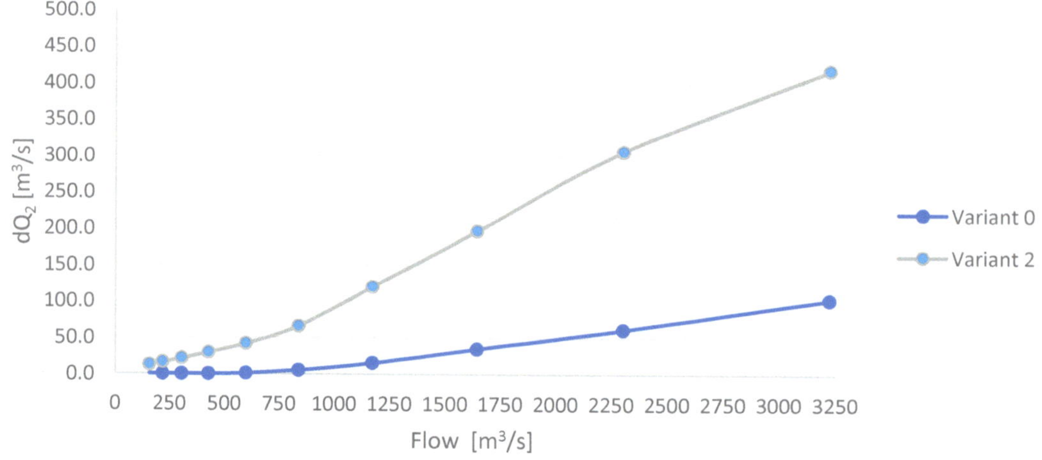

Figure 19. Changes in the value of dQ_2 at a steady water level in the outlet cross-section $Z_{R,OP} = 0.00$ [m a.s.l.].

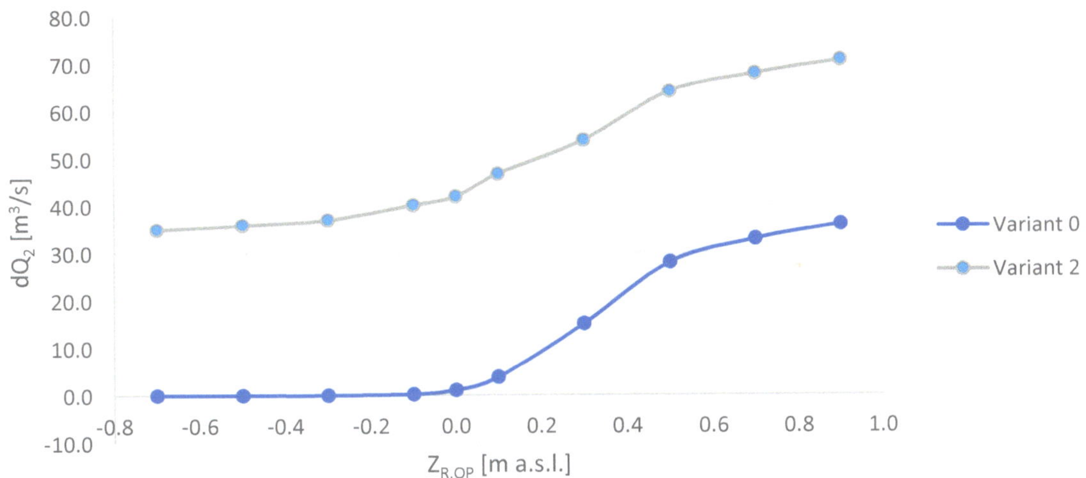

Figure 20. Changes in the value of dQ_2 at a steady water flow in the input cross-section $Q_O = 599 \, [\mathrm{m^3/s}]$.

The presented figures indicate that for Variant 2 of the lower Odra network model, there is a greater influence of the Miedzyodrze area on the water flow through the network. It is noteworthy that in the case of Variant 0, as shown in Figure 20, there is no influence of the Miedzyodrze area on the hydraulic flow until reaching approximately the average water level of 0.0 m above sea level.

The last group of steady-state flow analyses focuses on the flow values at the outlets of the models, specifically the ratio of flow in the Regalica to the flow in the Odra—Pucka part. This analysis is analogous to the first group of analyses, where the ratio of flow in the East Odra to the flow in the West Odra River was examined (Figures 21 and 22).

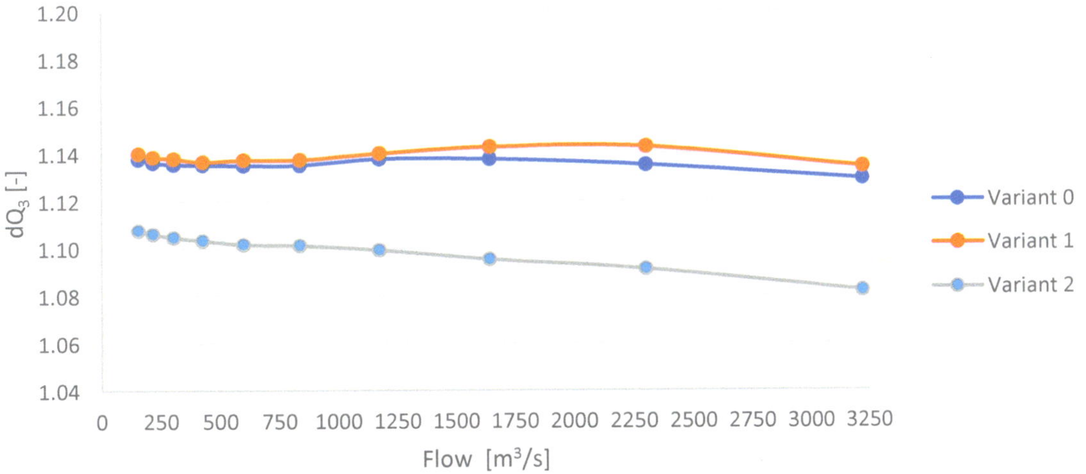

Figure 21. Changes in the value of dQ_3 at a steady water level in the outlet cross-section $Z_{R,OP} = 0.00 \, [\mathrm{m \ a.s.l.}]$.

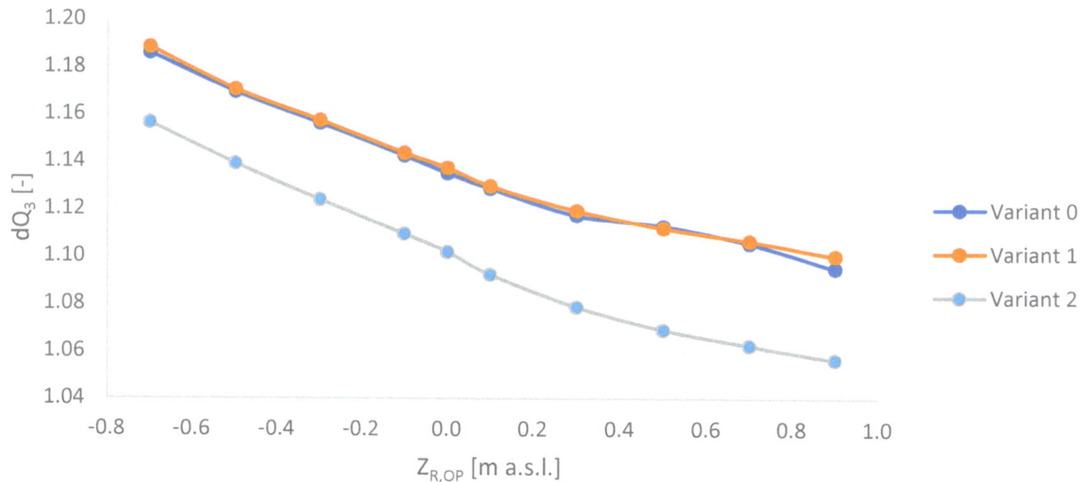

Figure 22. Changes in the value of dQ_3 at a steady water flow in the input cross-section $Q_O = 599$ [m^3/s].

From the presented graphs, it can be observed that the highest dQ_3 value occurs in variant 1, which corresponds to the complete closure of the Miedzyodrze area. However, compared to variant 0, which represents the current situation, the differences are not significant. This indicates that the Skosnica Channel serves as a compensatory feature in the distribution of flows in the downstream network of the Odra River. The simulation results of variant 2 aim to achieve a more balanced distribution of flows at the outlet of the model.

8. Discussion

Due to the rather innovative treatment of the Międzyodrze within the lower Odra River network system, a practical comparison of all the presented results in the study with prior research is not feasible. The values of characteristic flows presented in the study, which served as the basis for determining boundary conditions, align with those reported in [11,12].

In investigations conducted by Kurnatowski [1,52], it was identified that assuming a constant Manning coefficient value for the entire lower Odra River network area can yield erroneous outcomes. As demonstrated in this study, the calibration process successfully managed to identify Manning coefficient values as a function dependent on the incoming flow to the network.

Previous research regarding flow modeling in the lower Odra River network was predicated on the geometry of rivers alone without accounting for the adjacent floodplain areas associated with these channels, as reflected in Variant 1 within this study. Such models were developed by Kurnatowski [1] and Ewertowski [53]. However, earlier models exclusively operated under steady-state conditions, entailing substantial modeling limitations.

It is important to emphasize the limitations of this study. All investigations are grounded in historical measurements. While flow values tend to remain relatively stable, the bathymetry of lower Odra River channels exhibits variability. Additionally, Hec-Ras has its own constraints, including geometric limitations and numerical simplifications, which have been discussed in [54].

9. Conclusions

There is a possibility of constructing a complex model of the lower Odra River network in three computational variants using the Hec-Ras software, which also allows for the incorporation of controlled hydraulic structures such as weirs or locks into the river structure, as explicitly described in the program documentation.

The two-stage verification has demonstrated the accuracy of the constructed model. The verification phase yielded very good results, with errors occasionally not exceeding 1%. However, localized, significant discrepancies were observed between the measured and computed values.

The calculations confirmed the existence of a curvilinear relationship between the global roughness coefficient and the prevailing flow. The results of these calculations are crucial for understanding flow dynamics under various hydraulic conditions.

Dredging works on the Miedzyodrze channels have produced significant effects, such as an increase in the flow ratio at the Widuchowa node, which is of paramount importance for improving the overall functioning of the water network. Dredging the Miedzyodrze channels involves the removal of sediments, silt, and other pollutants from the riverbed to ensure better water flow. This investment is particularly important due to the strategic role of the Widuchowa node in the water flow system. Through dredging operations, it is possible to increase the channel's capacity, which affects the smoothness of water traffic and may reduce the risk of flooding.

The renovation of hydrotechnical structures, which is one aspect of these activities, aims to improve the stability and efficiency of the entire hydrotechnical infrastructure.

Dredging and hydrotechnical renovations require proper planning, financial resources, and collaboration between different institutions and services responsible for water management. It is also important to monitor the effects of these activities to assess their effectiveness and make further adjustments and improvements if necessary.

In summary, dredging works on the Miedzyodrze channels and the renovation of hydrotechnical structures aim to improve the functioning of the water network, increase the capacity of the channels, enhance infrastructure stability, and optimize water management. These investments bring benefits not only in terms of smoother water traffic but also have broader implications for overall water management.

Regrettably, the planned renovation of the Międzyodrze area has been indefinitely deferred due to financial constraints. Nevertheless, the possibility of implementing the renovation in the future remains contingent upon the amelioration of the budgetary situation and pertinent circumstances. However, during the forthcoming preparatory stages, careful consideration should be given to specific technical challenges in modeling, design, and execution.

The most noteworthy technical challenge arises from the absence of a contemporary bathymetric database, which may negatively impact the modeling and design process of the renovation. Undertaking costly surveying studies to obtain up-to-date bathymetric data becomes imperative to ensure the accuracy and reliability of the outcomes. This endeavor is time-consuming and necessitates substantial financial investment, thus potentially introducing formidable challenges in the planning phase of the renovation.

Despite these difficulties, there remains a firm commitment to undertake the renovation in the future to enhance the condition of the Międzyodrze area. This aspiration mandates meticulous planning and the identification of appropriate funding sources to fulfill the technical and budgetary requisites associated with this ambitious endeavor.

Due to the analysis and conclusions conducted, the following stages of model expansion should be included:

- Extending the study area directly to the Baltic Sea, considering the Szczecin Lagoon.
- Developing a broader set of hydrological scenarios, including the occurrence of low water levels and low sea states.
- Analyzing the influence of wind direction and magnitude on the formation of flows in the lower Odra River network.

Funding: This research received no external funding.

Data Availability Statement: Not applicable.

Conflicts of Interest: The authors declare no conflict of interest.

References

1. Kurnatowski, J. *Wybrane Problemy Obliczen Ruchu Wód w Sieciach Rzecznych o Strukturze Pierscieniowej na Przykladzie Dolnej Odry*; Wydawnictwo Uczelniane Zachodniopomorskiego Uniwersytetu Technologicznego w Szczecinie: Szczecin, Poland, 2011; ISBN 978-83-7663-096-0.
2. Pluta, M. Wplyw Stopnia Otwarcia Budowli Pietrzacej na Ksztaltowanie Sie Stanów i Przeplywów Wody w Sieci Ujscia Rzeki na Przykladzie Jazu w Widuchowej na Rzece Odrze. Ph.D. Thesis, Szczecin University of Technology, Szczecin, Poland, 1994.
3. Meyer, Z.; Pluta, M. Widuchowa weir Influence in Navigation Conditions in Lower Odra. In Proceedings of the 1st International Congress of Seas and Oceans, Szczecin/Miedzyzdroje, Poland, 18–22 September 2001.
4. Kaup, M.; Lozowicka, D.; Slaczka, W. A concept of an inland LNG barge designed for operation on the Odra waterway Scientific Journal of Silesian University of Technology. *Ser. Transp.* **2017**, *95*, 75–87. [CrossRef]
5. Khan, M.D.; Shakya, S.; Thi Vu, H.H.; Ahn, J.W.; Nam, G. Water Environment Policy and Climate Change: A Comparative Study of India and South Korea. *Sustainability* **2019**, *11*, 3284. [CrossRef]
6. Gucma, S.; Gucma, M.; Gralak, R.; Przywarty, M. Maximum safe parameters of Ships in Complex system of Port Waterways. *Appl. Sci.* **2022**, *12*, 7692. [CrossRef]
7. Durlik, I.; Gucma, L.; Miller, T. Statistical Model of Ship Delays on the Fairway in Terms of Restrictions Resulting from the Port Regulations: Case Study of Świnoujście-Szczecin Fairway. *Appl. Sci.* **2023**, *13*, 5271. [CrossRef]
8. Gucma, L.; Bąk, A.; Sokołowska, S. Stochastic Model of Ships Traffic Capacity and Congestion—Validation by Real Ships Traffic Data on Świnoujście—Szczecin Waterway. *Annu. Navig.* **2017**, *24*, 177–191. [CrossRef]
9. Kurnatowski, J. *Hydraulic Effects of Deepening the Świnoujście-Szczecin Fairway*; Gospodarka Wodna: Warszawa, Poland, 2021.
10. Kurnatowski, J. Inverse problem for looped river network—Lower Oder River case study. *Arch. Environ. Prot.* **2013**, *39*, 105–118. [CrossRef]
11. Buchholz, W. *Materialy Do Monografii Dolnej Odry*; IBW-PAN: Gdansk, Poland, 1990.
12. Buchholz, W. *Monografia Dolnej Odry*; IBW-PAN: Gdansk, Poland, 1991.
13. Kowalewska-Kalkowska, H. *Rola Wezbran Sztormowych w Ksztaltowaniu Ustroju Wodnego Ukladu Dolnej Odry i Zalewu Szczecinskiego*; Wydawnictwo Naukowe Uniwersytetu Szczecińskiego: Szczecin, Poland, 2012; ISBN 978-83-7241-882-1. ISSN 0860-2751.
14. Boczar, J.; Sajko, B.; Ziebro, J. *Okreslenie Optymalnego Rozdzialy Przeplywu Dolnej Odry ze Wzgledu na Ochrone wód Ochry Zachodniej*; Badania Hydrauliczne, Prace Naukowe Politechniki Szczecinskiej, Instytut Inzynierii Wodnej Politechniki Szczecinskiej nr 41: Szczecin, Poland, 1975.
15. Majewski, A. *Charakterystyka Hydrologiczna Estuariowych wód u Polskiego Wybrzeza*; Prace PIHM: Warszawa, Poland, 1972.
16. Borówka, K.R.; Osadczuk, A.; Osadczuk, K.; Witkowski, A.; Skowronek, A.; Łatałowa, M.; Mianowicz, K. *Postglacial Evolution of the Odra River Mouth, Poland-Germany, Coastline Changes of the Baltic Sea from South to East Past and Future Projection*; Springer: Berlin/Heidelberg, Germany, 2017; pp. 193–218. [CrossRef]
17. Nadudvari, A.; Czajka, A.; Wyżga, B.; Zygmunt, M.; Wdowikowski, M. Patterns of Recent Changes in Channel Morphology and Flows in the Upper and Middle Odra River. *Water* **2023**, *15*, 370. [CrossRef]
18. Orlewicz, S.; Mrozinski, Z. Hydrologia Dolnej Odry: Dolina Dolnej Odry. In *Monografia Przyrodnicza Parku Krajobrazowego*; Wydawnictwo Szczecinskie Towarzystwo Naukowe: Szczecin, Poland, 2002.
19. Wisniewski, B.; Wolski, T. Changes in Dabie lake bathymetry. *Limnol. Rev.* **2005**, *5*, 255–262.
20. Wolski, T.; Wisniewski, B. The analysis of Dabie Lake water balance. *Limnol. Rev.* **2005**, *5*, 263–270.
21. Kowalski, W.; Wróbel, M.; Jurzyk-Nordlow, S. The locality of *Trapa natans* L. within the region of Miedzyodrze—Dangers and protection perspective (the Lower Odra Valley, West Pomerania). *Biodivers. Res. Conserv.* **2018**, *49*, 29–36. [CrossRef]
22. Borkowski, M.; Mrozinska, G.; Szutowicz, J. *Historia Regulacji Dolnej Odry ze Szczegónym Uwzglednieniem Odcinka Widuchowa-Szczecin*; Wydawnictwo Uczelniane ZUT w Szczecinie: Szczecin, Poland, 2009.
23. Geographical Coordinates: 53.32081197425485, 14.494238288766322. Available online: https://www.google.com/maps (accessed on 12 April 2022).
24. Dąbkowski, S.L.; Wesołowski, P.; Brysewicz, A.; Humiczewski, M. Międzyodrze: An example of diverse economic and nature-related activities in the part of the Lower Odra Valley. *J. Water Land Dev.* **2017**, *34*, 117–129. [CrossRef]
25. Słogucki, Ł.; Czerniawski, R. Water Quality of the Odra (Oder) River before and during the Ecological Disaster in 2022: A Warning to Water Management. *Sustainability* **2023**, *15*, 8594. [CrossRef]
26. Indyk, W.; Mazon, S.; Potocki, A. *Instrukcja Gospodarowania Woda dla Jazu Usytuowanego w km 0,0 Rzeki Odry Zachodniej w Miejscowosci Widuchowa*; Kraków, Poland, 2011.
27. Szczepaniak-Kreft, A. *Jaz Widuchowa po Modernizacji. Regionalne Problemy Gospodarki Wodnej i Hydrotechniki*; Szczecin, Poland, 2007.
28. Jiang, H.Z. A new method of deducing the chasing coefficients in looped river networks with weirs. *Chin. J. Hydrodyn.* **2010**, *25*, 398–405.

29. Polish Institute of Meteorology and Water Management. Available online: https://danepubliczne.imgw.pl/data/dane_pomiarowo_obserwacyjne/ (accessed on 1 July 2021).
30. Kowalewska-Kalkowska, H. Storm-Surge Induced Water Level Changes in the Odra River Mouth Area (Southern Baltic Coast). *Atmosphere* **2021**, *12*, 1559. [CrossRef]
31. Kowalewska-Kalkowska, H. *Impacts of Storm Surges on the Water Level in the System of Downstream Reaches of the Odra River and the Szczecin Lagoon*; WNUS: Szczecin, Poland, 2012.
32. Wolski, T.; Wisniewski, B. Geographical diversity in the occurrence of extreme sea levels on the coasts of the Baltic Sea. *J. Sea Res.* **2020**, *159*, 101890. [CrossRef]
33. Dziaduszko, Z.; Jednoral, T. *Wahania Poziomów Morza na Polskim Bałtyku, Gdynia i Materiały Oceanologiczne*; Studia i Materiały Oceanologiczne, 52, Dynamika Morza, 6, Ossolineum: Szcecin, Poland, 1987; pp. 215–238.
34. Butts, M.; Dubicki, A.; Stronska, K.; Jørgensen, G.; Nalberczynski, A.; Lewandowski, A.; Van Kalken, T. Flood Forecasting for the Upper and Middle Odra River Basin. In *Flood Risk Management in Europe*; Springer: Dordrecht, The Nethlands, 2007. [CrossRef]
35. Gądek, W.; Baziak, B.; Tokarczyk, T.; Szalińska, W. A Novel Method of Design Flood Hydrographs Estimation for Flood Hazard Mapping. *Water* **2022**, *14*, 1856. [CrossRef]
36. Manko, R. The influence of weight coefficients in the Preissman scheme on flows in the lower Odra river network using the Hec-Ras software. *ITM Web Conf.* **2018**, *23*, 00024. [CrossRef]
37. Amaliah, R.; Minola, B. Investigating the Capability of HEC-RAS Model for Tsunami Simulation. *J. Civ. Eng. Forum* **2023**, *9*, 161–180. [CrossRef]
38. Tahmid, A.; Rahman, H.; Mounota, S.; Abid Ahsan, K. Relationship between Manning Roughness Coefficient and Flow Depth in Bangladesh Rivers. *Malays. J. Civ. Eng.* **2021**, *33*, 51–58. [CrossRef]
39. Hu, D.; Lu, C.; Yao, S.; Yuan, S.; Zhu, Y.; Duan, C.; Liu, Y. A Prediction–Correction Solver for Real-Time Simulation of Free-Surface Flows in River Networks. *Water* **2019**, *11*, 2525. [CrossRef]
40. Orlewicz, S.; Buchholz, W.; Kurnatowski, J.; Mrozinski, Z. *Metoda Obliczania Rozpływów w Ujsciowym Odcinku Odry na Przykładzie Wielkich Wód*; Prace Naukowe Politechniki Szczecinskiej: Szczecin, Poland, 1980.
41. Chow, V. *Open-Channel Hydraulics*; McGraw-Hill: New York, NY, USA, 1959; ISBN 07-0107769.
42. Orlewicz, S.; Kurnatowski, J.; Kreft, A.; Mrozinski, Z. *Wspólzaleznosc Dynamiki i Zasobów Wodnych Dolnej Odry*; Odra i Nadodrze, Instytut Slaski: Opole, Poland, 1983.
43. Kurnatowski, J. *Wspólczynniki Szorstkosci a Polozenie Geoidy*; Regionalne Problemy Gospodarki Wodnej i Hydrotechniki: Dziwnów, Poland, 2003.
44. Kurnatowski, J. *Wspólczynniki Szorstkosci Koryt Dolnej Odry*; Regionalne Problemy Gospodarki Wodnej i Hydrotechniki: Dziwnów, Poland, 2004.
45. Roszak, A. *Wplyw Rumowiska na Wspólczynnik Szorstkosci Dolnej Odry*; Regionalne Problemy Gospodarki Wodnej i Hydrotechniki: Dziwnów, Poland, 2007.
46. Arcement, G.; Schneider, V. *Guide for Selecting Manning's Roughness Coefficients for Natural Channels and Flood Plains*; Water-Supply Paper 2339; United States Geological Survey: Reston, FL, USA, 1989.
47. Elomaty, M.; Elsamman, T. Manning roughness coefficient in vegetated open channels. *Water Sci.* **2020**, *34*, 121–128. [CrossRef]
48. Cardenas, G.; Catalan, P.A. Accelerating Tsunami Modeling for Evacuation Studies through Modification of the Manning Roughness Values. *GeoHazards* **2022**, *3*, 492–507. [CrossRef]
49. Parhi, P.K.; Sankhua, R.N.; Roy, G.P. Calibration of Channel Roughness for Mahanadi River, (India) Using HEC-RAS Model. *J. Water Resour. Prot.* **2012**, *4*, 847–850. [CrossRef]
50. Philippus, D.; Wolfand, J.; Abdi, R.; Hogue, T.S. Raspy-Cal: A Genetic Algorithm-Based Automatic Calibration Tool for HEC-RAS Hydraulic Models. *Water* **2021**, *13*, 3061. [CrossRef]
51. Geographical Coordinates: 53.3098597, 14.6205277. Available online: https://www.google.com/maps (accessed on 12 April 2022).
52. Kurnatowski, J. Wrażliwe sieci rzeczne. *Gospod. Wodna* **2016**, *12*, 405–409.
53. Ewertowski, R. Mathematical model of the River Odra Estuary. *Bull. Perm. Int. Assoc. Navig. Congr.* **1988**, *60*, 95–114.
54. Armas, A.; Beilicci, R.; Beilicci, E. Numerical Limitations of 1D Hydraulic Models Using MIKE11 or HEC-RAS software—Case study of Baraolt River, Romania. *IOP Conf. Ser. Mater. Sci. Eng.* **2017**, *245*, 072010. [CrossRef]

Article

Detecting Annual and Seasonal Hydrological Change Using Marginal Distributions of Daily Flows

Borislava Blagojević [1,*], Vladislava Mihailović [2,*], Aleksandar Bogojević [1,3] and Jasna Plavšić [4]

1. Faculty of Civil Engineering and Architecture, University of Niš, 18000 Niš, Serbia
2. Faculty of Forestry, University of Belgrade, 11000 Belgrade, Serbia
3. Faculty of Civil and Environmental Engineering, Ruhr-Universität Bochum, 44780 Bochum, Germany; aleksandar.bogojevic@ruhr-uni-bochum.de
4. Faculty of Civil Engineering, University of Belgrade, 11000 Belgrade, Serbia; jplavsic@grf.bg.ac.rs
* Correspondence: borislava.blagojevic@gaf.ni.ac.rs (B.B.); vladislava.mihailovic@sfb.bg.ac.rs (V.M.)

Abstract: Changes in the hydrological regime are widely investigated using a variety of approaches. In this study, we assess changes in annual and seasonal flow characteristics based on a probabilistic representation of the seasonal runoff regime at the daily time scale. The probabilistic seasonal runoff pattern is constructed by determining quantiles from marginal distributions of daily flows for each day within the year. By applying Fourier transformation on the statistics of the daily flow partial series, we obtain smooth periodical functions of distribution parameters over the year and consequently of the quantiles. The main findings are based on the comparison of the dry, average, and wet hydrologic condition zones as defined by the daily flow quantiles of selected probabilities. This analysis was conducted for ten catchments in Serbia by considering changes between two 30-year nonoverlapping periods, 1961–1990 and 1991–2020. It was found that the relative change in runoff volume is the most pronounced in the extreme dry condition zone in the winter season (−33% to 34%). The annual time shift is the largest in the dry and average condition zones, ranging from −11 to 12 days. The applied methodology is not only applicable to the detection of hydrologic change, but could also be used in operational hydrology and extreme flow studies via drought indices such as the Standardized Streamflow Index.

Keywords: time series analysis; daily flow; Fourier series; annual periodicity; marginal distribution of daily flows; probabilistic thresholds; annual and seasonal regime change

Citation: Blagojević, B.; Mihailović, V.; Bogojević, A.; Plavšić, J. Detecting Annual and Seasonal Hydrological Change Using Marginal Distributions of Daily Flows. *Water* **2023**, *15*, 2919. https://doi.org/10.3390/w15162919

Academic Editors: Agnieszka Rutkowska and Katarzyna Baran-Gurgul

Received: 11 July 2023
Revised: 5 August 2023
Accepted: 9 August 2023
Published: 12 August 2023

1. Introduction

Hydrologic systems, and the water cycle in general, are subject to stresses and change caused by a range of drivers. The obvious or direct stressors on hydrologic systems include widespread land-cover change, urbanization, industrialization, and significant engineering interventions, while the indirect stressors are linked to the growing demands for drinking water, food, and energy for the population [1]. The major changes to the global hydrologic cycle over the last century are attributed to global warming and other influences of climate change in many regions of the world. The changes in frequency, intensity, and timing of precipitation directly contribute to modifications in the magnitude and timing of flow in rivers, including extreme floods and hydrologic droughts [1].

Assessment of expected change in the future hydrological regime is of interest for different projects and applications in water resources management and engineering, as well as in environmental studies. The aim and scope of the application determine the type of information needed from the assessment, as well as the baseline and future time horizons and spatial and temporal scales. The hydrologic change assessments may therefore differ significantly in aspects such as spatial extent (e.g., one location, catchment, region), assessment periods (past and/or future), timestep (e.g., daily, monthly, yearly), and hydrologic regime (mean-, high-, low-flow). In addition to the objectives of the assessment, data availability is a key factor in selecting the analysis technique.

Selection of the appropriate period for the assessment of any indicator of the hydrological regime may have a significant effect on the results because of the variability in hydrological regimes over longer periods. Climatological normal periods of 30 years often impose the use of 30-year-long periods for hydrological assessments in climate change impact studies. According to the World Meteorological Organization (WMO) [2], the 30-year normal period currently in use is acceptable for hydroclimatic applications. However, WMO [2] also warns that 30-year periods may not be acceptable for an analysis of extreme events (floods and low flows). The period of record is preferred for hydrologic and water resources engineering applications; for situational assessment and forecasting, a common period of record across the region of interest is preferred when using a record length less than 30 years is acceptable [2].

In Europe, the longest assessment period reported in the literature was related to the detection of long-term changes in the annual flow regime at the Ceatal Izmail hydrologic station on the Danube River over 1840–2015, which was the period of record for this station [3]. Renner et al. [4] used the 1930–2009 period for investigating 27 basins throughout Saxony/Germany, while the flood levels of the Danube at Novi Sad over the 1919–2007 period were analyzed in [5]. Stewart et al. [6] studied the period between 1948 and 2002 to detect changes toward earlier flow timing in a network of 302 western North American HSs, while the shortest period of twenty years (1977–1996) was used by Laaha and Blöschl [7] for analyzing hydrographs from 325 catchments in Austria. The studies that focus on future conditions and projections of hydrologic change are mostly based on 30-year baseline catchment behavior, such as 1981–2010 [8,9], but also 1975–2005 [10], 1980–2012 [11], and 1985–2008 [12]. Having in mind long-term persistence and oscillations in the hydrologic regime, shorter assessment periods in hydrologic change studies may lead to false signals about the tendency and signs of changes [13,14]. This is particularly important when trend assessment is used as a change detection technique. According to Kundzewicz and Robson [15], a hydrologic time series of at least 50 years in length is required to distinguish between trends and variability in the hydrologic regime.

Identifying and assessing the changes in the hydrological regime usually implies looking for different trends or other types of changes in the historical hydrological data. Application of various statistical tests prevails in this group of approaches, such as tests for trend detection, homogeneity, serial correlation, etc. The use of the nonparametric Mann–Kendall test for trends has been extremely popular, especially after it was modified by Hamed and Rao [16] for autocorrelated data. Tests for abrupt changes, such as Pettitt's test [17], are also used often (e.g., [18]) and mainly for detection of human-induced changes to river regimes. The use of statistical tests for detecting changes in hydrological time series has also been criticized, mainly because their null hypotheses and assumptions may be compromised by the correlation structure of the hydrological time series [19]. While these tests are always numerically feasible, their outcomes may not be truly informative if their basic assumptions are violated. Furthermore, the outcomes of the trend tests depend heavily on the period of record and may indicate false tendencies characteristic of a part of a long-range dependence such as long-term oscillations [14,20]. Moreover, trends are rarely identified from data with a time scale finer than annual (e.g., [21]).

Seasonal runoff variation is one of the most important indicators of the hydrological regime, i.e., one of the most important hydrological signatures. Information on seasonal runoff distribution over the annual cycle is essential for water resources management and hydroecology because it provides predictable patterns of water availability [22]. While the mean seasonal pattern over the annual cycle describes the average hydrological regime at the location of interest, seasonal runoff also exhibits interannual variability that is also important for other water- and environment-related management. Therefore, detecting the change in the seasonal runoff regime is important for water management and for adaptation and mitigation strategies.

Seasonal runoff variation is typically described over the annual cycle at the monthly time scale and less often at the daily time scale. Daily streamflow data have several advantages over monthly data: they provide a more functional representation of the runoff regime [23] by offering important insights for water resources management, including linking hydrological regimes to habitats and biotic communities [24], introducing variable hydrologic drought thresholds [25], and predicting streamflow characteristics in ungauged basins [12]. Most runoff indicators of interest for studying aquatic and riparian ecological response (magnitude, frequency, duration, timing, and rate of change [24]) can be assessed from the daily representation of the annual and seasonal runoff regimes. Daily flows also facilitate estimating different measures of runoff seasonality. The annual and seasonal changes in position of the hydrograph centroid are frequently considered [6,10,26], while the floods and droughts are investigated through onset date, end date [6,27], or duration [3].

Variability in the seasonal runoff pattern from year to year is typically described by computing and drawing lines around the mean (or median) pattern that correspond to seasonal runoff of certain non-exceedance probability. These lines may be considered as boundaries between the zones of different hydrological conditions such as average, wet, or dry conditions. There are no general recommendations in the literature about specific probabilities that could serve to define the boundaries between the hydrological conditions. These boundaries are frequently associated with various threshold levels that are used in water resources management to define certain design or reference quantities. In drought-related studies, the threshold levels used to identify drought events can either represent water demand or the boundary between normal and low-flow conditions [28]. Although some applications may require fixed threshold levels, the thresholds' variation throughout the year reflects seasonally different hydrological conditions, ecological requirements, or water demands. For instance, in [23] the 10th percentile of monthly flows delineates very dry and dry hydrological conditions, the 25th and 75th percentiles define the zone of average hydrological conditions, and wet conditions are those above the 75th percentile. Daily flows that exceed the 75th percentile are also classified as high flows by Pekarova et al. [3], but those below the 50th percentile are classified as low flows, with additional separation into extreme low flow and low flow categories.

The probabilistic approach to defining time-varying threshold levels is also embedded in the methodology for computing the Standardized Precipitation Index (SPI) [29], where monthly precipitation aggregated over scales of 1, 2, 3, ..., 12 months for each calendar month is transformed into a standard normal quantile, SPI, based on non-exceedance probability of aggregated precipitation. Typically, a threshold of SPI = 0 is used to define meteorological droughts as periods with less accumulated precipitation than expected in a long-term period. Similarly, the Standardized Streamflow Index (SSI) [30], as a probabilistic hydrologic drought index based on monthly flows, was introduced. Computation of both the SPI and the SSI involves an intensive procedure of determining the type of probability distribution for each aggregation scale [31,32]. The standardization of precipitation and flows in this probabilistic approach explicitly introduces fixed SPI or SSI thresholds (e.g., $-1, 0, 1$) corresponding to certain probabilities, which translate into variable thresholds for precipitation or runoff volume in each calendar month.

Representing the annual runoff cycle at the daily time scale has its advantages and disadvantages. The advantage is that the daily resolution provides more detail and smoother transitions compared with the monthly scale [33]. On the other hand, the disadvantage is that the calculation is much more demanding and extensive because it includes analyzing and fitting 365 probability distributions [33]. Furthermore, using the daily step requires smoothing to avoid jumps in consecutive days [25,33,34].

Considering all the above, the main issues in detecting annual and seasonal hydrological change are threefold. First, the analysis of changes, and particularly trend analysis, mainly deals with the annual scale while the changes in seasonal patterns are seldom addressed. In many cases, the change is not detectable on the annual scale, but is obvious on the seasonal (monthly) scale. Second, the seasonal pattern is typically described in terms of monthly flows, which may not be useful for some applications and creates an abrupt transition in flows from one month to another. Third, the change in the seasonal runoff pattern is usually identified from the average pattern, and the year-to-year variability in the pattern is typically neglected. In cases when this is considered, it is assessed from empirical distributions of monthly flows. Moreover, the change in hydrological conditions with the assessment period has not been analyzed to our best knowledge.

To address the above identified issues, the goals of this paper are as follows: (i) to investigate seasonal change despite the lack of significant change on the annual scale in our study area; (ii) to utilize probabilistic representation of the seasonal runoff cycle at the daily time scale; and (iii) to detect changes in zones of different hydrological conditions. To do so, we use a probabilistic approach for defining the seasonal runoff pattern at the daily time scale by fitting 365 marginal distributions of daily flows (MDDFs). The parameters for the 365 marginal distributions are modeled as periodic functions to avoid abrupt changes in quantiles in consecutive days. The estimated quantiles from the 365 MDDFs result in smooth lines in the plot of the annual runoff cycle and represent the boundaries of the zones with flows for the given probability of occurrence. To detect changes in the seasonal runoff pattern, we compare wetness (hydrological) conditions as defined by probabilistic thresholds between two nonoverlapping 30-year subperiods. The two subperiods are compared in terms of the overall seasonal runoff pattern, runoff volume, and timing of the annual and seasonal runoff. The analysis is performed for the selected catchments in Serbia for which the statistical tests do not reveal any change or trend at the annual scale in the 1961–2020 period. Therefore, the purpose of this research is to investigate potential alterations in the seasonal runoff pattern despite the lack of significant changes on the annual scale.

This paper is organized as follows. Section 2 describes the study area and data sets, while Section 3 presents the methodological framework used to construct a probabilistic annual runoff cycle using MDDFs, define hydrological condition zones, and evaluate annual and seasonal hydrologic changes per zones. Section 4 shows the results of distribution fitting, zoning, and the detected changes. Section 5 discusses the results from two aspects, first by comparing them with the previous findings on hydrologic change in the study area, and then by highlighting the benefits of the MDDF approach for hydrological applications. Section 6 gives the key conclusions of the presented work.

2. Study Area and Data

This study was conducted on daily flows in the period 1961–2020 at ten hydrological stations that belong to the Danube River basin in the Republic of Serbia (Figure 1) and are operated by the Republic Hydrometeorological Service of Serbia (RHMSS). Basic information about the ten stations and their drainage areas is given in Table 1.

The daily flow series were checked for data completeness and only minor gaps were found that were filled either through interpolation or through regression analysis using data from the neighboring stations as predictors. The series homogeneity and randomness, as well as the presence of trends, were tested on the annual flow series with a range of parametric and nonparametric tests. The results of all tests (p-values) are given in Table A1 in Appendix A, showing that the null hypothesis on homogeneity, randomness, and absence of trends cannot be rejected at a significance level of 0.05 for all stations.

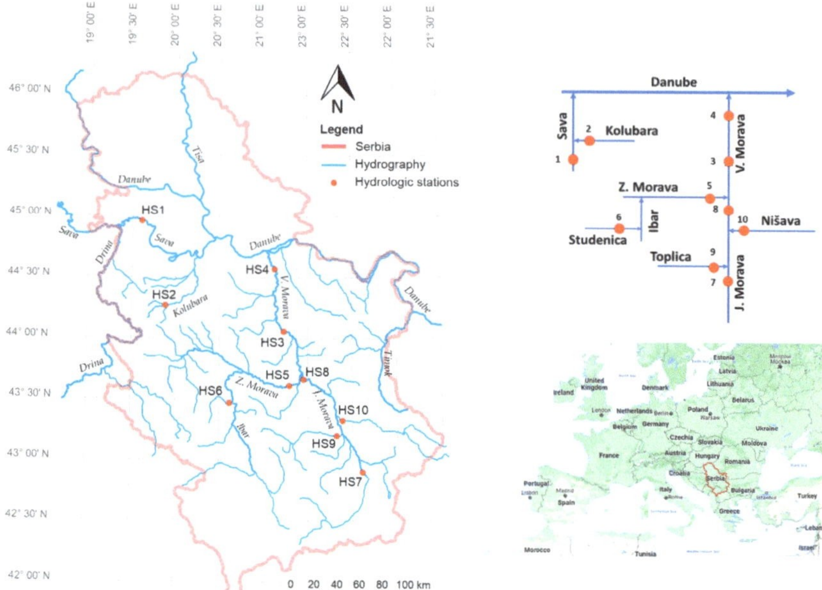

Figure 1. Locations of the hydrological stations used in this study on the hydrographic map of Serbia (**left**) and a schematic of the river network and stations (**upper right** panel).

Table 1. Hydrological stations (HS) used in this study.

HS#	Station Name	River	Catchment Area [km²]	Watershed
1	Sremska Mitrovica	Sava	87,996	Sava
2	Valjevo	Kolubara	340	Kolubara
3	Bagrdan	Velika Morava	33,446	Velika Morava
4	Ljubičevski most	Velika Morava	37,320	Velika Morava
5	Jasika	Zapadna Morava	14,721	Zapadna Morava
6	Ušće	Studenica	540	ZapadnaMorava
7	Grdelica	Južna Morava	3782	Južna Morava
8	Mojsinje	Južna Morava	15,390	Južna Morava
9	Doljevac	Toplica	2052	Južna Morava
10	Niš	Nišava	3870	Južna Morava

3. Methodology

Detection of changes in hydrological regime at selected stations follows the methodology organized in the workflow consisting of four steps and illustrated in Figure 2. The first three steps (a–c) result in probabilistic representation of the annual runoff cycle at daily time scale, while the fourth step (d) is dedicated to defining hydrologic condition zones, calculating indicators, and detecting the annual and seasonal change. The workflow in Figure 2 relates to the analysis for one station and is repeated for each one. The analysis is performed for the 1961–2020 period of the available data, and for two subperiods, 1961–1990 and 1991–2020, representing two standard 30-year periods used by the WMO [2].

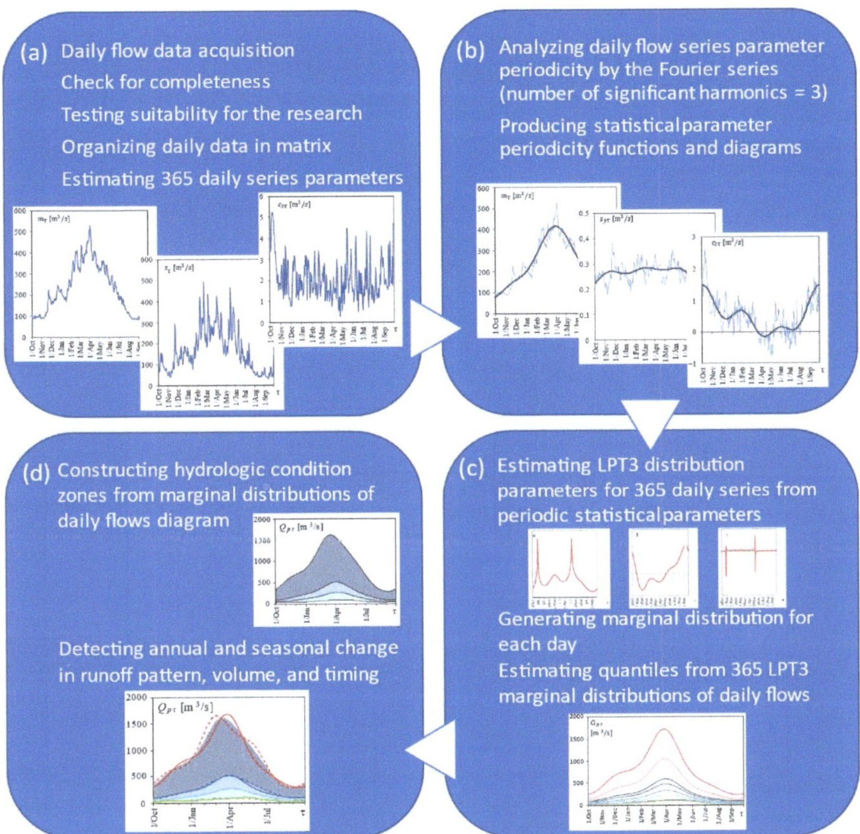

Figure 2. Illustration of methodological steps to detect changes in hydrological regime.

The core of the proposed methodological approach is the use of daily flows for construction of the probabilistic annual runoff cycle. In the first step (a), daily flows are prepared for analysis by defining the data series for each day within a year, resulting in a total of 365 data series for which the corresponding statistics are computed. This step is described in Section 3.1. In the second step (b), statistics of the partial daily series are modeled as the periodic functions to achieve their smooth representation over the year. This step is described in Section 3.2. In the third step (c), the periodic daily statistics are transformed into periodic parameters of marginal distributions of daily flows (MDDFs). The quantiles of MDDFs for different probabilities are then estimated for each day, and the diagrams of these quantiles are created to represent probabilistic seasonal runoff pattern. The lines of MDDF quantiles are smooth because an implicit, anterior smoothing is achieved by using periodic probability distribution parameters, thus avoiding sudden changes in flow between two successive days. The third step is described in Section 3.3. The fourth step (d) of the methodology comprises the definition of the zones of different hydrologic conditions (Section 3.4) and detection of changes in hydrological regime from one period to another (Section 3.5). The changes in the seasonal pattern, runoff volumes, and runoff timing at annual and seasonal scales are considered.

3.1. Daily Flows as a Stochastic Process

The continuous stochastic process $\{x_t; t \geq 0\}$ is mathematically formulated [35] by the time function $x_t = f(t; \alpha, \beta, \gamma, \ldots)$, where t is time, and α, β, γ, etc. are the parameters of a multidimensional distribution describing the temporal structure of the process. Hydrologic daily time series are a close approximation to the continuous stochastic processes, where the process x_t is represented by the observed daily flows $x_{\tau,i}$ (with $i = 1, 2, \ldots, N$—number of years, and $\tau = 1, 2, \ldots, 365$—number of days). Hence, the $x_{\tau,i}$ series is a set of N process realizations over the $[0, 365$ days$]$ interval.

The observed daily flow series $x_{\tau,i}$ at each station is arranged in a matrix, in which each row represents one date (i.e., the ordinal number of a day) within a year, and each column represents one year:

$$\begin{bmatrix} x_{1,\,1} & \cdots & x_{1,\,N} \\ \vdots & \ddots & \vdots \\ x_{365,\,1} & \cdots & x_{365,\,N} \end{bmatrix}. \tag{1}$$

The data are arranged in the matrix (1) by hydrologic years, starting on 1 October ($\tau = 1$) and ending on 30 September ($\tau = 365$). In the leap years, 29 February is omitted, similar to many studies (e.g., [25,36]).

For each of the 365 dates, i.e., for each row in the matrix (1), the sample statistics are estimated using the method of moments for both the original series (x) and its logarithmic transformation ($y = \log x$). In this way, column vectors of the statistics with 365 rows are obtained for each day in the year. We will refer to these series of statistics as daily statistics. They include the mean m_τ, standard deviation s_τ, and coefficient of skewness $C_{s\tau}$ ($\tau = 1, 2, \ldots, 365$).

3.2. Periodicity Analysis

The series of daily means m_τ, $\tau = 1, 2, \ldots, 365$, represents the mean seasonal runoff pattern. Due to variability in daily flows, the annual cycle of m_τ may not resemble a periodic-like function that would be obtained in a similar process with monthly data. The same is valid for the other daily statistics as well. Therefore, the seasonal runoff pattern can be described by introducing the periodic functions of the mean, standard deviation, or other parameters. These periodic functions are estimated from the column vectors of the statistics of the observed daily flow series. For any parameter v, its periodic component $v_{\tau,per}$ may be expressed using the Fourier series [35,37]:

$$v_{\tau,per} = \overline{v}_\tau + \sum_{j=1}^{h}\left(A_j \cos\frac{2\pi j\tau}{\omega} + B_j \sin\frac{2\pi j\tau}{\omega}\right) \tag{2}$$

where $\overline{v}_\tau$ is the mean of v_τ, h is the number of significant harmonics, A_j and B_j are the Fourier coefficients, $f_j = j/\omega$ is the frequency of the j-th harmonic, ω is the base period of 365 days, and $\tau = 1, 2, \ldots, 365$. The Fourier coefficients are calculated as follows:

$$A_j = \frac{2}{\omega}\sum_{\tau=1}^{\omega}(v_\tau - \overline{v}_\tau)\cos\frac{2\pi j\tau}{\omega} \tag{3}$$

$$B_j = \frac{2}{\omega}\sum_{\tau=1}^{\omega}(v_\tau - \overline{v}_\tau)\sin\frac{2\pi j\tau}{\omega} \tag{4}$$

while the amplitude of the j-th harmonic is computed from the Fourier coefficients:

$$C_j^2 = A_j^2 + B_j^2. \tag{5}$$

Modeling the periodicity of the daily flow series includes finding significant harmonics for the periodic component shown in Equation (2) [35]. The number of significant harmonics adopted in this study is three, as identified in the previous study for the same stations [38] based on the periodicity analysis for the mean via the periodograms, as suggested by Yevjevich [35].

3.3. Marginal Distributions

The row vectors of the process x_t in matrix (1), i.e., the vectors

$$[x_{\tau,1} \quad x_{\tau,2} \quad \ldots \quad x_{\tau,N}], \quad \tau = 1, 2, \ldots 365, \tag{6}$$

represent N realizations of the process for day τ in the annual cycle. The process in matrix (1) can therefore be seen as a 365-dimensional random variable. For each day, based on vector (6), a one-dimensional or marginal distribution is identified representing the distribution law of the process at the point τ in time. A family of 365 distributions defines a multidimensional distribution of the observed stochastic process.

The series of daily flows is nonstationary because of the intra-annual periodicity of its parameters. Therefore, the parameters of the 365 marginal distributions also change periodically during the year because they are estimated from the periodic daily statistics.

There are several approaches for identifying marginal distribution functions [35]. The simplest approach that is suitable for extensive multisite computations at regional level is to apply a unique theoretical distribution type for all marginal distributions, while allowing its parameters to change periodically during the year. Radić and Mihailović [38] have shown that the log-Pearson type 3 (LPT3) distribution is the most appropriate choice among the two- and three-parameter distributions in the studied region of Serbia not only according to the statistical tests, but also because it satisfies two conditions related to the nature of the modeled physical process: (i) the upper bound of the marginal distribution should be higher than the observed maximum flows; and (ii) the lower bound of the marginal distribution should not be less than zero.

The flow quantiles $Qp\tau$ of a non-exceedance probability $p = p(x)$ are estimated from the 365 fitted LPT3 marginal distributions. The probability density function of LPT3 distribution is given with

$$f(x; a, b, c) = \frac{k}{x|b|\Gamma(a)} \left[\frac{log_n x - c}{b}\right]^{a-1} \cdot e^{\frac{(log_n x - c)}{b}}, \quad a > 0 \tag{7}$$

where a, b, and c are the shape, scale, and location parameters of LPT3 distribution, respectively, and $\Gamma(a)$ denotes a gamma function. If the natural logarithm of data is used ($y = \ln x$), then $n = e$ and $k = 1$. In this study, we use the common (decadic) logarithm with $n = 10$ and $k = 0.434$. The LPT3 parameters are estimated using the method of moments with log-transformed data as

$$a = 4/C_{sy}^2, \; b = S_y \cdot C_{sy}/2, \; c = \bar{y} - ab \tag{8}$$

where $\bar{y}$, S_y, and C_{sy} are the mean, standard deviation, and coefficient of skewness of the log-transformed observed data $y = log_{10}$, respectively. The bounds of LP3 distribution depend on the sign of the shape parameter b:

$$\begin{aligned} b > 0: &\quad e^{c/k} \leq x < +\infty \\ b < 0: &\quad 0 \leq x < e^{c/k} \end{aligned} \tag{9}$$

3.4. Hydrological Condition Zones

The family of 365 MDDFs allows us to understand how the daily flows at some locations can change within the annual cycle. It also provides the basis for constructing the intervals in which the daily flows can occur with certain probability, and thus it could be used to define an interval, or a zone, of typical, "normal" conditions at different times in the year. The zone is bounded by the specific lower and upper quantiles of MDDFs. The zones outside the "normal" condition zone are characterized by deviations of daily flows from the typical regime for a given day [39].

Hydrological condition zones are defined here for specific probabilities of occurrence of daily flows in the annual cycle. The zone thresholds, i.e., boundaries, are quantiles of specific probabilities (Figure 3a), selected to represent dry, average, and wet conditions, based on the previous research [3,7,23,36,40]. The average condition zone is the interval between the 0.3 and 0.7 quantiles. This zone is divided into two subzones by the median flow, $p(x) = 0.5$, an important hydrological indicator recommended by WMO as a measure of central tendency when undertaking assessments that focus on characterizing typical conditions [2]. The 0.05 quantile delineates the dry condition zone from the lowest zone representing extremely dry conditions, while the 0.99 quantile delineates the wet condition zone from the upper zone representing extremely wet conditions.

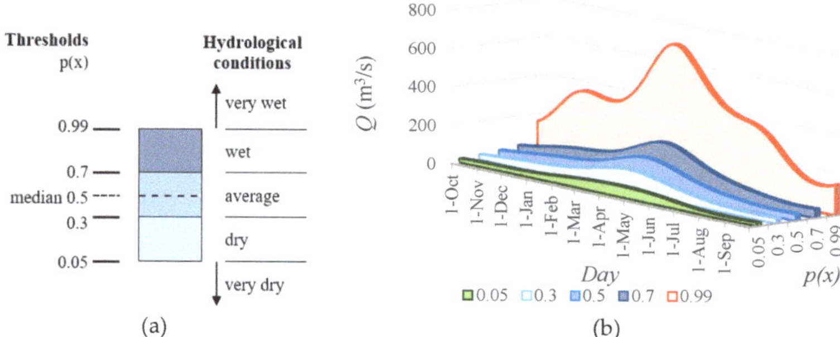

Figure 3. Hydrological condition zones based on probabilistic threshold levels for (**a**) flows, (**b**) runoff volumes.

Runoff volumes corresponding to different hydrologic conditions are computed by integrating the area under the MDDF quantile line representing the upper boundary of that zone, as shown in Figure 3b.

3.5. Estimating Change in Hydrological Regime

The changes in seasonal runoff patterns are first examined visually by comparing the average zone and the median threshold (Figure 3a), as well as the periodical function of the mean daily flows. The changes in the annual and seasonal runoff volumes and timing between the two periods are computed by examining the area below the upper threshold of the hydrologic condition zone. The runoff volume is obtained by integrating this area (i.e., the flows representing specific MDDF quantiles), while the timing is described by the time coordinate of the centroid of this area [6,26]. The following two indicators of change are computed at annual and seasonal scales:

1. Relative change in runoff volume, ΔVp (%):

$$\Delta Vp = \frac{Vp_{1991-2020} - Vp_{1961-1991}}{Vp_{1961-1991}} \cdot 100 \tag{10}$$

where Vp is runoff volume below the p-th MDDF quantile representing the upper threshold of the zone for the given period.

2. Time shift of the centroid of the area below the MDDF quantile line, Δt (in days):

$$\Delta t = Cp_{1991-2020} - Cp_{1961-1990} \tag{11}$$

where Cp is the time coordinate (ordinal number of day within year) of the centroid of the area under the p-th MDDF quantile for the given period.

The seasons within the hydrological year used for evaluation of the changes are defined as follows: October, November, and December represent autumn, January, February, and March—winter, April, May, and June—spring, and July, August, and September—summer.

4. Results

4.1. Daily Flow Statistics and Their Periodicity

The daily statistics computed for each day within a year include the mean (m_τ), standard deviation (s_τ), and coefficient of skewness ($c_{s\tau}$). Figure 4 shows the seasonal variation in the daily statistics for HS3 in all three considered periods, together with their corresponding periodic functions (smooth lines). The mean and standard deviation both exhibit distinct seasonal patterns which agree in phase. The highs occur in the spring and the lows in the summer (Figure 4a,b). This pattern is characteristic for all but two considered stations (HS1 and HS6). The skewness does not exhibit a distinct seasonal pattern like the mean and standard deviation, but a complex one with more highs and lows. It is also worth noting that there is a significant positive skewness over the entire annual cycle at all stations, as shown for HS3 in Figure 4c.

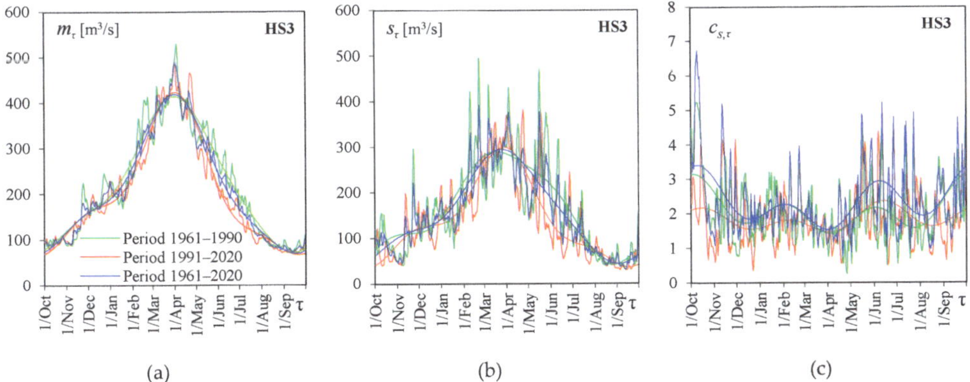

Figure 4. Seasonal variation in daily flow statistics (peaky lines) and their corresponding periodic functions (smooth lines) at HS3 estimated for the three periods: (**a**) mean, (**b**) standard deviation, and (**c**) coefficient of skewness. Modified from [41].

Figure 5 shows the daily statistics of the log-transformed data at HS3: mean ($m_{y\tau}$), standard deviation ($s_{y\tau}$), and skewness ($c_{sy\tau}$), as well as their corresponding periodic functions. Expectedly, logarithmic transformation suppresses the variability in the daily statistics compared with the ones in the original space. The distinct seasonal pattern is still visible for the mean, but not for the standard deviation, for which the amplitude of the intra-annual oscillation is significantly smaller. The logarithmic transformation reduced the skewness and even produced negative values in both the daily skewness series and the estimated periodic function during certain parts of the year. This was noticed at all stations in some part of the year for at least one period considered. This further affects the parameters and the shape of the LPT3 marginal distributions. The statistics and their periodic functions for all stations are given in the Supplementary Materials (Figures S1 and S2).

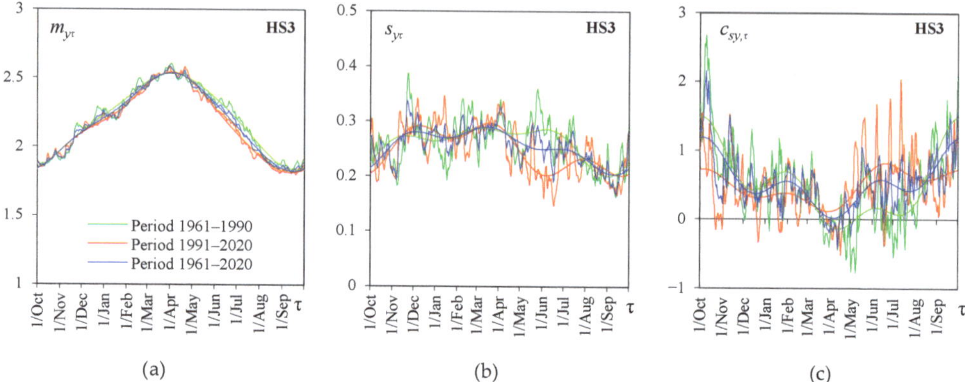

Figure 5. Seasonal variation in log-transformed daily flow statistics (peaky lines) and their corresponding periodic functions at HS3 estimated for the three periods: (**a**) mean, (**b**) standard deviation, and (**c**) coefficient of skewness. Modified from [41].

4.2. Periodic Parameters of the Marginal LPT3 Distributions

The parameters of marginal LPT3 distributions are periodic functions, resulting from the periodic statistics $m_{y\tau,per}$, $s_{y\tau,per}$, and $c_{sy\tau,per}$. Figure 6 shows the seasonal variation in the three LPT3 distribution parameters at HS3. The LPT3 parameters highly depend on the value and sign of the skew, c_{sy}. Therefore, a comparison of periodical functions of statistics (Figure 5) with LPT3 distribution parameters (Figure 6) expectedly shows that the differences between studied periods are transferred from the statistics (especially the skew) to the distribution parameters in accordance with the expressions in Equation (8). The results for all stations are given in the Supplementary Materials (Figure S3).

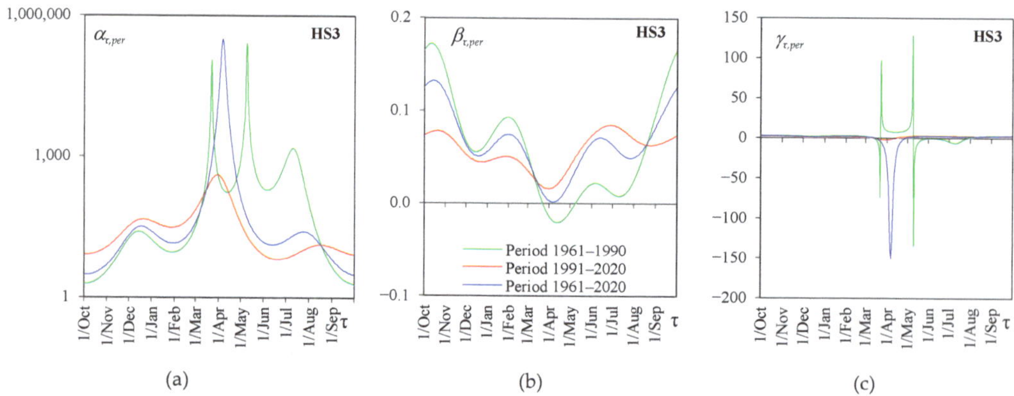

Figure 6. Periodical parameters of LPT3 marginal distribution of daily flows at HS3: (**a**) shape parameter $\alpha_{\tau,per}$, (**b**) scale parameter $\beta_{\tau,per}$, (**c**) location parameter $\gamma_{\tau,per}$. Modified from [41].

4.3. Marginal Distributions of Daily Flows

The flow quantiles $Q_{p\tau}$ of a non-exceedance probability $p = p(x)$ are estimated from the 365 fitted LPT3 distributions, one per each day within a year, with its periodic parameters. The smooth lines shown in Figure 7 represent the quantiles of MDDFs for the three periods considered at HS3. The results for all stations are given in the Supplementary Materials (Figure S4).

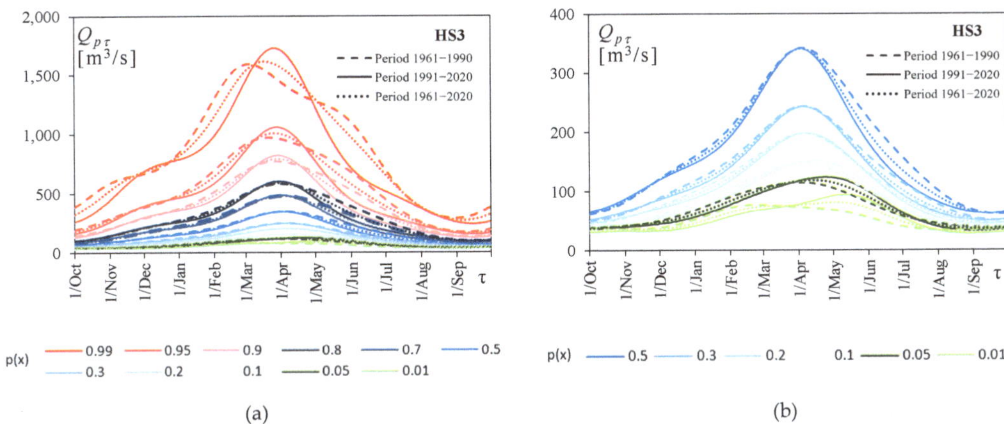

Figure 7. The diagrams of MDDFs at HS3 for the three periods showing the quantiles (**a**) for a full range of probabilities $p(x)$, from 0.01 to 0.99, and (**b**) for low to median probabilities. The solid lines represent the later 1991–2020 period, the dashed lines represent the earlier 1961–1990 period, and the dotted lines represent the whole 1961–2020 period.

In general, the patterns of MDDFs in Figure 7 are defined by the periodic function of the mean (Figure 4a). The periodic functions of standard deviation (Figure 4b) and skew (Figure 4c) influence the spread and skewness of the annual MDDFs, respectively. Despite the peaky pattern for $\alpha_{\tau,per}$ and $\gamma_{\tau,per}$ (Figure 6), the MDDF quantiles are represented by smooth lines. For days with a negative skew of log-transformed data ($c_{sy\tau,per} < 0$) for 1961–1990 (green line in Figure 5c), marginal LPT3 distributions are bounded from above (Equation (9), also, e.g., [42]). The values of the upper bound were computed for all such cases and it was found that they do not cause underestimation of daily flow upper-tail quantiles for any non-exceedance probability of interest in hydrologic applications.

Visual inspection of the MDDFs in Figure 7 reveals that the differences between the three periods are the smallest for the median daily flows and in the average zone, and grow larger for the upper and lower quantiles. For this station, HS3, the maximum quantiles for 1991–2020 are higher and occur later compared with those for 1961–1990. For the joint 1961–2020 period, the quantile lines lie expectedly between the lines of the two subperiods.

4.4. The Zones of Hydrological Conditions

The zones of hydrological conditions for all stations are shown in Figure 8 in accordance with the adopted classification.

The patterns of thresholds (quantile lines) in Figure 8 obviously change with the period. The changes occurring in the wet condition zone at all stations are more visible due to the scale, but the changes in the dry condition zone are also significant, as can be seen in Figure 7b. At all stations, the pattern remains the same for quantiles 0.3, 0.5, and 0.7, which define the normal hydrological conditions. The seasonal pattern is not changed at any station from a simple unimodal shape to a mixed bimodal shape, or vice versa. The bimodal shape of the seasonal pattern is the most visible at HS1, where the most interesting change is for the 0.99 quantiles. Here, the primary peak in 1961–1990 was in the autumn while the secondary one was in the spring; in 1991–2020, the primary peak was in spring, while the secondary was in winter.

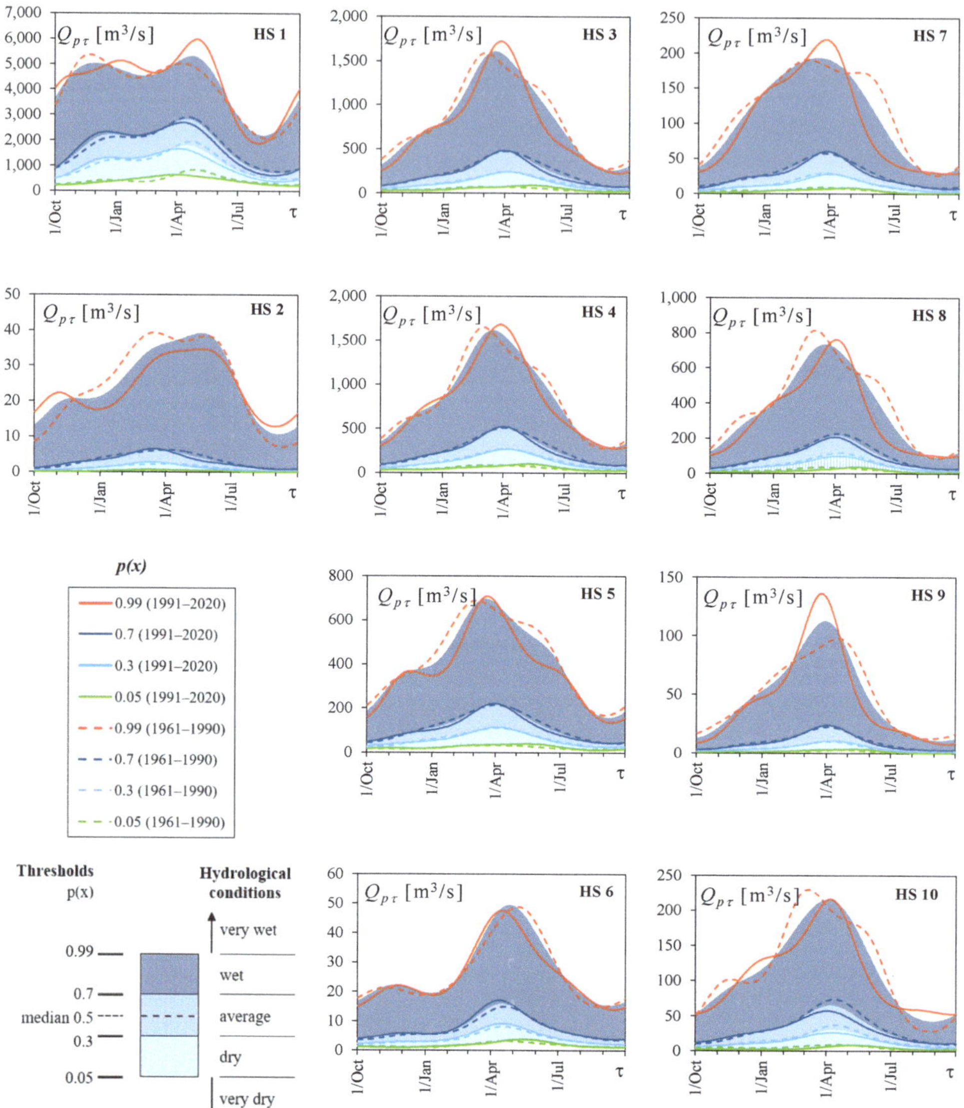

Figure 8. Hydrological condition zones for all stations defined by the quantiles of marginal daily flow distributions for selected probabilities. The zones for the 1961–2020 period are shaded, while the zones for two subperiods are indicated by their upper threshold lines.

4.5. Annual and Seasonal Hydrologic Condition Changes

The graphical presentation of hydrologic condition zones defined in Figure 3a is helpful for detecting changes in daily flow patterns, while the quantification of change in respect to runoff timing and volume (the areas defined in Figure 3b) is required for practical purposes. Change in runoff timing is assessed from the shifts of the centroids of the areas below the upper thresholds of the hydrologic condition zones, as shown in Figure 9a for the annual scale and in Figure 9b for the seasonal scale at HS3.

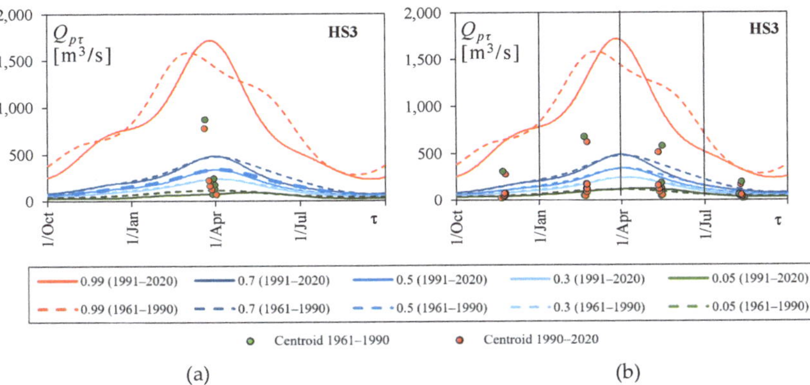

Figure 9. The centroids of the area below the upper thresholds of the hydrologic condition zones for (**a**) annual and (**b**) seasonal scales. Red dots refer to the recent 1991–2020 period, and green ones to the earlier 1961–1990 period.

Relative changes in runoff volume and time shifts of the centroids, calculated from Equations (10) and (11), are shown in Table 2 and Table 3, respectively, for all stations.

Table 2. Relative change in runoff volume (%) in the recent 1991–2020 period compared with the 1961–1990 period. The first column indicates the probabilities $p(x)$ of the upper thresholds of the hydrologic condition zones. Changes are shown for annual (Ann.) and seasonal scales (Autumn, Winter, Spring, and Summer), and colored in blue (increase) and red (decrease).

$p(x)$	Season	HS1	HS2	HS3	HS4	HS5	HS6	HS7	HS8	HS9	HS10
	Ann.	−8	11	−1	−2	5	7	−8	−15	9	−23
	Aut.	−3	−9	−10	−3	7	−2	−24	−25	0	−37
0.05	Win.	14	34	−9	−13	−1	6	−16	−28	2	−33
	Spr.	−22	2	16	11	16	17	8	3	28	−12
	Sum.	−16	−13	−10	−7	−4	−2	0	−17	−15	−13
	Ann.	−5	−1	−6	−5	−2	10	−5	−11	1	−18
	Aut.	12	4	−2	3	5	8	−7	−3	−5	−8
0.3	Win.	5	15	−5	−6	−4	13	−11	−14	−2	−16
	Spr.	−19	−16	−5	−4	−1	12	2	−12	10	−23
	Sum.	−19	−13	−14	−10	−7	2	1	−8	−10	−15
	Ann.	−4	−5	−7	−5	−5	10	−4	−10	−2	−15
	Aut.	15	9	0	4	3	11	−2	3	−8	4
0.5	Win.	3	6	−4	−4	−6	15	−9	−9	−2	−11
	Spr.	−16	−20	−11	−8	−7	9	−2	−16	3	−25
	Sum.	−19	−11	−15	−10	−8	3	−1	−6	−9	−13
	Ann.	−2	−8	−8	−5	−7	9	−5	−10	−4	−12
	Aut.	15	11	1	5	1	12	−1	6	−11	12
0.7	Win.	2	−2	−4	−3	−7	15	−6	−7	−1	−7
	Spr.	−11	−23	−14	−10	−10	6	−6	−19	−3	−26
	Sum.	−17	−7	−16	−10	−9	3	−4	−6	−9	−8

Table 2. *Cont.*

$p(x)$	Season	HS1	HS2	HS3	HS4	HS5	HS6	HS7	HS8	HS9	HS10
	Ann.	5	−6	−11	−6	−12	−2	−15	−19	−5	−8
	Aut.	−3	12	−11	−2	−8	−1	−21	−20	−19	−4
0.99	Win.	5	−26	−8	−9	−13	5	3	−17	14	−14
	Spr.	15	−9	−12	−5	−14	−10	−28	−21	−18	−15
	Sum.	3	29	−13	−2	−7	2	−24	−21	−13	38

Legend: ΔV (%) −30 −20 −10 −5 −1 1 5 10 20 30

Table 3. The changes in runoff timing expressed as the shift in the centroid date (in days) in the recent 1991–2020 period compared with 1961–1990. The first column indicates the probabilities $p(x)$ of the upper threshold of the hydrologic condition zones. Changes are shown for annual (Ann.) and seasonal scales (Autumn, Winter, Spring, and Summer), and colored in blue (later date) and red (earlier date).

$p(x)$	Season	HS1	HS2	HS3	HS4	HS5	HS6	HS7	HS8	HS9	HS10
	Ann.	−8	−4	4	3	0	1	9	8	3	12
	Aut.	0	−1	−1	−1	−2	1	2	−1	0	−1
0.05	Win.	0	2	2	1	1	−1	−1	1	2	1
	Spr.	0	−2	1	1	1	1	0	1	2	3
	Sum.	0	2	−2	−1	−1	−1	−1	−2	−5	−4
	Ann.	−11	−7	−3	−2	−2	−1	4	−1	1	−3
	Aut.	0	−2	−1	−1	−3	−1	2	−1	−2	−2
0.3	Win.	−1	2	1	1	2	1	1	0	1	−1
	Spr.	0	−2	−1	−1	−1	0	0	1	0	2
	Sum.	1	1	2	2	2	0	−3	0	1	0
	Ann.	−11	−8	−4	−4	−3	−2	1	−3	1	−6
	Aut.	0	−2	−1	−1	−3	−1	2	−1	−2	−2
0.5	Win.	−1	2	1	1	2	1	1	0	1	−2
	Spr.	−1	−1	−2	−2	−1	−1	0	0	−1	1
	Sum.	1	1	3	2	2	1	−3	1	3	2
	Ann.	−10	−7	−5	−4	−3	−3	−1	−5	0	−8
	Aut.	0	−3	−1	−1	−2	−1	2	0	−2	−2
0.7	Win.	−1	2	1	1	2	2	2	0	2	−2
	Spr.	−1	−1	−3	−2	−2	−1	−1	−1	−2	0
	Sum.	2	2	3	3	2	1	−2	2	3	3
	Ann.	4	4	−1	0	0	−2	−6	−2	−2	4
	Aut.	−1	−8	4	3	1	2	4	5	4	4
0.99	Win.	−1	2	3	2	3	2	2	1	6	−1
	Spr.	−1	0	−5	−4	−2	−2	−9	−8	−4	−5
	Sum.	4	7	−1	0	−3	−2	7	6	−5	4

Legend: Δt (days) −10 −7 −3 −1 1 3 7 10

A general impression from Table 2 is that there is less runoff volume in the recent 30-year period compared with the previous one. HS6 is the only station at which runoff volume has increased in the recent period. HS2 and HS9 have distinct increases in extremely dry conditions in the winter and spring seasons, respectively. Additionally, HS2 exhibits a significant runoff volume increase in the wet condition zone in summer, as does HS10. The most unfavorable changes are found in HS10, with significant volume reductions in dry condition zones both annually and seasonally, and in all hydrologic condition zones in the spring. This station is located approximately 90 km downstream of the outflow of the Pirot hydropower plant (HPP) that started operation in 1990, at the very end of the first 30-year study period, as a peaking HPP [43]. It should be noted that we did not detect a change in the mean annual flow series of HS10 in the data preparation phase. This is most likely due to the small effect of this HPP and its reservoir on the regime of the downstream HS10 for several reasons: the small drainage area of the reservoir compared with the drainage area of HS10, the annual water balancing mode of the reservoir, and subdaily disturbance of the regime at HS10 by HPP outflows.

The shift in runoff timing shown in Table 3 is less pronounced compared with the relative volume change. The changes in more than 7 or 10 days might be significant for some applications in the water sector such as irrigation and hydropower operation planning. In that respect, the most significant changes are found at the annual scale in the dry and average condition zones in HS1 and HS2 of the Sava River basin, where the runoff centroids occur up to 11 and 8 days earlier, respectively. A later occurrence of dry conditions is found at the annual scale in HS7, HS8, and HS10, which are all situated in the upper and middle parts of the Južna Morava River basin. Wet hydrologic conditions occur 7 days later in HS2 and HS7 in summer, and 9 and 8 days earlier in HS7 and HS8, respectively, in the spring season.

The data from Tables 2 and 3 are shown as boxplots in Figure 10 and Figure 11, respectively. The results are organized in such a way that annual and seasonal changes in the hydrologic condition zones are easily grasped for the whole study area.

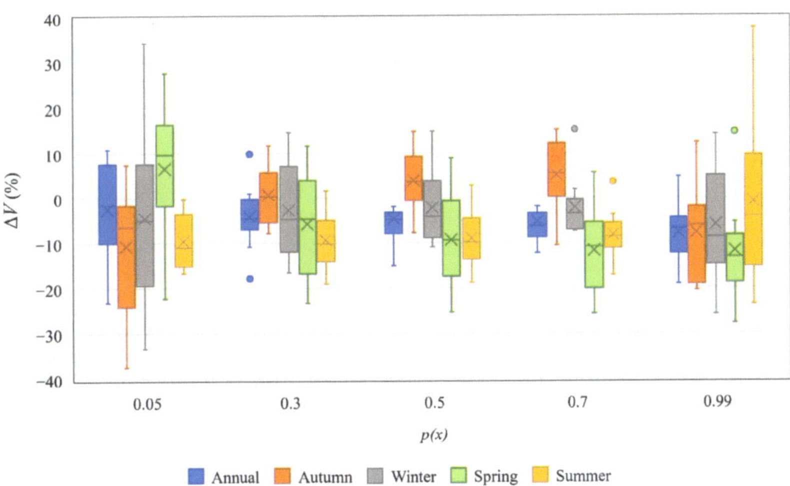

Figure 10. The boxplots of changes in the annual and seasonal runoff volumes in 1991–2020 compared with 1961–1990 at all stations for the wetness condition zones defined using the upper quantiles of probability $p(x)$. The lines in the boxplots indicate the median value and x indicates the mean, while the whiskers correspond to $1.5 \times$ IQR, where IQR is the interquartile range, with the outliers outside IQR displayed.

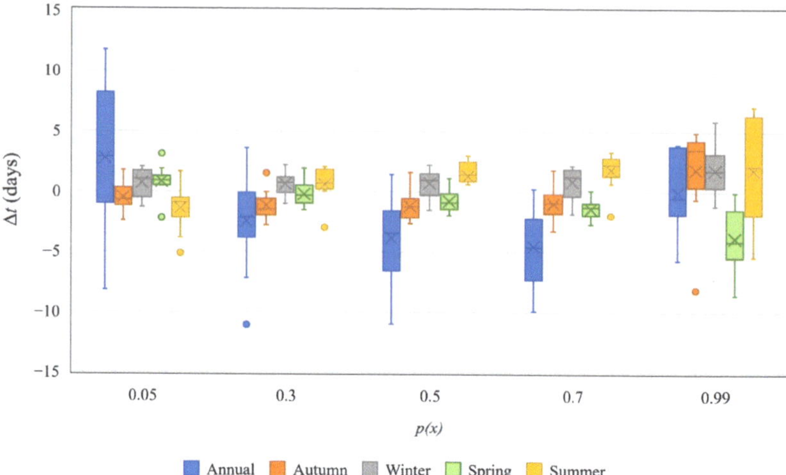

Figure 11. The boxplots of changes in the annual and seasonal runoff timing in 1991–2020 compared with 1961–1990 at all stations for the wetness condition zones defined by the upper quantiles of probability $p(x)$. The lines in the boxplot indicate the median value and x indicates the mean, while the whiskers correspond to $1.5 \times$ IRQ, where IRQ is the interquartile range, with the outliers outside IQR displayed.

A general decreasing tendency in annual runoff volume in 1991–2020 compared with 1961–1990 at all stations is clearly visible from the boxplot in Figure 10. The most pronounced decrease is for the wet condition zone. Runoff volume has also decreased in all seasons and for all hydrological condition zones, except the autumn runoff in the average zone (for the 0.5 and 0.7 quantiles), and the spring runoff in the extremely dry zone. There is less water available in wet zones (0.7 to 0.99 quantiles) during the high-flow season in winter and spring. Similarly, in dry zones (0.3 and 0.05 quantiles) there is less water available in the summer and autumn seasons in the recent period 1991–2020.

In terms of runoff timing at the annual level (Figure 11), the centroids for dry and average conditions occur earlier in 1991–2020 compared with 1961–1990, while the centroids for wet conditions occur somewhat later. The seasonal changes in timing are not pronounced for all zones, but the centroids for wet conditions appear earlier in spring at all stations, and later in winter at most stations.

The annual changes in both runoff volume and runoff timing at all stations are mapped for spatial analysis in Figure 12 and Figure 13, respectively.

The spatial distribution of the relative volume change between the two periods shown in Figure 12 reveals that the catchments in western Serbia have experienced an increase in runoff volume in the recent period for the dry condition zone. It should be noted that these catchments are small to medium areas in size. The decrease in annual runoff volume is more pronounced at the transition from the average to wet condition zone in all catchments. The most western station, HS1, behaves differently compared with the other stations. It has the largest catchment area and represents the only station with runoff generated almost completely out of the country. This difference in behavior is pronounced in respect to runoff timing, as shown in Figure 13. The runoff timing for extreme dry conditions is later in the recent period at the annual level for all other stations. For the average wetness conditions, runoff generally occurs earlier, but this change is insignificant, while it generally occurs a bit later for wet conditions. The most southern station, HS7, shows earlier runoff timing for wet conditions. Expectedly, the timing shifts at the annual level are more pronounced compared with the shifts within seasons.

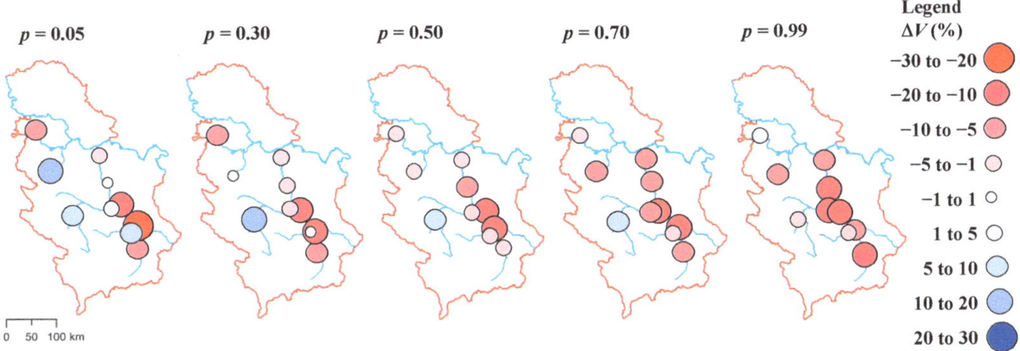

Figure 12. The maps of relative change in annual runoff volume at ten stations for dry, average, and wet condition zones, represented by upper quantiles of probability $p(x)$. Red circles denote volume reduction in 1991–2020 compared with 1961–1990, and blue circles indicate volume increase. The circle size corresponds to the amount of change.

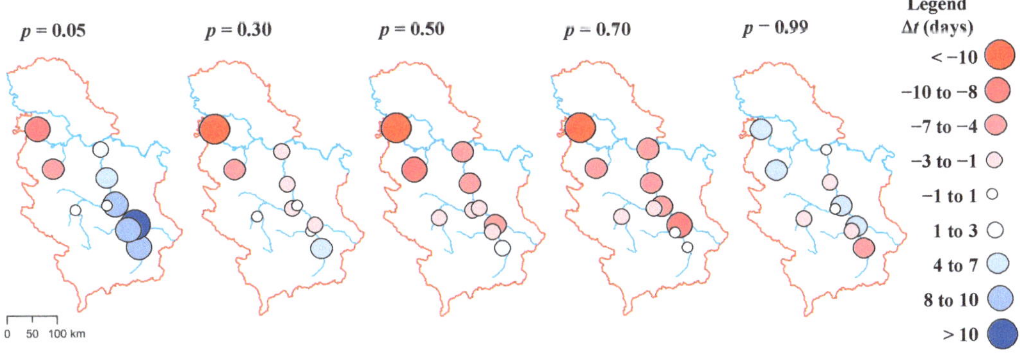

Figure 13. The maps of changes in timing of annual runoff at ten stations for dry, average, and wet condition zones, represented by upper quantiles of probability $p(x)$. Red circles denote earlier runoff occurrence in 1991–2020 compared with 1961–1990, and blue circles indicate later timing. The circle size corresponds to the amount of change.

5. Discussion

5.1. Long-Term Changes in Hydrological Regime

The majority of previous studies considering the same area used trend detection of annual/seasonal/extreme flows to study hydrologic change in the past or in the future, or the "climate-hydrology-assessment chain" approach [44], in which a selected climate scenario plays an important role. Comparison of the approach presented here with other studies can therefore only be made indirectly, i.e., based on the detected changes. The methodology presented here can be used to analyze the seasonal runoff pattern, but it is less convenient for the analyses of phenomena at short, event-based time scales (such as flow flashiness) or at time scales longer than one year (such as interannual variability), as underlined by Blum et al. [36]. The probabilistic seasonal runoff pattern obtained from the MDDFs is used here to define zones of different hydrologic conditions and to detect a change in this pattern by looking into the changes in runoff timing and volume between the two 30-year periods. Our study is therefore aimed at detecting a structural change in the hydrological regime instead of the simple detection of trends or projections via climate change scenarios. The proposed approach is also not aimed at detecting the changes in intensity and frequency of flood events, which are often considered important consequences of climate change. However, our approach can be useful for a general understanding of the hydrological regime at a given location and can provide indications of the changes in the high-water regime, which is related to the floods.

In general, our results are consistent with the findings of Blöschl et al. [45], who reported decreasing flows in the Balkan region, and with Lobanova et al. [46], who used hydrologic simulations with a daily timestep under future climate conditions and concluded that shifts in runoff seasonality, particularly earlier occurrence of high flows in the snowmelt-driven catchments, can be expected. Also, according to [46], winter and early spring runoff in Central and Eastern Europe (in the high-flow season) are expected to increase. This is fairly consistent with our results (Figure 9), which show an increase in runoff in the wet season over 1991–2020 that is not significant (yet), but later timing of winter runoff and earlier timing of spring runoff (Figure 10) are detected, indicating the shifts in the wet period projected for the future in [46].

In a comprehensive analysis of trends in mean annual and seasonal flows in Serbia over 1961–2010 [47], no trend in mean annual flows was detected at all stations used in our study except HS8 and HS10, for which moderate decreasing trends were found. This is not in agreement with our findings from the data preparation phase (Table A1 in Appendix A) on the absence of trends over the 1961–2020 period at all stations considered. On the other hand, we found the greatest negative changes in annual runoff volume at HS8 and HS10 for the 0.5 quantile zone (−10% and −15%, respectively; see Table 2). In the same zone, HS6 exhibits the highest positive change in annual runoff volume of 10%, while there was no trend detected at this station in [47]. These discrepancies may stem from the difference in the periods for analysis (1961–2010 in [47] vs. 1961–2020 in our study), because the additional 10 years in our study cover a wetter period in Serbia, including an extremely wet year, 2014 (also known for extreme floods). On the seasonal level, a low-significance positive trend is found in [47] in autumn and winter runoff at HS6, whereas we found changes in seasonal runoff volume of 11% and 15% for autumn and winter, respectively. Additionally, a moderate negative trend in spring runoff at HS10 was found in [47], while our results show a change in spring runoff volume of −25%. However, we detected several considerable seasonal changes exceeding 10%, both positive and negative, in the average condition zone, while these were not detected as trends in [47].

A variety of studies about different aspects of hydrologic change for the Sava River (e.g., [14,48–50]) are in line with our results for HS1, showing that the mean flow is slightly declining. Hydrologic simulations in [49] for the near future (2011–2041), which partially overlaps with the second period in our study, indicate changes consistent with our results about increases in winter runoff volume in all hydrologic condition zones. An increase in winter runoff at HS1 is also found as a weak (statistically insignificant) positive trend over the long period of 1928–2017 [48]. A substantial decrease in river flows expected in the spring and summer seasons [49] is also visible in our results for all zones except for wet condition zones. The analysis of long-term trends of climate drivers and assessment of runoff in [50] also shows declining runoff in the Sava River basin as a consequence of increased air temperature and evapotranspiration. General findings in [13] for HS1 and HS4 indicate decreasing trends in summer and autumn runoff, and mostly increasing trends in winter and spring runoff, which are partially in accordance with our results. Our study indicates that HS1 and HS4 exhibit decreases in spring runoff volume and an increase in autumn runoff volume for the 0.5 quantile zone.

The hydrologic projections for the Kolubara (HS2) and Toplica (HS9) catchments indicate a decrease in snow storage and a substantial decrease in runoff in 2001–2030 (near future) compared with 1961–1990 (baseline period) [51]. Our results do not show a substantial but rather a slight decrease in runoff volume in the average and wet zones, which is closer to the findings of Idrizović et al. [52] for HS9. In [51], the largest decrease in runoff at HS9 is associated with the spring season, which is in accordance with our results for the wet zone only. The flood flow series simulated in [51] indicate little change for HS9 and a small increase for HS2 in median annual maximum floods. Our results for the wet zone at the annual scale for HS9 are consistent with these results, and also with the results in [52]. The analyses in [51] additionally show that the future droughts are expected to be more frequent, to last longer, and to start earlier in the summer for both the HS2 and HS9 catchments. Judging by the combination of change in runoff volume and time shifts in the dry and extremely dry zones (Tables 2 and 3, 0.05 and 0.3 quantile), runoff reduction in these zones is already evident, but the time shift toward an earlier date can be seen in the extremely dry zone at HS9 only.

5.2. Probabilistic Annual Runoff Cycle as an Indicator of Hydrological Conditions

The probabilistic annual runoff cycle obtained from the MDDFs is used in this study to assess the changes over time in hydrological regime at selected rivers and locations in Serbia. It can also be used for other purposes, including operational monitoring of the hydrological regime aimed at early warnings and water management, and especially in operational reservoir management and management of droughts. The diagram with the MDDF of the annual runoff cycle (Figure 7) shows the zones of different hydrological conditions. Plotting actual hydrological data over this diagram clearly indicates current conditions in the catchment, but it can also reveal departures from normal conditions and anomalies. It can therefore be a support for responding to unfavorable situations and their mitigation.

The zones of hydrological conditions in the probabilistic annual runoff cycle are delimited by different probabilistic thresholds, which vary over the annual cycle because they represent the quantiles of MDDFs. This type of zoning based on probabilistic thresholds is very similar to the application of the SPI and SSI as drought indicators [29,30]. The main rationale behind the SPI and SSI as nondimensional (standardized) probabilistic indicators is to enable comparative analyses of precipitation or streamflow at different locations and at different times. Application of these two indices in monitoring hydrological conditions therefore requires the transformation of current precipitation or streamflow observations into standardized values. Contrary to this, the probabilistic annual runoff cycle allows direct evaluation of current streamflow observations for a given location.

A common characteristic of the methodologies for assessing the probabilistic annual runoff cycle and the SPI/SSI is related to fitting probability distributions to the observations. The SPI and SSI are usually calculated from monthly data over different accumulation periods from 1 to 12 months, whereas precipitation or streamflow accumulations for each month in a year are fitted by separate parametric or nonparametric distributions. Stagge et al. [33] attempted to fit a set of 365 parametric probability distributions for daily SPI estimation, but they retained the monthly averaging period. When compared with the SPI/SSI applications, using the daily time scale for the probabilistic annual runoff cycle provides information on runoff conditions with finer resolution and helps to avoid discontinuities when using the monthly scale and transiting from one month to another.

Regardless of temporal resolution, finding appropriate distribution type(s) for precipitation or streamflow is a crucial issue because it can considerably affect the computed SPI/SSI series and the derived drought characteristics [30,53]. Inappropriate distribution can lead to inconsistencies in the SPI or SSI such as the average not being equal to 0 and the variance not being equal to 1 [30]. When assessing the probabilistic annual runoff cycle for a given location, finding the most appropriate distribution is still a great challenge, but the targets of this process are less demanding regarding spatial comparability and in terms of constraints related to statistical properties.

Furthermore, the probabilistic annual runoff cycle is obtained here using the periodic functions for statistical properties of daily streamflow. This approach leads to smooth periodic-like lines of streamflow quantiles within a year and additionally ensures continuity of the thresholds that delineate certain hydrological condition zones.

As the SPI and SSI are devised as indices that should facilitate comparability of conditions in space and time, relevant thresholds that define specific conditions (e.g., droughts) are obtained via the equiprobability transformation from a selected distribution into the standardized normal quantiles. Typical thresholds are SPI = 0 or SSI = 0, and they are constant over the annual cycle. Conversely, the thresholds in the probabilistic annual runoff cycle are expressed directly in the physical units of streamflow at a given location and are therefore variable throughout a year. This provides a more intuitive presentation of the hydrological conditions in the annual cycle, and the zones delimited by these thresholds readily show deficits and surpluses in available water quantities.

6. Conclusions

Hydrologic change detection involves different identification approaches, temporal scales, and spatial scales. This study presents an analysis of changes in the annual and seasonal runoff of ten catchments in Serbia in the 1961–2020 period at the daily time scale. The analysis is based on the probabilistic representation of the seasonal runoff pattern that is obtained from marginal distributions of daily flows (MDDFs) with periodic distribution parameters inherited from the periodic statistical parameters of daily flow series. The hydrologic (wetness) conditions are defined as the zones between the MDDF quantiles of specific probabilities, selected to depict dry, average, and wet normal conditions in the catchment.

Based on a comparison of these zones between the two 30-year periods at the annual and seasonal scales, the following may be concluded:

1. The seasonal runoff pattern changed from one period to another in terms of temporal shift and the occurrence of more extreme flows. However, the general pattern of seasonal runoff remained the same. The prevailing pattern is simple and unimodal, while the less present mixed regime is bimodal.
2. In most of the catchments, runoff volume has decreased in the recent 1991–2020 period at both the annual and seasonal scales. The critical season is summer for dry and average conditions, with volume reduction in all catchments.
3. The most pronounced shift in runoff timing is found on the annual scale. Dry and average conditions occur earlier at this scale. The change in runoff timing is found to be insignificant for all seasons and zones, except for wet conditions, which occur earlier in spring.

The proposed zoning of hydrologic conditions based on probabilistic variable daily flow thresholds enables more precise analysis of runoff volume and timing between the selected periods. It also enables more detailed analysis of zones that are critical for some applications in a certain period of the year. For instance, such an analysis may be of interest in drought-related studies in the seasons when water deficit conflicts with water demand.

The results of this study are generally in line with the results of the previous hydrologic change analyses in the study area, in terms of both past changes and those predicted for the near-future climate, but our results may be considered more detailed due to the applied approach based on MDDFs with periodic distribution parameters. It is demonstrated that the proposed approach allows for detecting a structural change in the hydrological regime, and detecting changes in the regulated flow regime when statistical tests cannot detect the change. The daily time scale of the seasonal runoff pattern representation makes this approach more useful for monitoring actual catchment wetness and runoff conditions and therefore for operational water management and early warning systems. In situational assessment and forecasting, the wetness condition zones may be used for flow forecasting by focusing on the zone to which current daily flows belong. It is worth noting that the boundaries of the wetness zones could be changed easily depending on the purpose required. Furthermore, the zone boundaries, i.e., the probabilistic thresholds, can also be converted to hydrologic drought indicators such as the SSI.

Supplementary Materials: The following supporting information can be downloaded at https:// www.mdpi.com/article/10.3390/w15162919/s1. Figure S1: Seasonal variation of daily flow statistics (peaky lines) and their corresponding periodic functions (smooth lines) at all HS estimated for the three periods: (a) mean, (b) standard deviation, and (c) coefficient of skewness; Figure S2: Seasonal variation of log-transformed daily flow statistics (peaky lines) and their corresponding periodic functions (smooth lines) at all HS estimated for the three periods: (a) mean, (b) standard deviation, and (c) coefficient of skewness; Figure S3: Periodical parameters of LPT3 marginal distribution of daily flows at all HS: (a) shape parameter $\alpha_{\tau,per}$, (b) scale parameter $\beta_{\tau,per}$, (c) location parameter $\gamma_{\tau,per}$; Figure S4: The diagrams of MDDF at all HS for the three periods showing the quantiles (a) for a full range of probabilities p(x), from 0.01 to 0.99, and (b) for low to median probabilities. Solid lines represent the latest 1991–2020 period, the dashed lines represent the earlier 1961–1990 period, and the dotted lines represent the whole 1961–2020 period.

Author Contributions: Conceptualization, B.B. and V.M.; methodology, B.B. and V.M.; validation, B.B., V.M. and A.B.; formal analysis, B.B., V.M., A.B. and J.P.; investigation, B.B., V.M. and A.B.; data curation, V.M. and A.B.; writing—original draft preparation, B.B., V.M. and J.P.; visualization, V.M. and B.B.; funding acquisition, B.B., V.M. and J.P. All authors have read and agreed to the published version of the manuscript.

Funding: This research was partially funded by the Ministry of Education, Science and Techno-logical Development of the Republic of Serbia, grant numbers 451-03-47/2023-01/200095, 451-03-47/2023-01/200169, and 451-03-47/2023-01/200092.

Data Availability Statement: The data are not publicly available because they are owned by the Republic Hydrometeorological Service of Serbia.

Acknowledgments: The authors would like to thank the Republic Hydrometeorological Service of Serbia for the provision of data. The authors would like to acknowledge the initial support to the research of the late Ass. Prof. Dr. Zoran M. Radić from the University of Belgrade—Faculty of Civil Engineering, Serbia.

Conflicts of Interest: The authors declare no conflict of interest.

Appendix A

Table A1 shows the results of statistical tests for the homogeneity, randomness, and trends of annual flows at the 10 hydrological stations used in this study over the 1961–2020 period. The table shows p-values of the test statistics in the following tests: a test on the difference between two means (Z-test), two tests on the difference between two variances (Fisher's F-test and Levene's test), the Mann–Whitney test for homogeneity, the Wald–Wolfowitz run test for independence, and the Mann–Kendall test for trends (with the null hypothesis that there is no trend in the data).

Table A1. Results (p-values) of the tests applied to the annual flow series for the period 1961–2020.

HS#	Z [1]	F [2]	L [3]	M-W [4]	W-W [5]	M-K [6]
1	0.792	0.888	0.946	0.792	0.520	0.444
2	0.507	0.469	0.732	0.584	0.367	0.329
3	0.150	0.430	0.563	0.092	0.899	0.211
4	0.402	0.344	0.442	0.382	0.700	0.506
5	0.206	0.809	0.420	0.152	0.896	0.367
6	0.374	0.673	0.619	0.393	0.053	0.415
7	0.358	0.089	0.215	0.184	0.520	0.255
8	0.111	0.729	0.793	0.084	0.367	0.154
9	0.254	0.968	0.588	0.262	0.520	0.293
10	0.601	0.654	0.390	0.516	0.700	0.354

Note(s): [1] Z-test; [2] Fisher's F-test; [3] Levene's test; [4] Mann–Whitney test; [5] Wald–Wolfowitz test; [6] Mann–Kendall test.

References

1. Bridgewater, P.; Guarino, E.; Thompson, R.M. Hydrology in the Anthropocene. In *Encyclopedia of the Anthropocene*; Elsevier: Amsterdam, The Netherlands, 2017; Volume 1–5, pp. 87–92, ISBN 9780128096659.
2. WMO Hydrological Coordination Panel. Statement on the Term Hydrological Normal. Available online: https://community.wmo.int/en/activity-areas/hydrology-and-water-resources/hydrological-coordination-panel (accessed on 13 June 2023).

3. Pekarova, P.; Gorbachova, L.; Mitkova, V.B.; Pekar, J.; Miklanek, P. Statistical Analysis of Hydrological Regime of the Danube River at Ceatal Izmail Station. *IOP Conf. Ser. Earth Environ. Sci.* **2019**, *221*, 012035. [CrossRef]
4. Renner, M.; Bernhofer, C. Long Term Variability of the Annual Hydrological Regime and Sensitivity to Temperature Phase Shifts in Saxony/Germany. *Hydrol. Earth Syst. Sci.* **2011**, *15*, 1819–1833. [CrossRef]
5. Plavšić, J.; Milutinović, R. O računskim nivoima vode za zaštitu od poplava na Dunavu kod Novog Sada. *Vodoprivreda* **2010**, *42*, 69–78.
6. Stewart, I.T.; Cayan, D.R.; Dettinger, M.D. Changes toward Earlier Streamflow Timing across Western North America. *J. Clim.* **2005**, *18*, 1136–1155. [CrossRef]
7. Laaha, G.; Blöschl, G. Seasonality Indices for Regionalizing Low Flows. *Hydrol. Process.* **2006**, *20*, 3851–3878. [CrossRef]
8. Eisner, S.; Flörke, M.; Chamorro, A.; Daggupati, P.; Donnelly, C.; Huang, J.; Hundecha, Y.; Koch, H.; Kalugin, A.; Krylenko, I.; et al. An Ensemble Analysis of Climate Change Impacts on Streamflow Seasonality across 11 Large River Basins. *Clim. Chang.* **2017**, *141*, 401–417. [CrossRef]
9. Poschlod, B.; Willkofer, F.; Ludwig, R. Impact of Climate Change on the Hydrological Regimes in Bavaria. *Water* **2020**, *12*, 1599. [CrossRef]
10. Yeste, P.; Rosa-Cánovas, J.J.; Romero-Jiménez, E.; García-Valdecasas Ojeda, M.; Gámiz-Fortis, S.R.; Castro-Díez, Y.; Esteban-Parra, M.J. Projected Hydrologic Changes over the North of the Iberian Peninsula Using a Euro-CORDEX Multi-Model Ensemble. *Sci. Total Environ.* **2021**, *777*, 146126. [CrossRef]
11. Hobeichi, S.; Abramowitz, G.; Ukkola, A.M.; De Kauwe, M.; Pitman, A.; Evans, J.P.; Beck, H. Reconciling Historical Changes in the Hydrological Cycle over Land. *NPJ Clim. Atmos. Sci.* **2022**, *5*, 17. [CrossRef]
12. Brunner, M.I.; Melsen, L.A.; Newman, A.J.; Wood, A.W.; Clark, M.P. Future Streamflow Regime Changes in the United States: Assessment Using Functional Classification. *Hydrol. Earth Syst. Sci.* **2020**, *24*, 3951–3966. [CrossRef]
13. Stojković, M.; Plavšić, J.; Prohaska, S. Dugoročne promene godišnjih i sezonskih proticaja: Primer reke Save. *Vodoprivreda* **2014**, *46*, 39–48.
14. Stojković, M.; Ilić, A.; Prohaska, S.; Plavšić, J. Multi-Temporal Analysis of Mean Annual and Seasonal Stream Flow Trends, Including Periodicity and Multiple Non-Linear Regression. *Water Resour. Manag.* **2014**, *28*, 4319–4335. [CrossRef]
15. Kundzewicz, Z.W.; Robson, A. (Eds.) *Detecting Trend and Other Changes in Hydrological Data*; WMO/TD-No. 1013; World Meteorological Organization: Geneva, Switzerland, 2000.
16. Hamed, K.H.; Ramachandra Rao, A. A Modified Mann-Kendall Trend Test for Autocorrelated Data. *J. Hydrol.* **1998**, *204*, 182–196. [CrossRef]
17. Pettitt, A.N. A Non-Parametric Approach to the Change-Point Problem. *Appl. Stat.* **1979**, *28*, 126. [CrossRef]
18. Zhang, A.; Zheng, C.; Wang, S.; Yao, Y. Analysis of Streamflow Variations in the Heihe River Basin, Northwest China: Trends, Abrupt Changes, Driving Factors and Ecological Influences. *J. Hydrol. Reg. Stud.* **2015**, *3*, 106–124. [CrossRef]
19. Serinaldi, F.; Kilsby, C.G.; Lombardo, F. Untenable Nonstationarity: An Assessment of the Fitness for Purpose of Trend Tests in Hydrology. *Adv. Water Resour.* **2018**, *111*, 132–155. [CrossRef]
20. Hannaford, J.; Buys, G.; Stahl, K.; Tallaksen, L.M. The Influence of Decadal-Scale Variability on Trends in Long European Streamflow Records. *Hydrol. Earth Syst. Sci.* **2013**, *17*, 2717–2733. [CrossRef]
21. Stojković, M.; Prohaska, S.; Plavšić, J. Stochastic Structure of Annual Discharges of Large European Rivers. *J. Hydrol. Hydromech.* **2015**, *63*, 63–70. [CrossRef]
22. Blöschl, G.; Sivapalan, M.; Wagener, T.; Viglione, A.; Savenije, H. (Eds.) *Runoff Prediction in Ungauged Basins: Synthesis across Processes, Places and Scales*; Cambridge University Press: Cambridge, UK, 2013; ISBN 978-1-107-02818-0.
23. Salinas-Rodríguez, S.A.; Sánchez-Navarro, R.; Barrios-Ordóñez, J.E. Frequency of Occurrence of Flow Regime Components: A Hydrology-Based Approach for Environmental Flow Assessments and Water Allocation for the Environment. *Hydrol. Sci. J.* **2021**, *66*, 193–213. [CrossRef]
24. Zeiringer, B.; Seliger, C.; Greimel, F.; Schmutz, S. River Hydrology, Flow Alteration, and Environmental Flow. In *Riverine Ecosystem Management*; Springer International Publishing: Cham, Switzerland, 2018; pp. 67–89.
25. Cammalleri, C.; Vogt, J.; Salamon, P. Development of an Operational Low-Flow Index for Hydrological Drought Monitoring over Europe. *Hydrol. Sci. J.* **2016**, *62*, 346–358. [CrossRef]
26. Mendoza, P.A.; Clark, M.P.; Mizukami, N.; Newman, A.J.; Barlage, M.; Gutmann, E.D.; Rasmussen, R.M.; Rajagopalan, B.; Brekke, L.D.; Arnold, J.R. Effects of Hydrologic Model Choice and Calibration on the Portrayal of Climate Change Impacts. *J. Hydrometeorol.* **2015**, *16*, 762–780. [CrossRef]
27. Abdul Aziz, O.I.; Burn, D.H. Trends and Variability in the Hydrological Regime of the Mackenzie River Basin. *J. Hydrol.* **2006**, *319*, 282–294. [CrossRef]
28. Fleig, A.K.; Tallaksen, L.M.; Hisdal, H.; Demuth, S. A Global Evaluation of Streamflow Drought Characteristics. *Hydrol. Earth Syst. Sci.* **2006**, *10*, 535–552. [CrossRef]
29. McKee, T.B.; Doesken, N.J.; Kleist, J. The Relationship of Drought Frequency and Duration to Time Scales. In Proceedings of the 8th Conference on Applied Climatology, Anaheim, CA, USA, 17–22 January 1993; American Meteorological Society: Boston, MA, USA, 1993; pp. 179–183.
30. Vicente-Serrano, S.M.; López-Moreno, J.I.; Beguería, S.; Lorenzo-Lacruz, J.; Azorin-Molina, C.; Morán-Tejeda, E. Accurate Computation of a Streamflow Drought Index. *J. Hydrol. Eng.* **2012**, *17*, 318–332. [CrossRef]

31. Mihailović, V.; Blagojević, B. Izbor teorijske raspodele verovatnoća na osnovu L-momenata u proračunu pokazatelja meteorološke i hidrološke suše. *Vodoprivreda* **2015**, *47*, 273–278.
32. Ozkaya, A.; Zerberg, Y. A 40-Year Analysis of the Hydrological Drought Index for the Tigris Basin, Turkey. *Water* **2019**, *11*, 657. [CrossRef]
33. Stagge, J.H.; Tallaksen, L.M.; Gudmundsson, L.; Van Loon, A.F.; Stahl, K. Candidate Distributions for Climatological Drought Indices (SPI and SPEI). *Int. J. Climatol.* **2015**, *35*, 4027–4040. [CrossRef]
34. Sutanto, S.J.; Van Lanen, H.A.J. Streamflow Drought: Implication of Drought Definitions and Its Application for Drought Forecasting. *Hydrol. Earth Syst. Sci.* **2021**, *25*, 3991–4023. [CrossRef]
35. Yevjevich, V. *Structure of Daily Hydrologic Series*; Water Resources Publications: Littleton, CO, USA, 1984.
36. Blum, A.G.; Archfield, S.A.; Vogel, R.M. On the Probability Distribution of Daily Streamflow in the United States. *Hydrol. Earth Syst. Sci.* **2017**, *21*, 3093–3103. [CrossRef]
37. Yevjevich, V. *Stochastic Processes in Hydrology*; Water Resources Publications: Littleton, CO, USA, 1972.
38. Radić, Z.; Mihailović, V. Marginalne raspodele dnevnih proticaja na reprezentativnim profilima u Srbiji. *Vodoprivreda* **2010**, *42*, 17–38.
39. Mihailović, V. Složena Analiza Hidroloških Vremenskih Serija za Potrebe Modeliranja Ekstremnih Događaja (A Complex Analysis of Hydrological Time Series for Modelling Extreme Events). Doctoral Thesis, University of Belgrade—Faculty of Civil Engineering, Belgrade, Serbia, 2012.
40. Hayes, D.S.; Brändle, J.M.; Seliger, C.; Zeiringer, B.; Ferreira, T.; Schmutz, S. Advancing towards Functional Environmental Flows for Temperate Floodplain Rivers. *Sci. Total Environ.* **2018**, *633*, 1089–1104. [CrossRef] [PubMed]
41. Bogojević, A. Strukturna Analiza Serija Dnevnih Protoka u Profilima Hidroloških Stanica na Teritoriji Srbije (Structural Analysis of Daily Flow Series at Hydrological Stations in the Territory of Serbia). Bachelor's Thesis, University of Niš, Faculty of Civil Engineering and Architecture, Niš, Serbia, 2022.
42. Bobée, B.; Ashkar, F. *The Gamma Family and Derived Distributions Applied in Hydrology*; Water Resources Publications: Littleton, CO, USA, 1991; ISBN 978-0918334688.
43. Elektroprivreda Srbije, HE Pirot. Available online: https://www.eps.rs/lat/djerdap/Stranice/he_pirot.aspx (accessed on 4 August 2023).
44. Olsson, J.; Arheimer, B.; Borris, M.; Donnelly, C.; Foster, K.; Nikulin, G.; Persson, M.; Perttu, A.-M.; Uvo, C.; Viklander, M.; et al. Hydrological Climate Change Impact Assessment at Small and Large Scales: Key Messages from Recent Progress in Sweden. *Climate* **2016**, *4*, 39. [CrossRef]
45. Blöschl, G.; Hall, J.; Viglione, A.; Perdigão, R.A.P.; Parajka, J.; Merz, B.; Lun, D.; Arheimer, B.; Aronica, G.T.; Bilibashi, A.; et al. Changing Climate Both Increases and Decreases European River Floods. *Nature* **2019**, *573*, 108–111. [CrossRef] [PubMed]
46. Lobanova, A.; Liersch, S.; Nunes, J.P.; Didovets, I.; Stagl, J.; Huang, S.; Koch, H.; López, M.D.R.R.; Maule, C.F.; Hattermann, F.; et al. Hydrological Impacts of Moderate and High-End Climate Change across European River Basins. *J. Hydrol. Reg. Stud.* **2018**, *18*, 15–30. [CrossRef]
47. Kovacevic-Majkic, J.; Urosev, M. Trends of Mean Annual and Seasonal Discharges of Rivers in Serbia. *J. Geogr. Inst. Jovan Cvijic SASA* **2014**, *64*, 143–160. [CrossRef]
48. Leščešen, I.; Šraj, M.; Pantelić, M.; Dolinaj, D. Assessing the Impact of Climate on Annual and Seasonal Discharges at the Sremska Mitrovica Station on the Sava River, Serbia. *Water Supply* **2022**, *22*, 195–207. [CrossRef]
49. The World Bank. *Water and Climate Adaptation Plan for the Sava River Basin: Main Report*; World Bank Group: Washington, DC, USA, 2015. Available online: https://www.savacommission.org/documents-and-publications/technical-and-project-reports/water-and-climate-adaptation-plan-for-the-sava-river-basin-watcap/245 (accessed on 13 June 2023).
50. Pandžić, K.; Trninić, D.; Likso, T.; Bošnjak, T. Long-Term Variations in Water Balance Components for Croatia. *Theor. Appl. Climatol.* **2009**, *95*, 39–51. [CrossRef]
51. Langsholt, E.; Lawrence, D.; Wong, W.K.; Andjelic, M.; Ivkovic, M.; Vujadinovic, M. *Effects of Climate Change in The Kolubara and Toplica Catchments, Serbia*; Haddeland, I., Ed.; Norwegian Water Resources and Energy Directorate: Oslo, Norway, 2013.
52. Idrizovic, D.; Pocuca, V.; Vujadinovic Mandic, M.; Djurovic, N.; Matovic, G.; Gregoric, E. Impact of Climate Change on Water Resource Availability in a Mountainous Catchment: A Case Study of the Toplica River Catchment, Serbia. *J. Hydrol.* **2020**, *587*, 124992. [CrossRef]
53. Tijdeman, E.; Stahl, K.; Tallaksen, L.M. Drought Characteristics Derived Based on the Standardized Streamflow Index: A Large Sample Comparison for Parametric and Nonparametric Methods. *Water Resour. Res.* **2020**, *56*, e2019WR026315. [CrossRef]

Article

Study on a Hybrid Hydrological Forecasting Model SCE-GUH by Coupling SCE-UA Optimization Algorithm and General Unit Hydrograph

Yingying Xu, Chengshuai Liu *, Qiying Yu, Chenchen Zhao, Liyu Quan and Caihong Hu *

Yellow River Laboratory, Zhengzhou University, Zhengzhou 450001, China; xyy13271673906@gs.zzu.edu.cn (Y.X.); yqy258369@gs.zzu.edu.cn (Q.Y.); zcczcc@gs.zzu.edu.cn (C.Z.); quanliyv3014@outlook.com (L.Q.)
* Correspondence: liucs@gs.zzu.edu.cn (C.L.); hucaihong@zzu.edu.cn (C.H.)

Abstract: Implementing real-time prediction and warning systems is an effective approach for mitigating flash flood disasters. However, there is still a challenge in improving the accuracy and reliability of flood prediction models. This study develops a hydrological prediction model named SCE-GUH, which combines the Shuffled Complex Evolution-University of Arizona optimization algorithm with the general unit hydrograph routing method. Our aims were to investigate the applicability of the general unit hydrograph in runoff calculations and its performance in predicting flash flood events. Furthermore, we examined the influence of parameter variations in the general unit hydrograph on flood simulations and conducted a comparative analysis with the conventional Nash unit hydrograph. The research findings demonstrate that the utilization of the general unit hydrograph method can considerably decrease computational errors and enhance prediction accuracy. The flood peak detection rate was found to be 100% in all four study watersheds. The average Nash–Sutcliffe efficiency coefficients were 0.83, 0.83, 0.84, and 0.87, while the corresponding coefficients of determination were 0.86, 0.85, 0.86, and 0.94, and the absolute errors of peak present time were 0.19 h, 0.40 h, 0.91 h, and 0.82 h, respectively. Moreover, the utilization of the general unit hydrograph method was found to significantly reduce the peak-to-current time difference, thereby enhancing simulation accuracy. Parameter variations have a substantial influence on peak flow characteristics. The SCE-GUH model, which incorporates the topographic and geomorphological features of the watershed along with the optimization algorithm, is capable of effectively characterizing the catchment properties of the watershed and offers valuable insights for enhancing the early warning and prediction of hydrological forecasting.

Keywords: general unit hydrograph; rainfall–runoff relationship; optimization algorithm; flash flood simulation; application test; surface confluence

check for updates

Citation: Xu, Y.; Liu, C.; Yu, Q.; Zhao, C.; Quan, L.; Hu, C. Study on a Hybrid Hydrological Forecasting Model SCE-GUH by Coupling SCE-UA Optimization Algorithm and General Unit Hydrograph. *Water* **2023**, *15*, 2783. https://doi.org/10.3390/w15152783

Academic Editors: Agnieszka Rutkowska and Katarzyna Baran-Gurgul

Received: 5 July 2023
Revised: 22 July 2023
Accepted: 27 July 2023
Published: 1 August 2023

1. Introduction

In the context of global climate change, extreme rainfall occurs frequently, with flash floods contributing significantly to natural disasters [1]. Flash floods pose significant challenges for flood control and disaster management, given their abruptness, destructive impact, and the difficulty of providing early warnings and forecasts [2]. One of the central topics in flood forecasting research is the theory of watershed confluence, which analyzes the interaction between various factors of natural phenomena [3]. Nevertheless, the existing concentration calculation methods have several unknown parameters, leading to a commonly observed low accuracy in predicting peak discharge [4].

The primary methods used for responding to flash flood forecast and early warning are the dynamic critical rainfall method and the constructed hydrological model. The dynamic critical rainfall method aims to determine the relationship between rainfall and runoff using hydrological techniques [5]. This involves analyzing the measured soil moisture and rainfall data in the study area to establish the correlation between rainfall and soil moisture;

subsequently, a correlation model is developed for rainfall–runoff and soil moisture [6]. The outlet flow of the flash flood watershed is then derived based on the soil moisture, and a specific method is employed to calculate the rainfall, which corresponds to the early warning flow. This calculated rainfall value is defined as the dynamic critical rainfall. The decision to issue early warning information for flash flood disasters is made based on real-time synchronous or predicted rainfall. If the rainfall reaches or exceeds the specified threshold, immediate early warning information is sent to the threatened area. The Flash Flood Guidance System (FFGS) has been designed and developed by the Hydrologic Research Center (HRC) of the United States [7,8]; The Japan International Cooperation Agency has developed a community-based flash flood early warning system [9]. This system determines early warning indicators by establishing a correlation between rainfall intensity and effective cumulative rainfall, drawing from both experiential knowledge and statistical data [10]; Europe has developed its own flood awareness system called the European Flood Awareness System (EFAS) [11,12]; Malaysia has provided assistance in the development of the internet-based Geospatial Data Exchange System (GEOREX FLOOD), which is specifically designed to serve the local area [13]. A constructed hydrological model is another method used for predicting and issuing early warnings for flash flood disasters. This model involves creating a mathematical representation of the hydrological processes in a specific area, such as rainfall–runoff relationships, soil moisture dynamics, and channel flow. By inputting real-time or forecasted rainfall data into the model, it can simulate the response of the watershed and provide estimates of potential flood events, enabling early warnings to be issued to at-risk areas [14]. Progea, an Italian company, has implemented the TOPKAPI distributed hydrological model for conducting research on forecasting and early warning systems. Furthermore, they have successfully established a flood forecasting system specifically designed for small and medium-sized watersheds in the region [15]; The National Hydrological and Meteorological Administration of the United States employs the HL-RMS hydrological model to establish flood forecasting systems specifically tailored for the Red River watershed in Arkansas and the Colorado watershed [16].

The watershed catchment theory forms the fundamental basis for flood forecasting, and the analysis of slope catchments has emerged as a prominent research focus [17]. The core of catchment calculation involves the convergence of net rain droplets toward the outlet section of the watershed. Building upon this physical foundation, numerous scholars had put forth a multitude of methods to calculate the watershed catchment process utilizing the unit hydrograph principle. In 1945, Clark examined the calculation method of the instantaneous unit hydrograph, which was based on research conducted by Sherman, Zoch, and other researchers; however, an accurate mathematical expression was not obtained at that time [18–20]. It was not until 1957 that Nash summarized the method of time conversion and derived the equation for the unit hydrograph [21,22]. In 1997, Gupat and other researchers integrated the topographic distribution characteristics of the watershed to develop the geomorphic unit hydrograph [23,24]. In 1989, Lai Peiying and other researchers introduced the concept of the variable speed geomorphic unit hydrograph, recognizing the nonlinear characteristics of catchment processes [25]. Subsequently, researchers successively developed the applications of fractional instantaneous unit hydrograph and time-variant distributed unit hydrograph for calculating river watershed concentration [4,26,27]. In 2021, Guo Junke proposed a physical perspective challenging the assumption made in most current unit hydrograph concentration calculations, where the watershed is considered as being in series with linear reservoirs. However, Guo argued that the actual watershed concentration should predominantly occur in parallel [28]. Guo Junke developed a GUH model based on a negative exponential function distribution and utilized this model to simulate 10 actual watersheds located in the United States and the United Kingdom, and the simulation results demonstrated good performance [29–31].

While the existing flash flood prediction models come in various types and offer a rich assortment of runoff structures, a significant number of these models are structurally complex and involve a multitude of parameters. Consequently, these models are unable to satisfy the demands for swift and accurate early warning and prediction of flash floods [32]. Accurate prediction becomes difficult in areas with inadequate data because of the demanding requirements for fundamental data on flash flood occurrence. Given the complex underlying surfaces commonly found in flash-flood-prone areas, it is crucial to introduce prediction models featuring simple structures and a reduced number of parameters. This approach enables rapid predictions to effectively respond to the swift onset of flash floods [33]. Choosing the parameters for the flash flood model can be challenging. Consequently, it is essential to leverage computer technology to optimize these parameters, resulting in streamlined and simplified model parameters. This optimization process facilitates accurate predictions while reducing reliance on extensive basic data [34]. Duan Qingyun, a professor at the University of Arizona, proposed the Shuffled Complex Evolution-University of Arizona (SCE-UA) algorithm in 1992 [35]. This algorithm is based on the simplex algorithm developed by Nelder and Mead in 1965, integrating concepts from both natural biological competition principles and the fundamental principle of genetic algorithms [36]. SCE-UA is an effective approach for addressing nonlinear constrained optimization problems. It possesses the ability to consistently, efficiently, and swiftly explore the global optimal solution of hydrological model parameters [37]. The SCE-UA algorithm is applicable for parameter estimation in hydrological models, optimization of decision variables in water resource management models, evaluation of flood risk, and support of water resource planning [34]. Utilizing the SCE-UA algorithm for parameter estimation enhances the accuracy and predictive ability of hydrological models. This aspect is particularly crucial for rainfall–runoff models, evapotranspiration models, and other hydrological models [38]. Through thorough exploration of the parameter space and identification of the optimal parameter combination, the model can accurately reflect real-world conditions. The SCE-UA algorithm is widely regarded as the most effective approach for parameter optimization in watershed hydrological models, and it finds extensive application in this area [39].

Our study strives to develop a streamlined and precise flood prediction model, which tackles the intricacy and data dependency of the current prediction models. To accomplish this objective, we introduce a hybrid hydrological prediction model named SCE-GUH. By integrating the SCE-UA algorithm with the calculation of a general unit hydrograph, this model aims to enhance prediction accuracy while simultaneously streamlining the complexity. In this study, the SCE-UA mixed complex evolution theory was employed to optimize the parameters of the general unit hydrograph. Furthermore, the general unit hydrograph was extended to flash flood watersheds where data scarcity exists, bringing forth fresh perspectives for international flash flood warning and forecasting. As an example, we conducted rainfall–runoff simulations in four watersheds. In these simulations, we utilized the Nash unit hydrograph method to re-calculate the surface concentration. By doing so, we aimed to compare and validate the predictive effectiveness of the general unit hydrograph principle in simulating flash floods.

2. Materials and Methods

2.1. Study Area

The applicability of the SCE-GUH model was tested by selecting control watersheds (Lixin and Xiagushan watersheds) from the Lixin and Xiagushan hydrological stations in the Huaihe River in China, as well as control watersheds (Liqingdian and Miping watersheds) from the Liqingdian and Miping hydrological stations in the Yangtze River in China, with similar climatic characteristics. The geographical distribution of the study area is presented in Figure 1. These study areas are characterized by hilly terrain and complex and changeable weather patterns. The spatial and temporal distribution of annual precipitation is uneven, and the instability of precipitation often results in frequent flash flood disasters, which can cause significant impacts and damages. The fundamental information of these four watersheds is summarized below:

- Lixin watershed: Located within the Huaihe River Basin of China, it covers an area of 79 km^2 (113°36′–113°46′ E, 32°86′–32°98′ N). The watershed experiences a continental monsoon climate, characterized by hot and rainy summers, and a humid climate. The average annual precipitation is 960 mm. Land use analysis reveals that farmland and grassland dominate the area, accounting for 39.51% each, followed by forest land at 14.81%.

- Xiagushan watershed: Located within the Huaihe River Basin of China, it covers an area of 383.5 km^2 (112°28′–112°43′ E, 33°48′–34°00′ N). The watershed experiences a warm temperate continental monsoon climate with four distinct seasons and abundant rainfall, averaging 1000 mm annually. Land use analysis reveals that the largest proportion of land is dedicated to farmland (45.50%), followed by grassland (41.64%) and forest land (9.51%).

- Liqingdian watershed: Located within the Yangtze River Basin of China, it covers an area of 634 km^2 (112°06′–112°31′ E, 33°27′–33°44′ N). The watershed exhibits distinct features of transitioning from a subtropical to warm temperate zone, with precipitation concentrated in the summer. It has an average annual precipitation of 868.8 mm. The largest proportion of land use in the watershed is grassland (73.63%), followed by farmland (18.97%) and forest land (4.01%).

- Miping watershed: Located within the Yangtze River Basin of China, it covers an area of 1402.8 km^2 (110°49′–111°29′ E, 33°34′–33°59′ N). The watershed experiences a warm temperate continental monsoon climate with mild weather conditions, four distinct seasons, and moderate rainfall. It has an average annual rainfall of 830 mm. Land use analysis shows that grassland accounts for the largest proportion (90.52%), followed by farmland (7.86%) and forest land (2.41%).

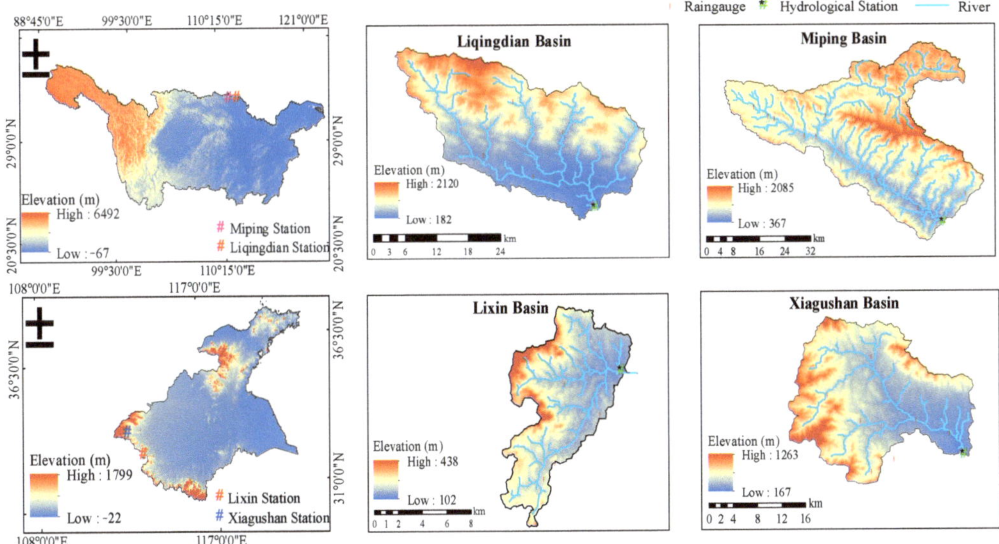

Figure 1. Map showing the geographical distribution of the study area.

2.2. Data Processing

Rainfall and discharge data were obtained from China's Hydrological Statistical Yearbook (1980–2013), ensuring a temporal resolution of 1 h. DEM data were obtained from China's geospatial cloud data (https://www.gscloud.cn/) with a spatial resolution of 30 m. The access date was 6 May 2023. The land use data were obtained from the 1:100,000 land use dataset provided by the National Tibetan Plateau Scientific Data

Center (http://data.tpdc.ac.cn) and the Global Geographic Information Public Product (http://www.globallandcover.com). The access date was 8 May 2023.

2.3. Methods

2.3.1. General Unit Hydrograph

The rainfall–runoff simulation approximates the watershed as a time-invariant linear hydrological system. The simulation assumes that the regulation and storage of the watershed's net rain on the surface can be represented by the regulation of M parallel linear reservoirs, as illustrated in Figure 2. Each linear reservoir within the confluence adheres to the watershed's water storage equation and the conservation law of mass [28]. The unit hydrograph is divided into three sections. The rising and recessing sections follow the "exponential" pattern of growth and decline. As time approaches infinity, the unit hydrograph tends to zero [30]. The equation for the instantaneous unit hydrograph is obtained through the superposition of multiple negative exponential functions [29,31]. It can be expressed as

$$u_{(t)} = \frac{\mu}{t_p} e^{\frac{\mu}{t_p}(t-t_p)} \left(1 + m e^{\frac{\mu}{t_p}(t-t_p)}\right)^{-(1+1/m)} \tag{1}$$

where $u_{(t)}$ represents the instantaneous unit hydrograph; μ represents the rising coefficient determined by watershed characteristics (s); t_p represents the time interval from the peak of net rainfall to the peak of flood (s); m represents the recessing coefficient affected by the downstream water surface conditions.

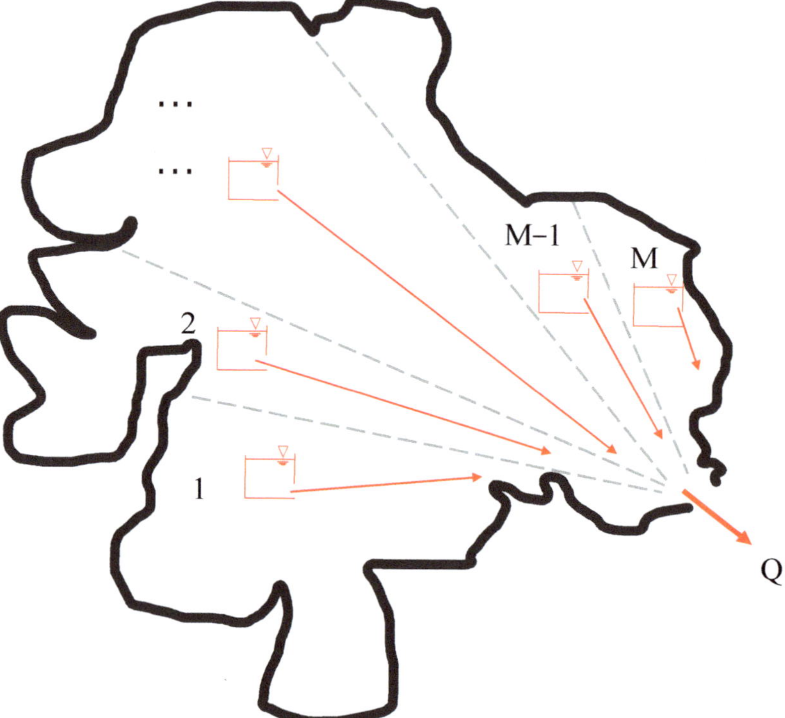

Figure 2. Schematic diagram illustrating the calculation of watershed concentration using a general unit hydrograph.

To make μ dimensionless, we can perform a transformation by letting $\frac{\mu}{t_p} \to \mu$. This transformation leads to the general unit hydrograph $g_{(T)}$:

$$g_{(T)} = \int_0^T u_{(t)} dt = 1 - \left\{ 1 + me^{[\mu(T-t_p)]} \right\}^{-1/m} \tag{2}$$

Given the knowledge of the surface runoff $R_{S(t)}$ for different durations $\Delta\tau$ (starting at $t = \tau$ and ending at $t = \tau + \Delta\tau$), the general unit hydrograph can be employed to deduce the corresponding surface runoff $Q_{s,t}$:

$$\begin{aligned} Q_{s,t} &= \sum_\tau F \times R_{S(\tau)} \frac{1}{\Delta\tau} \left[g_{(t-\tau)} - g_{(t-\tau-\Delta\tau)} \right] \\ &= \sum_\tau F \times R_{S(\tau)} \frac{1}{\Delta\tau} \left\{ \left[1 + me^{\mu(t-\tau-\Delta\tau-t_p)} \right]^{-1/m} - \left[1 + me^{\mu(t-\tau-t_p)} \right]^{-1/m} \right\} \end{aligned} \tag{3}$$

where $Q_{s,t}$ represents the surface runoff at time t ($m^3 \cdot s^{-1}$); F represents the watershed area (m^2); $\Delta\tau$ represents rainfall duration starting at $t = \tau$ and ending at $t = \tau + \Delta\tau$; $R_{S(\tau)}$ represents the surface runoff depth in duration $\Delta\tau$ (m); τ represents a dummy variable in terms of time starting at $\tau = 0$ and ending at $\tau = t$ (s). For the selected flood data, we assume that τ changes with a magnitude equal to the selected time step, which is 1 in this case.

2.3.2. SCE-UA Algorithm

The SCE-UA algorithm is an efficient and robust global optimization method, known for its ability to effectively exploit population information, thereby enhancing algorithm convergence speed. The method integrates four key concepts: (1) a combination of deterministic and probabilistic approaches; (2) systematic evolution across point groups in the parameter space to achieve global improvement; (3) competitive evolution; (4) complex recombination [35]. Complex recombination improves survival capability by enabling the sharing of independently acquired information about the search space among the complexes. While the SCE-UA algorithm involves numerous parameters, most of them can adopt default values based on existing research findings. The determination of the complex number "p" is the only parameter, which needs to be tuned based on the specific problem at hand [40]. Based on the recommendation in the literature, "p" represents the number of complexes; "n" represents the number of parameters; "m_1" represents the number of vertices in each complex; "q" represents the number of vertices in each subcomplex; "s" represents the population size; while "α" and "β" correspond to the number and algebraic characteristics of offspring generated from the parent generation; the values of the parameters are defined as follows: $m_1 = 2n + 1$, $q = n + 1$, $s = pm_1$, $\alpha = 1$, and $\beta = 2n + 1$ [35]. Figure 3 illustrates the flowchart framework of the SCE-UA method employed in this study. Here, we provide a detailed description of the method:

1. Initialization: Set the dimensionality of the problem, "p" (value = 2); the value of "n" is 3 in the calculation process using the general unit hydrograph method, which involves only three parameters: μ, m, t_p; "m_1" (value = 7); "s" (value = 14).
2. Sample generation: Randomly generate s sample points within the feasible parameter space and compute the criterion value F for each point. F comprises four components: f_{NSE} denotes the Nash efficiency coefficient function; f_{R2} represents the determination coefficient function; f_{RE} corresponds to the absolute error function; and $f_{\Delta t}$ symbolizes the peak time difference function.

$$F = 0.6 \times (f_{NSE} + f_{R2}) + 0.2 \times (1 - f_{RE}) + 0.2 \times (1 - f_{\Delta t}) \tag{4}$$

3. Sorting points: Arrange the s points in descending order based on their criterion values, with the first point corresponding to the maximum F value and the last

point representing the F value (in the study, the maximum value is considered as the objective function).

4. Partition into complexes: Divide the s points into p complexes, where each complex contains m points.
5. Complex evolution: Evolve each complex using the competitive complex evolution (CCE) algorithm.
6. Complex recombination: Merge the points from the evolved complexes into a sample population; sort this population in ascending order of F.
7. Criterion evaluation: If the termination criteria are satisfied, stop; otherwise, go to Step 4. The convergence criteria for the optimization process are as follows: the iteration terminates when the objective function F attains the maximum allowable value or when the rate of change satisfies the specified minimum ratio (0.01%), indicating convergence.

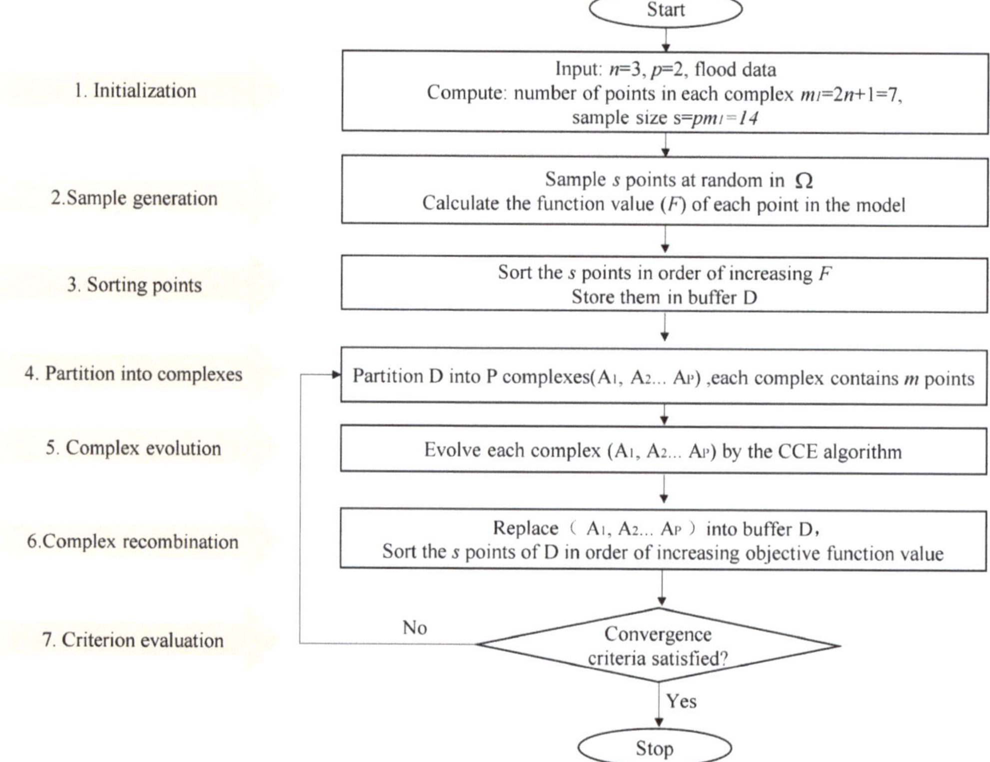

Figure 3. Flowchart of the SCE-UA optimization algorithm.

2.3.3. SCE-GUH Model

This study investigates the suitability of the general unit hydrograph routing calculation for hydrological forecasting and flood disaster defense by analyzing the characteristics of its flow routing calculation. A relatively simple two-source model is implemented, where the routing module is divided into surface runoff and groundwater runoff components. The linear reservoir method, a well-established approach, is utilized to compute the groundwater runoff by simulating the processes of groundwater storage and release. To compute the surface runoff, the general unit hydrograph recession calculation, investigated in this study, is utilized, with parameter values being optimized using the SCE-UA algorithm. The SCE-GUH model integrates the SCE-UA optimization algorithm with the general unit hydrograph routing method. Specifically, the coupling process involves iterative computation. Initially, the SCE-UA algorithm generates a set of parameters for optimization, determined by the specified parameter range and initial values. Subsequently, these parameter values are utilized to compute the output of the general unit hydrograph method, which is compared with observed data in order to obtain the objective function, serving as an indicator for model fitting. Subsequently, the SCE-UA algorithm employs specific strategies to update or adjust the parameters, considering the current parameter values and the objective function. These changes may include parameter increases, decreases, or substitutions. The updated parameter values are subsequently employed to calculate the output of the general unit hydrograph routing method, repeating the process for the subsequent round of computing the statistic and adjusting the parameters. This process continues until the statistic achieves its maximum value or meets other specified stopping criteria. In summary, the aim of integrating the optimization algorithm with the model is to iteratively update the parameters and identify the parameter combination, which optimally aligns the general unit hydrograph routing method with real-world observed data, ultimately enhancing the SCE-GUH model's applicability and predictive accuracy.

2.3.4. Model Benchmarks and Methods

Multiple benchmarks were employed to assess the performance of the SCE-GUH model. These benchmarks comprise Nash's instantaneous unit hydrograph (NIUH), Nash's instantaneous unit hydrograph coupled with the SCE-UA algorithm model (SCE-NIUH), and traditional general unit hydrograph (GUH). The model construction framework is depicted in Figure 4. The 53 flood scenarios employed for model calibration and verification are consistent, which is the premise for comparing the performance of different benchmarks. Among the four watersheds, we select flood events with strong data availability, where the first 70% of the data is used as the calibration period, and the remaining 30% is used as the validation period. Both the calibration and validation periods encompass different hydrological conditions, such as drought periods, normal periods, and high-flow periods, to ensure the model performs well under various hydrological conditions. We concurrently calculated and compared the performance evaluation metrics of all benchmarks.

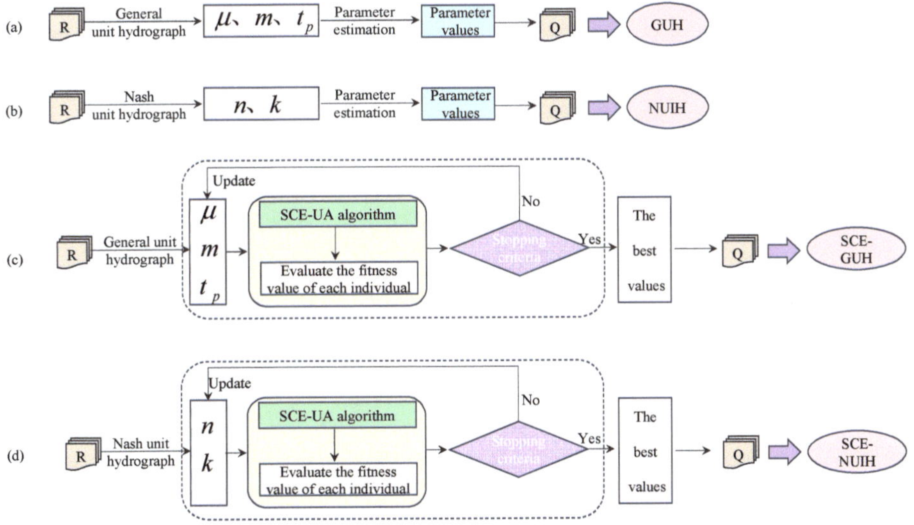

Figure 4. Diagram illustrating the structure of the model: (**a**) GUH; (**b**) NIUH; (**c**) SCE-GUH; (**d**) SCE-NIUH.

The NIUH and SCE-NIUH models were calculated using the Nash unit hydrograph method, which assumes linearity and time invariance in the watershed's response, treating the watershed as a sequence of linear reservoirs connected in series, depicted in Figure 5. The Nash unit hydrograph calculates the confluence using the gamma function distribution, which has emerged as one of the most frequently utilized methods for confluence calculation [22,41].

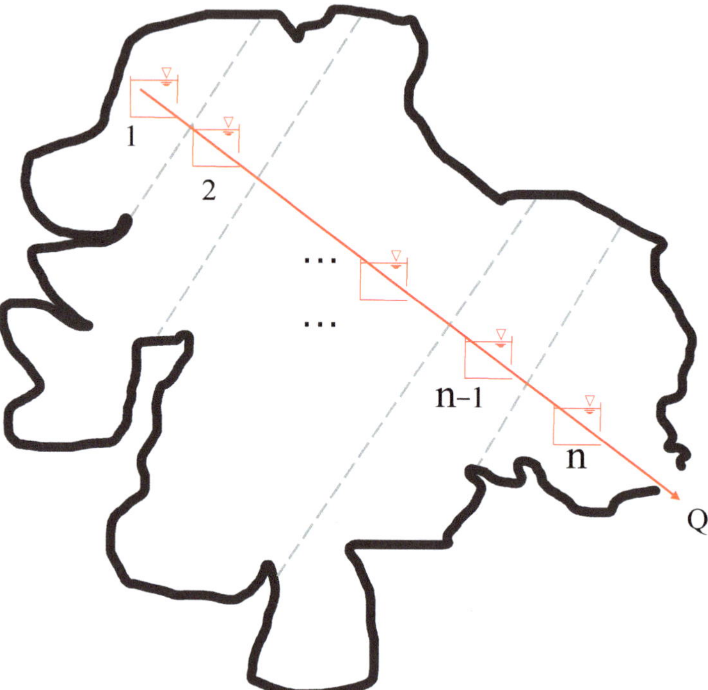

Figure 5. Schematic diagram illustrating the calculation of watershed concentration using the Nash unit hydrograph method.

2.3.5. Selection of Evaluation Indicators

The mathematical expressions for these metrics can be described as follows.

The average Nash–Sutcliffe efficiency coefficient (*NSE*) quantifies the model's ability to predict variables deviating from the mean. It indicates the proportion of the initial variance explained by the model and varies from $-\infty$ to 1, where 1 represents a perfect fit. Higher values closer to 1 indicate more accurate predictions:

$$NSE = 1 - \frac{\sum_{i=1}^{n} \left(Q_S^i - Q_O^i\right)^2}{\sum_{i=1}^{n} \left(Q_S^i - \overline{Q_O}\right)^2} \tag{5}$$

where n represents the total number of measured data; Q_S^i and Q_O^i denote the simulated and observed discharge, respectively; $\overline{Q_0}$ represents the average observed discharge.

The coefficient of determination (R^2) is commonly employed to quantify the level of fit between data. A higher R^2 value indicates a stronger association with the reference equation, while a lower R^2 value (closer to 0) implies a weaker association, as illustrated in Equation (5):

$$R^2 = \frac{\left[\sum_{i=1}^{n}(Q_S^i - \overline{Q_S})(Q_O^i - \overline{Q_O})\right]^2}{\sum_{i=1}^{n}(Q_S^i - \overline{Q_S})^2(Q_O^i - \overline{Q_O})^2} \tag{6}$$

where $\overline{Q}_s$ represents the average simulated discharge.

The relative error (RE) is calculated by multiplying the ratio of the absolute error of a measurement to the actual value by 100%. Equation (6) shows the calculation. In the evaluation of flash flood models, RE is commonly employed to assess the reliability of the simulated flood peak discharge:

$$RE = \frac{Q_s^{max} - Q_0^{max}}{Q_0^{max}} \times 100\% \tag{7}$$

where Q_s^{max}, Q_0^{max} represent the maximum simulated peak discharge and observed peak discharge, respectively.

The absolute error of peak present time (Δt) can be calculated as the difference between the moment when the maximum flood discharge appears in the forecast process and the moment when it appears in the actual flood process. The formula to calculate Δt is as follows:

$$\Delta t = t_s^{max} - t_0^{min} \tag{8}$$

where t_s^{max}, t_0^{min} represent the occurrence time of simulated and observed flood peaks, respectively.

3. Results

3.1. Model Calibration and Validation Results

The evaluation employs multiple discriminant indicators, including NSE, R^2, RE, and Δt. Considering the small drainage area and short flood duration in the study area, the allowable accuracy for predicting the flood peak is set at 20% of the measured flood peak. This implies that the predicted flood peak should be within $\pm 20\%$ of the actual observed value. Furthermore, a permissible deviation of up to 3 h is set for the peak time to avoid significant differences between the predicted and actual peak times. Table 1 presents the optimal parameter values utilized in computing surface flow concentration using the SCE-UA algorithm. These parameters are determined during the calibration process of the SCE-GUH model and the SCE-NUIH model.

Table 1. The parameter values were derived using the SCE-UA algorithm.

Watershed	SCE-GUH			SCE-NIUH	
	m	tp/h	μ	n	k
Lixin	1.60	0.96	2.18	1.61	3.32
Xiagushan	1.90	0.60	0.20	1.86	3.74
Liqingdian	6.70	2.10	2.50	2.53	3.11
Miping	1.00	2.00	1.50	2.79	4.40

A total of 39 representative floods were selected from 53 flood events in four watersheds: 12 from the Lixin watershed, 11 from the Xiagushan watershed, 8 from the Liqingdian watershed, and 8 from the Miping watershed. The selection process aimed to capture a range of different flood characteristics. The simulation results of the four models during the parameter calibration period are presented in Table 2, and Figure 6 illustrates the simulated hydrograph. The hydrographs generated by the NIUH, SCE-NIUH, GUH, and SCE-GUH models exhibit a close agreement with the observed hydrographs. These representative floods were utilized for model parameter calibration.

Table 2. Summary table of simulated results for flood events during the calibration period.

Basin	RE/%				Nse				R2				△t/h			
	NUIH	SCE-NUIH	GUH	SCE-GUH	NUIH	SCE-NUIH	GUH	SCE-GUH	NUIH	SCE-NUIH	GUH	SCE-GUH	NUIH	SCE-NUIH	GUH	SCE-GUH
Lixin	14.21	17.30	9.70	11.35	0.69	0.73	0.75	0.82	0.73	0.74	0.79	0.85	0.67	0.50	0.42	0.17
Xiagushan	14.33	15.72	8.55	8.40	0.73	0.76	0.80	0.82	0.75	0.83	0.84	0.85	0.82	1.00	0.73	0.45
Liqingdian	12.13	12.52	13.35	12.60	0.71	0.76	0.82	0.84	0.79	0.86	0.87	0.87	1.13	1.00	0.75	0.75
Miping	13.70	16.78	8.21	5.54	0.70	0.72	0.83	0.87	0.75	0.79	0.93	0.95	2.38	2.38	0.63	0.63

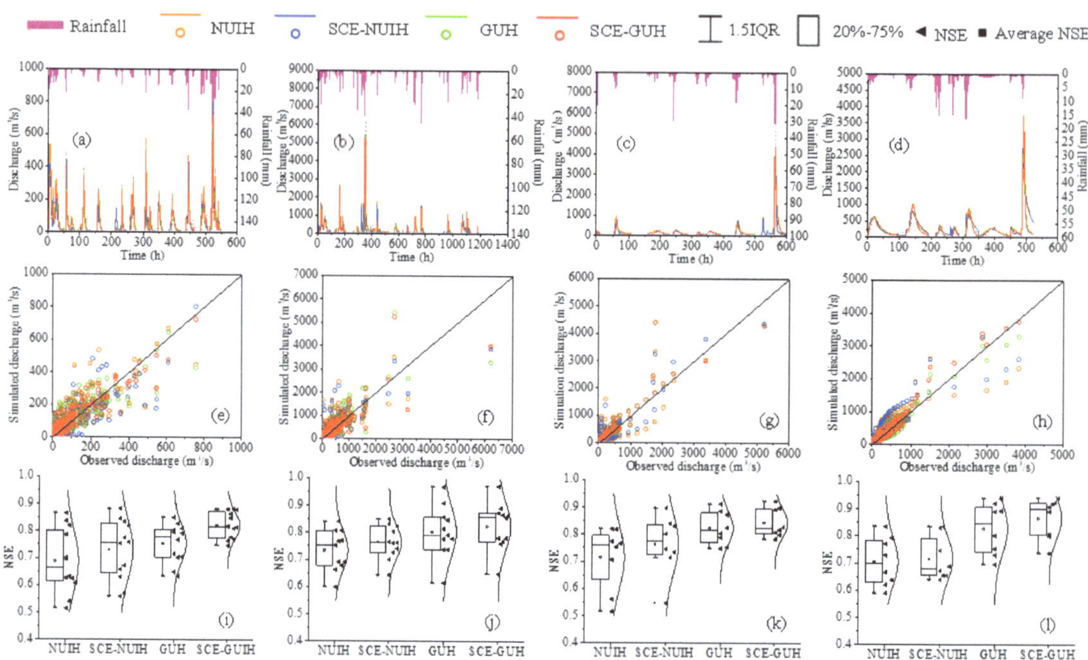

Figure 6. Calibration period flood events simulation results. Among them, (**a–d**) represents the process diagram of observed and simulated runoff in Lixin, Xiagushan, Liqingdian, and Miping watersheds; (**e–h**) is a scatter plot of observed and simulated runoff in Lixin, Xiagushan, Liqingdian, and Miping watersheds; (**i–l**) is the *NSE* map of Lixin, Xiagushan, Liqingdian, and Miping watersheds.

We conducted 12 typical flood simulations in the Lixin watershed. The average *RE* calculated for these simulations were 14.21%, 17.30%, 9.70%, and 11.35% for the NIUH, SCE-NIUH, GUH, and SCE-GUH models, respectively. The respective models achieved peak flow rates with qualification rates of 83.33%, 83.33%, 100%, and 100%. Δt across all four models were determined as 0.67 h, 0.50 h, 0.42 h, and 0.17 h, respectively. Each model achieved a peak flow time qualification rate of 100%. The average *NSE* values for the NIUH, SCE-NIUH, GUH, and SCE-GUH models were 0.69, 0.73, 0.75, and 0.82, respectively. Additionally, the corresponding R^2 values were 0.73, 0.74, 0.79, and 0.85. However, when performing calculations with the NIUH model, the relative errors of peak flow rates for the floods on 14 June 1984 and 20 July 1987 exceeded 20%, violating the allowable error threshold. In contrast, the flood simulations conducted using the GUH model produced peak flow rates, which closely matched the measured flow rates within the acceptable error limit for both floods. It is worth noting that the simulation results obtained from the GUH model exhibited superior performance compared to the NIUH model in terms of *RE*, peak flow rate qualification, R^2, Δt, and *NSE*. We simulated 11 typical flood events in the Xiagushan watershed using four models: NIUH, SCE-NIUH, GUH, and SCE-GUH.

The *RE* of the average simulation results for these models were obtained and found to be 14.33%, 15.72%, 8.55%, and 8.40%, respectively. The flood peak qualification rates for each model were calculated as 90.91%, 90.91%, 100%, and 100%, respectively. The Δt were 0.82 h, 1 h, 0.73 h, and 0.45 h, respectively. Furthermore, all four models achieved a peak timing qualification rate of 100%. Moreover, the average *NSE* values were 0.73, 0.76, 0.80, and 0.82, and the average R^2 values were 0.75, 0.83, 0.84, and 0.85, respectively. It is worth noting that applying the SCE-NIUH model in this watershed resulted in an overall improvement of 0.03 in *NSE* and 0.08 in R^2 compared to the NIUH model. However, the simulations of average *RE* and Δt using the SCE-NIUH model slightly underperformed compared to the NIUH model. This discrepancy can be attributed to the SCE-UA algorithm, which assigns an *NSE* proportion of up to 0.6 during parameter calibration for the objective function and prioritizes improving the model's *NSE* [42]. Additionally, the simulation results using GUH for flow routing showed significant improvements over the NIUH. Moreover, the simulation results using the SCE-GUH model outperformed the GUH model. Specifically, the average *RE* and Δt values were reduced by 0.15% and 0.28 h, respectively, and the average *NSE* and R^2 improved by 0.02 and 0.01, respectively, compared to the GUH model.

We simulated eight typical flood events in the Liqingdian watershed using four models: NIUH, SCE-NIUH, GUH, and SCE-GUH. The resulting average *RE* values for the models were 12.13%, 12.52%, 13.35%, and 12.60%, respectively. While the *RE* values of other models decreased compared to the NIUH model, all models (except NIUH, with a peak flow rate passing the rate of 87.5%) achieved a 100% pass rate. The average Δt were obtained as 1.13 h, 1 h, 0.75 h, and 0.75 h, corresponding to peak time pass rates of 87.50%, 90.91%, 100%, and 100%, respectively. Moreover, the NIUH, SCE-NIUH, GUH, and SCE-GUH models achieved average *NSE* values of 0.71, 0.76, 0.82, and 0.84, and average R^2 values of 0.79, 0.86, 0.87, and 0.87, respectively. A case analysis of the flood event, which occurred on 23 July 2005, showed that the NUIH model had *RE*, Δt, and *NSE* values of 15.38%, 5 h, and 0.71, respectively, while for the SCE-NIUH model, the corresponding values were 16.7%, 4 h, and 0.71, respectively. In contrast, the GUH model produced *RE*, Δt, and *NSE* values of 13.36%, 1 h, and 0.79, and the SCE-GUH model had values of 13.38%, 1 h, and 0.78, respectively. It is important to note that utilizing the NIUH model led to a Δt value for the flood event, which exceeded the acceptable error range. While the SCE-UA algorithm reduced the error by 1 h, it still exceeded the forecast accuracy standards. However, by utilizing the GUH model, enhancements in *NSE* were observed, and Δt was significantly reduced to 1 h, effectively controlling it within the allowable error range.

We conducted eight typical flood simulations in the Miping watershed using four models, namely NIUH, SCE-NIUH, GUH, and SCE-GUH. The average *RE* values for these models were determined as 13.70%, 16.78%, 8.21%, and 5.54%, respectively. The qualification rates for flood peak accuracy were recorded as 87.50%, 87.50%, 100%, and 100% for these models, respectively. The average Δt values were determined as 2.38 h, 2.38 h, 0.63 h, and 0.63 h for the four models, respectively, while achieving qualification rates of 75%, 75%, 100%, and 100%. The average *NSE* values were calculated as 0.70, 0.72, 0.83, and 0.87 for the NIUH, SCE-NIUH, GUH, and SCE-GUH models, respectively. The average R^2 values were 0.75, 0.79, 0.93, and 0.95 for NIUH, SCE-NIUH, GUH, and SCE-GUH, respectively. Notably, the NIUH and SCE-NIUH models displayed significant errors in simulating the floods, which occurred on 4 September 2003 and 1 October 2005, with average Δt of 4 h and 6 h, respectively. In contrast, the GUH and SCE-GUH models precisely reproduced the timing of flood peaks observed in the actual hydrographs. Consequently, there was a significant improvement in the *NSE* values, with the NIUH model transitioning from 0.67 and 0.62 to 0.75 and 0.73, and the SCE-NIUH model transitioning from 0.67 and 0.65 to 0.80 and 0.81 for the GUH and SCE-GUH models, respectively.

The simulation results of the four models, based on the application practices during the calibration period in four watersheds, demonstrate that the general unit hydrograph algorithm significantly outperforms the NIUH algorithm in terms of controlling peak timing error and relative peak flow error. Figure 7 illustrates the error in peak present time. Additionally, calibrating unit hydrograph parameters using the SCE-UA algorithm can enhance simulation accuracy. This demonstrates that the optimized parameter values are reasonably close to the true values, making them suitable for broader application. In summary, the performance of the four models can be ranked as follows: SCE-GUH > GUH > SCE-NIUH > NIUH.

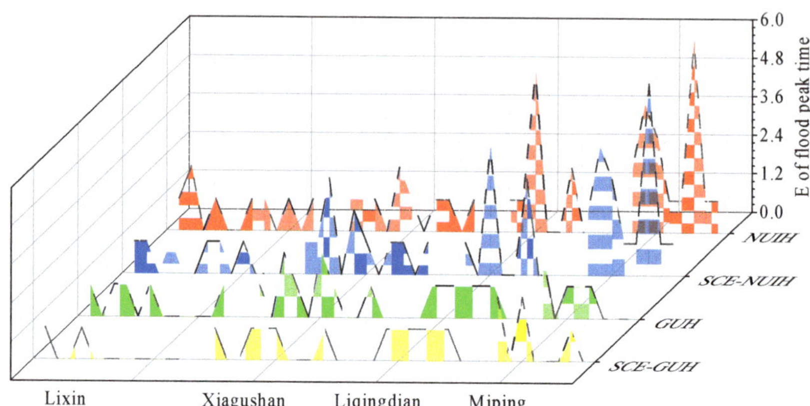

Figure 7. Diagram showing the time difference of simulated flood peaks during the calibration period.

The verification period involved four typical flood events in the Lixin and Xiagushan watersheds, respectively, as well as three events in the Liqingdian and Miping watersheds, respectively. Table 3 provides the simulated results for flood events during the verification period, whereas Figure 8 illustrates the simulated rainfall and runoff, along with the performance evaluation chart of the four models throughout this period.

Table 3. Table of simulated results during the validation period for flood events. In the relative error, a negative sign (−) indicates that the simulated peak flow is lower than the observed flow. In the peak timing error, a negative sign (−) indicates that the simulated peak occurs earlier than the actual peak.

Number	RE/%				Nse				R2				△t/h			
	NUIH	SCE-NUIH	GUH	SCE-GUH	NUIH	SCE-NUIH	GUH	SCE-GUH	NUIH	SCE-NUIH	GUH	SCE-GUH	NUIH	SCE-NUIH	GUH	SCE-GUH
Lixin watershed																
20030830	−2.22	−15.76	−5.83	1.19	0.89	0.93	0.92	0.94	0.93	0.95	0.95	0.96	0	0	0	0
20050708	10.47	13.33	8.19	11.42	0.79	0.94	0.82	0.87	0.84	0.88	0.88	0.90	0	0	0	0
20080722	10.39	−15.23	−12.23	−5.54	0.54	0.72	0.77	0.83	0.70	0.77	0.82	0.85	−1	−1	−1	−1
2010019	7.71	−14.30	−16.21	−16.14	0.73	0.78	0.81	0.86	0.81	0.85	0.88	0.89	0	0	0	0
Average	7.70	14.66	10.62	8.57	0.73	0.82	0.83	0.87	0.82	0.86	0.88	0.90	0.25	0.25	0.25	0.25
Xiagushan watershed																
20000714	−85.90	−57.85	−25.88	−18.66	0.52	0.61	0.59	0.60	0.62	0.65	0.62	0.63	0	0	−1	1
20020626	−4.30	−16.26	0.20	6.99	0.62	0.78	0.91	0.95	0.65	0.79	0.92	0.96	−1	−1	0	0
20100718	−0.84	−17.95	−7.24	−11.38	0.66	0.67	0.89	0.91	0.70	0.69	0.89	0.92	0	0	0	0
20130525	−1.62	3.50	1.87	5.53	0.71	0.79	0.86	0.89	0.72	0.79	0.86	0.89	3	3	0	0
Average	23.17	23.89	8.80	10.64	0.63	0.71	0.81	0.84	0.67	0.73	0.82	0.85	1	1	−0.25	0.25
Liqingdian watershed																
20100819	−23.99	−17.58	0.77	1.28	0.62	0.59	0.72	0.73	0.72	0.63	0.76	0.76	1	1	−1	1
20100823	4.46	10.23	3.95	6.23	0.70	0.76	0.77	0.81	0.79	0.85	0.82	0.84	3	0	−3	3
20110914	1.02	8.92	−8.55	−3.35	0.80	0.88	0.93	0.94	0.88	0.94	0.95	0.95	0	1	0	0
Average	9.82	12.24	4.42	3.62	0.71	0.74	0.80	0.83	0.8	0.81	0.84	0.85	1.33	0.67	1.33	1.33
Miping watershed																
20090816	−17.62	−19.04	−12.74	−7.06	0.54	0.55	0.86	0.85	0.73	0.75	0.86	0.87	3	1	−1	−1
20100724	−7.81	−2.60	−7.69	−4.86	0.79	0.83	0.88	0.88	0.89	0.89	0.89	0.89	−1	−1	2	2
20110913	0.81	4.62	−3.13	1.59	0.75	0.82	0.92	0.91	0.84	0.87	0.93	0.93	1	1	1	−1
Average	8.75	8.75	7.85	4.50	0.69	0.73	0.89	0.88	0.82	0.84	0.89	0.90	1.67	1	1.33	1.33

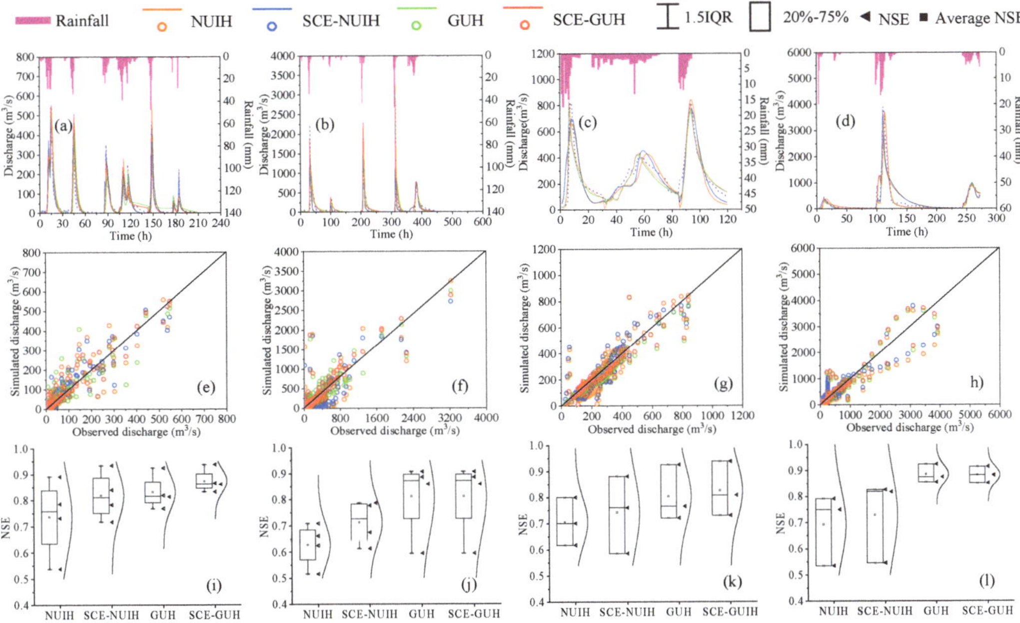

Figure 8. Verification period flood events simulation results. Among them, (**a–d**) represents the process diagram of observed and simulated runoff in Lixin, Xiagushan, Liqingdian, and Miping watersheds; (**e–h**) is a scatter plot of observed and simulated runoff in Lixin, Xiagushan, Liqingdian, and Miping watersheds; (**i–l**) is the *NSE* map of Lixin, Xiagushan, Liqingdian, and Miping watersheds.

The four models in the Lixin watershed all achieved a total qualification rate of 100%. There was a 1 h time lag for the flood event, which occurred on 22 July 2008, while no time lag was observed for other events. The average *NSE* values for the four models, namely NIUH, SCE-NIUH, GUH, and SCE-GUH, were 0.73, 0.82, 0.83, 0.87, while the average R^2 values were 0.82, 0.86, 0.88, 0.90, respectively. In the Xiagushan watershed, the *RE* values for simulating the typical flood event on 14 July 2000 were 85.90%, 57.85%, 25.88%, and 18.66% for the NIUH, SCE-NIUH, GUH, and SCE-GUH models, respectively. The total qualification rates for the NUIH, SCE-NUIH, GUH, SCE-GUH models were 75%, 75%, 75%, and 100%, respectively, with average time lags of 1 h for the NIUH and SCE-NIUH models, and 0.25 h for the GUH and SCE-GUH models. The average *NSE* values for the models were 0.63, 0.71, 0.81, and 0.84, while the average R^2 values were 0.67, 0.73, 0.82, and 0.85, respectively. In the Liqingdian watershed, the *RE* values for the four models during the flood event on 19 August 2010 were 23.99%, 17.58%, 0.77%, and 1.28% for NUIH, SCE-NUIH, GUH, and SCE-GUH, respectively. Consequently, the NUIH model had a total qualification rate of only 66.67%. However, the simulation results of the other three models were within the acceptable prediction error range, resulting in a total qualification rate of 100%. The SCE-NIUH model had an average time lag of 0.67 h, while the other three models had an average time lag of 1.33 h. The average *NSE* values for the models were 0.71, 0.74, 0.80, 0.83, respectively. Similarly, the average R^2 values were 0.80, 0.81, 0.84, and 0.85, respectively. The four simulated models in the Miping watershed all achieved a total qualification rate of 100%. The average time lags for the NUIH, SCE-NIUH, GUH, SCE-GUH models were 1.67 h, 1 h, 1.33 h, and 1.33 h, respectively. The average *NSE* values for the models were 0.69, 0.73, 0.89, and 0.88, while the average R^2 values were 0.82, 0.84, 0.89, and 0.90, respectively. Figure 9 shows the errors in peak time presentation for the four watersheds based on the four methods.

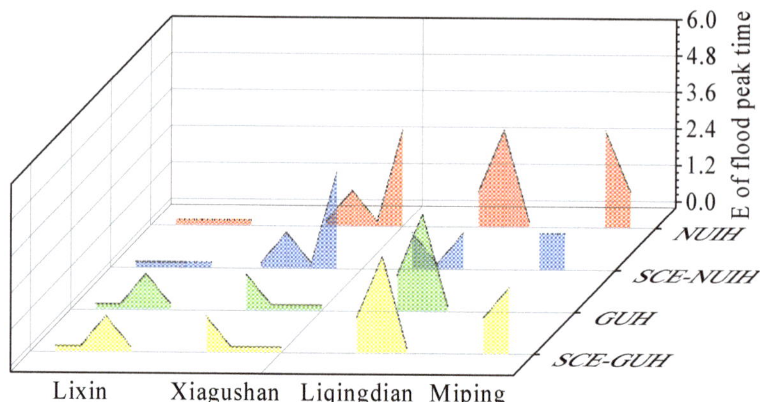

Figure 9. Diagram showing the time difference of simulated flood peaks during the validation period.

We conducted a comprehensive performance analysis of the four models across the selected watersheds. The SCE-GUH model was calibrated, and the average values of NSE, R^2, RE were found to be 0.82, 0.85, 11.35%, respectively, with Δt of 1.67 h in the Lixin watershed. The calibration period yielded results of 0.82 for NSE, 0.85 for R^2, 8.40% for RE, and Δt of 0.45 h in the Xiagushan watershed. Average values of 0.84 (NSE), 0.87 (R^2), 12.60% (RE), and 0.75 h (Δt) were obtained in the Liqingdian watershed. The values during the calibration period were 0.87 for NSE, 0.95 for R^2, 5.54% for RE, and Δt of 0.63 h in the Miping watershed. The SCE-GUH model was validated, and the average values of NSE, R^2, RE were found to be 0.87, 0.90, 8.57%, respectively, with Δt of 0.25 h in the Lixin watershed. The validation period yielded results of 0.84 for NSE, 0.85 for R^2, 10.64% for RE, and Δt of 0.25 h in the Xiagushan watershed. Average values of 0.83 (NSE), 0.85 (R^2), 3.62% (RE), and 1.33 h (Δt) were obtained in the Liqingdian watershed. The values during the validation period were 0.88 for NSE, 0.90 for R^2, 4.50% for RE, and Δt of 1.33 h in the Miping watershed. Thus, it can be concluded that the SCE-GUH model demonstrates favorable applicability in simulating rainfall and runoff in the four watersheds.

The peak occurrence time errors for Lixin, Xiagushan, Liqingdian, and Miping in the SCE-GUH model were 0.19 h, 0.40 h, 0.91 h, and 0.82 h, respectively. In the GUH model, the errors were 0.38 h, 0.60 h, 0.91 h, 0.82 h. In the SCE-NUIH model, the errors were 0.38 h, 1.00 h, 0.91 h, 2.00 h. In the NUIH model, the errors were 0.51 h, 0.87 h, 1.18 h, 2.19 h. Compared with the other three models, the SCE-GUH model showed a significant reduction in peak time error. Among the four studied watersheds, the peak time qualified rate was 100% in both the GUH and SCE-GUH models, while in the NUIH and SCE-NUIH models, the qualified rates were only 81.80% in the Miping watershed and 90.91% in the Liqingdian watershed. The NSE values for the four watersheds in the SCE-GUH model were 0.83, 0.83, 0.84, 0.87; in the GUH model, the values were 0.77, 0.81, 0.82, 0.84; in the SCE-NUIH model, the values were 0.75, 0.75, 0.76, 0.72; in the NUIH model, the values were 0.70, 0.70, 0.71, 0.70. By comparing the results, it can be observed that the SCE-GUH model improved the fitting degree of the flood hydrograph. Analyzing the overall qualification rate of the four watersheds, the SCE-GUH model achieved a 100% qualification rate. The GUH model had the second highest qualification rate, with all watersheds except Xiagushan achieving a 100% qualification rate. It is worth noting that the overall qualification rate of the Xiagushan watershed is 93.33%. However, in the SCE-NUIH model, only the Liqingdian watershed achieved a 100% qualification rate, while the other three watersheds had qualification rates of 87.5% for Lixin, 86.67% for Xiagushan, and 72.73% for Miping. In the NUIH model, the highest qualification rate was 87.50% in the Lixin watershed, while the lowest qualification rate was only 72.73% in the Liqingdian and Miping watersheds. Fluctuations occurred between the observed and simulated data during the simulation process of the four models. The main reasons for these fluctuations can be summarized as follows. First, there is an error in calculating the runoff phase. Second, in some cases, the rainfall during flood events exhibits scattered patterns, characterized by multiple rainfall peaks and downstream accumulation. The centralized model employed in this experiment utilized Thiessen polygons for processing rainfall, but it neglected the uneven spatial distribution of rainfall, leading to an error in simulating the flow rate. Lastly, the time intervals for the measured flow data are inconsistent. The study utilized a 1 h interval obtained through interpolation, which might compromise the accuracy of the simulation results.

The comparison among the four models reveals that the model constructed with the general unit hydrograph produces superior simulation results in comparison to the model, which uses the same conditions but relies on the Nash unit hydrograph. Consequently, the general unit hydrograph model, by utilizing runoff calculations, provides a more accurate depiction of the hydrological processes within the watershed compared to the Nash unit hydrograph model. This finding significantly contributes to enhancing our understanding and facilitating efficient management of water resources within the watershed.

3.2. Typical Site Flood Analysis

Figure 10a presents the simulation results of the flood event of 4 August 1995 in the Lixin watershed using four models, NIUH, SCE-NIUH, GUH, and SCE-NIUH, with corresponding Δt of 1 h, 1 h, 0, and 0. The *NSE* values for these models were 0.62, 0.63, 0.79, and 0.88, respectively. The comparison clearly shows that applying the general unit hydrograph method significantly enhances the accuracy of flood forecasting and reduces errors in predicting the timing of peak flow. Furthermore, optimizing the parameters of the SCE-UA algorithm led to additional enhancements in the *NSE*. The floods, which occurred on 20 August 1995, were studied in the Xiagushan watershed. The *RE* for this flood, as predicted by the four models, were 12.79%, 3.90%, 5.90%, and 0.61%. The Δt used in the models were 1 h, 1 h, 0, and 0, while the *NSE* values were 0.81, 0.82, 0.97, and 0.97, correspondingly. Figure 10b demonstrates that the rising and falling water process, specifically in the rising and falling water section, based on the GUH model closely resembles the observed process. The peak time is in perfect agreement with the measured process. In contrast, the total duration of the confluence as predicted by the NIUH differs from the observed process and is considerably shorter in duration. The results indicate that the SCE-GUH model demonstrates superior simulation performance. Moreover, the application of the general unit hydrograph method significantly enhanced the accuracy of flood forecasting while reducing the discrepancy between the simulated time of peak occurrence and the actual observed time of peak occurrence. By optimizing the parameters of the SCE-UA algorithm, the *NSE* and the rate of prediction qualification were further improved to a certain extent. The simulation of the flood event, which occurred on 14 September 2011 in the Liqingdian watershed, indicates that the *RE* and Δt values obtained from the four models fall within an acceptable range. Moreover, the *NSE* values exceed 0.80, indicating satisfactory simulation results. Please refer to Figure 10c for the simulation hydrograph. The simulation results of the four models exhibit a high degree of similarity during the rising stage of the flood. However, during the initial phase of recession, the two models based on NIUH and SCE-NIUH demonstrate a noticeably slower decline in the simulated flood compared to the observed flood. Nevertheless, the parameter *m* utilized in the calculations for both the GUH and SCE-GUH models partly accounts for the regression rate by incorporating the hydrological characteristics of the watershed. This integration helps align the simulated regression process more closely with the observed regression process. Figure 10d presents the simulation results of the flood event, which occurred on 29 July 2007 in the Miping watershed, using four models: NIUH, SCE-NIUH, GUH, and SCE-NIUH. The *RE* for these models were 19.00%, 16.82%, 17.01%, and 2.20%, respectively. The Δt were 1 h, 1 h, 0, and 0, respectively. Additionally, the *NSE* values were 0.64, 0.69, 0.94, and 0.94, respectively. The simulation results for the rising and falling stages of this flood exhibit considerably higher accuracy in the GUH and SCE-GUH models compared to the NUIH and SCE-NUIH models. This flood is categorized as a major flood; due to the NUIH and SCE-NUIH models' sensitivity to early rainfall, within the first 10 h of rainfall, the water levels rise rapidly, and the flood peak recedes swiftly after reaching its maximum. These processes occur at a shorter duration compared to the observed measurements. Moreover, the flood peak emerges earlier than recorded in the observation period. This suggests that the SCE-GUH model exhibits superior simulation performance, followed by the GUH model in the second place. Conversely, the SCE-NUIH and NUIH models perform notably worse than the previously mentioned two models in simulating this flood event.

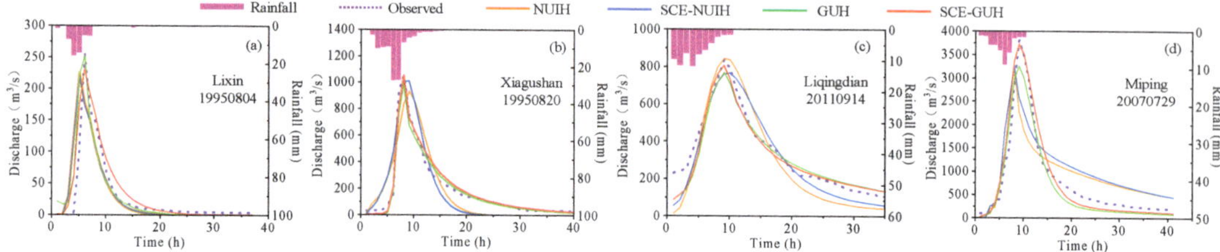

Figure 10. Results of a typical flood simulation. Subfigures (**a–d**) show the comparison between the simulated results and the measured results of the selected typical flood events in four basins: Lixin, Xiagushan, Liqingdian, and Miping, respectively, under four different models.

The combined use of the general unit hydrograph and the SCE-UA algorithm holds significant practical implications. The ranking order of flood simulation performance for the four models in the four watersheds is as follows: SCE-GUH > GUH > SCE-NIUH > NIUH. The concentration calculation equation of the general unit hydrograph is based on a negative exponential function, according to the calculation principle. The parameters, such as μ and m, partially reflect the watershed characteristics and control the amplitude of flood rise and fall [31]. The convergence of the Nash unit hydrograph is typically calculated using a gamma function with parameter n. When calculating runoff from the instantaneous unit hydrograph S curve, a table lookup is commonly used [43]. Consequently, the general unit hydrograph convergence model exhibits higher flexibility and is more aligned with the actual confluence process. The Nash unit hydrograph is based on the fundamental assumption of a linear reservoir series within the watershed [17]. Higher sensitivity to net rainfall is observed when the current soil moisture content is relatively high, resulting in a greater change in the hydrograph amplitude compared to the measured hydrograph. The three parameters of the general unit hydrograph integrate rainfall characteristics and watershed features, including water system shape and watershed slope, based on the analysis of parameter characteristics [28]. For instance, in the case of the Liqingdian watershed, which possesses a parallel water system with an elevation difference close to 2000 m, incorporating these factors into the confluence calculation leads to improved simulation results for rainfall–runoff. In conclusion, the application of the general unit hydrograph principle in the simulation of rainfall–runoff provides a more accurate characterization of watershed catchment characteristics compared to the Nash unit hydrograph.

3.3. Analysis of the Influence of Parameters on Unit Hydrograph

Extensive research had been conducted on the influence of Nash unit hydrograph parameters in the existing literature. Previous studies have revealed that two parameters in the Nash unit hydrograph exert a certain influence on the three elements of the unit hydrograph, namely peak flow, peak lag time, and total duration [20,44]. Nonetheless, further discussion is needed regarding the influence of the three parameters of the general unit hydrograph on the three elements of the unit hydrograph [28]. For instance, as an illustration, a net rainfall input of 10 mm per unit time period (1 h) in the Liqingdian watershed was considered. In the experiment, only one parameter was debugged at a time, allowing for an analysis of the effects of parameter changes on the three elements of the unit hydrograph. The results are depicted in Figure 11.

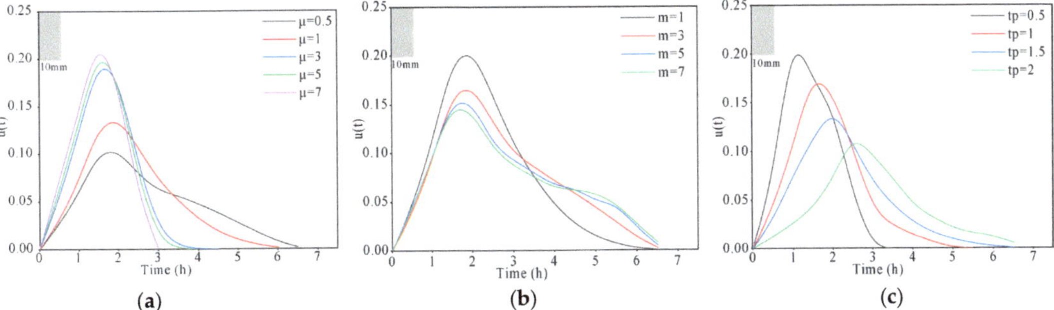

Figure 11. Illustration of how variations in parameters affect the general unit hydrograph. Subfigures (**a–c**) respectively represent the effects of parameter μ, m, t_p variations on the general unit hydrograph.

Figure 11a illustrates that as μ increases, the unit hydrograph exhibits a "tall and thin" shape, resulting in an increase in the flood peak, a decrease in the delay time of the flood peak, and a reduction in the total duration. The results presented in Figure 11b demonstrate that increasing the value of m leads to a "squat" shape of the unit hydrograph, which in turn causes a decrease in the flood peak. However, the change in m has minimal impact on the delay time and total duration of the flood peak. It is worth noting that at the end of the confluence, the flow will increase alongside the increase in m. The observation depicted in Figure 11c indicates that an increase in t_p results in a "plump and flat" shape of the unit hydrograph. This change leads to a decrease in the flood peak, a significant increase in the delay time of the flood peak, and a significant extension of the total duration.

4. Discussion

The performance of the four constructed test models in the watershed is satisfactory. These models effectively capture the variations and attenuation of flood events, maintaining the simulation errors of peak flow and peak time within acceptable limits. Nevertheless, all models typically underestimate the magnitude of flood peaks, a common observation in other lumped models [6]. The performance of the four models is ranked as follows: SCE-GUH > GUH > SCE-NIUH > NIUH. This ranking result aligns with the expectations and validates the applicability of the SCE-GUH model, which outperforms the other three models in simulating episodic floods. The GUH model exhibited slightly better overall error control and simulations of flood fluctuation and dissipation compared to the NIUH model. Consistent with previous studies, this article's research confirms that the general unit hydrograph method is more applicable than the Nash unit hydrograph method [29,30]. Furthermore, the SCE-GUH model exhibited superior overall performance compared to the GUH model. This is due to the effective reduction in errors in predicting peak flow and flood volume achieved by utilizing the complex evolutionary theory within the SCE-UA algorithm for parameter optimization. The model is also automatically calibrated to improve the NSE and R^2 based on the constraints of the objective function [38,42]. The SCE-GUH model demonstrates superior compliance in terms of the relative error of peak flow and enhances the correlation coefficient compared to other lumped models [45]. Overall, the SCE-GUH model demonstrates superior performance, notably in accurately simulating the peak flow and improving the correlation coefficient. This distinction sets it apart from other models in the field, making it a valuable tool for hydrological and flood prediction research. Regrettably, we did not consider the problem of target function divergence during the model construction process [46]. Further discussion is required to address the divergence issue, enhance convergence speed, and ensure the target function achieves the desired effect. Our experiments were conducted exclusively in small watershed areas, which are susceptible to flash floods in the Yangtze River and Huaihe River regions of China. The performance of this model may be influenced by variations in climate change, terrain, hydrological processes, and data availability across different regions. Consequently, further research is necessary to investigate the model's applicability in other regions worldwide based on this theory, as well as to explore its practical utilization in decision support and disaster management.

5. Conclusions

This study introduces a new flood forecasting model, SCE-GUH, which combines the general unit hydrograph with the SCE-UA optimization algorithm. The SCE-GUH model is founded on a robust theoretical basis and substantiated by scientific evidence. The applicability of SCE-GUH in simulating flash flood was examined using data from 53 observed flash flood events in the Lixin, Xiagushan, Liqingdian, and Miping watersheds. Moreover, a comparison was made between the simulation results of SCE-GUH and those of the NIUH, SCE-NIUH, and traditional GUH models. The key findings of this study are summarized as follows.

The performance ranking of these four models across the four watersheds is as follows: SCE-GUH > GUH > SCE-NIUH > NIUH. The SCE-GUH model exhibits stability and robustness across various flood scenarios. The model's structure is succinct, optimizing a mere three parameters. This reduces the risk of overfitting, lowers computational and storage costs, and enhances overall efficiency. The calculation formula of the general unit hydrograph method is relatively simple. The application of the general unit hydrograph enhances the description of surface runoff processes and introduces a new perspective into flood prediction research. Additionally, the application of the negative exponential function in the confluence is simpler compared to the gamma function (Nash unit hydrograph). This simplification reduces error transmission and the cumulative effect. The parameters involved in the calculation include the topographic and geomorphic characteristics of the watershed and the rainfall characteristics. Changes in these parameters have a significant impact on the three elements of the general unit hydrograph, with the flood peak being the most sensitive. The utilization of the SCE-UA composite optimization strategy for parameter optimization allows for an improved characterization of the catchment characteristics in the watershed, resulting in enhanced timeliness and prediction accuracy. Therefore, the SCE-UA model is well suited for regions prone to flash floods, where data are limited. By considering factors such as rainfall, soil moisture content, and topography comprehensively, it enables precise simulation of water flow and accumulation processes in a watershed, thus facilitating accurate prediction of flood propagation and evolution trends. Furthermore, this serves as a vital scientific basis for flood forecasting and prevention within the watershed while also providing a fresh perspective on international flash flood early warning and prediction. However, it is important to note that this experiment was only conducted in four selected flow domains, and its applicability should be tested further in other watersheds.

Author Contributions: Conceptualization, Y.X. and C.L.; methodology, Y.X. and C.H.; software, Q.Y.; validation, C.Z.; formal analysis, Y.X. and L.Q.; investigation, C.L.; resources, C.H.; data curation, Y.X. and C.L.; writing—original draft preparation, Y.X.; writing—review and editing, C.H.; supervision, Q.Y.; project administration, C.H.; funding acquisition, C.H. All authors have read and agreed to the published version of the manuscript.

Funding: Innovative research projects of the first-class project special fund of the Yellow River Laboratory (Zhengzhou University), grant number YRL22IR02; Projects of National Natural Science Foundation of China, grant number U2243219; Projects of National Natural Science Foundation of China, grant number 51979250.

Data Availability Statement: Not applicable.

Acknowledgments: We would like to thank the potential reviewer very much for their valuable comments and suggestions. We also thank our other colleagues for their valuable comments and suggestions, which have helped improve the manuscript.

Conflicts of Interest: The authors declare no conflict of interest.

References

1. Zhang, Z. Characteristics of flash floods in China and their prevention and control ideas. *China Water Resour.* **2007**, *14*, 14–15.
2. Liu, C.; Guo, L.; Ye, L.; Zhang, S.; Zhao, Y.; Song, T. A review of advances in China's flash flood early-warning system. *Nat. Hazards* **2018**, *92*, 619–634. [CrossRef]
3. Tamm, O.; Saaremäe, E.; Rahkema, K.; Jaagus, J.; Tamm, T. The intensification of short-duration rainfall extremes due to climate change—Need for a frequent update of intensity–duration–frequency curves. *Clim. Serv.* **2023**, *30*, 100349. [CrossRef]
4. Li, B.; Liang, Z.; Fu, Y.; Wang, J.; Hu, Y. A model for production and sink flow based on variable saturation zone and its parameter determination method. *Adv. Water Sci.* **2022**, *33*, 208–218.
5. Corral, C.; Berenguer, M.; Sempere-Torres, D.; Poletti, L.; Silvestro, F.; Rebora, N. Comparison of two early warning systems for regional flash flood hazard forecasting. *J. Hydrol.* **2019**, *572*, 603–619. [CrossRef]
6. Usman, M.; Ndehedehe, C.E.; Farah, H.; Ahmad, B.; Wong, Y.; Adeyeri, O.E. Application of a Conceptual Hydrological Model for Streamflow Prediction Using Multi-Source Precipitation Products in a Semi-Arid River Basin. *Water* **2022**, *14*, 1260. [CrossRef]
7. Georgakakos, K.P. Analytical results for operational flash flood guidance. *J. Hydrol.* **2006**, *317*, 81–103. [CrossRef]
8. Ntelekos, A.A.; Georgakakos, K.P.; Krajewski, W.F. On the uncertainties of flash flood guidance: Toward probabilistic forecasting of flash floods. *J. Hydrometeorol.* **2006**, *7*, 896–915. [CrossRef]
9. Tamm, O.; Kokkonen, T.; Warsta, L.; Dubovik, M.; Koivusalo, H. Modelling urban stormwater management changes using SWMM and convection-permitting climate simulations in cold areas. *J. Hydrol.* **2023**, *622*, 129656. [CrossRef]
10. He, B. Overview of flash flood warning methods of the Japan Meteorological Agency. *China Flood Control Drought Relief* **2020**, *30*, 149–152.
11. Bartholmes, J.C.; Thielen, J.; Ramos, M.H.; Gentilini, S. The european flood alert system EFAS—Part 2: Statistical skill assessment of probabilistic and deterministic operational forecasts. *Hydrol. Earth Syst. Sci.* **2009**, *13*, 141–153. [CrossRef]
12. Thielen, J.; Bartholmes, J.; Ramos, M.H.; de Roo, A. The European Flood Alert System—Part 1: Concept and development. *Hydrol. Earth Syst. Sci.* **2009**, *13*, 125–140. [CrossRef]
13. Cheng, W. A review of critical rainfall research on flash floods. *Adv. Water Sci.* **2013**, *24*, 901–908.
14. Guo, L.; Tang, X.; Kong, F. Research and application of flash flood early warning and forecasting system based on distributed hydrological model. *China Water Resour.* **2007**, *14*, 38–41.
15. Liu, Z.; Martina, M.L.; Todini, E. Flood forecasting using a fully distributed model: Application of the TOPKAPI model to the Upper Xixian catchment. *Hydrol. Earth Syst. Sci.* **2005**, *9*, 347–364. [CrossRef]
16. Koren, V.; Reed, S.; Smith, M.; Zhang, Z.; Seo, D.J. Hydrology Laboratory Research Modeling System (HL-RMS) of the US National Weather Service. *J. Hydrol.* **2004**, *291*, 297–318. [CrossRef]
17. Rui, X. Stochastic production and convergence theory. *Adv. Water Resour. Hydropower Sci. Technol.* **2016**, *36*, 8–12+39.
18. Ghumman, A.R.; Khan, Q.U.; Hashmi, H.N.; Ahmad, M.M. Comparison of Clark, Nash Geographical Instantaneous Unit Hydrograph Models for Semi Arid Regions. *Water Resour.* **2014**, *41*, 364–371. [CrossRef]
19. Zhu, C.; Wang, X.; Dai, X. Calculate Sherman unit hydrograph by stepwise filter optimization method. *J. Hydroelectr. Eng.* **2015**, *34*, 1–6.
20. Ge, W.; Chen, C. A review of unit line-based confluence studies: A review of Huang Wanli's "transient process line". *Water Sci. Econ.* **2021**, *27*, 24–27.
21. Agirre, U.; Goni, M.; Lopez, J.J.; Gimena, F.N. Application of a unit hydrograph based on subwatershed division and comparison with Nash's instantaneous unit hydrograph. *Catena* **2005**, *64*, 321–332. [CrossRef]
22. Babaali, H.R.; Sabzevari, T.; Ghafari, S. Development of the Nash instantaneous unit hydrograph to predict subsurface flow in catchments. *Acta Geophys.* **2021**, *69*, 1877–1886. [CrossRef]
23. Lee, K.T. Generating design hydrographs by DEM assisted geomorphic runoff simulation: A case study. *J. Am. Water Resour. Assoc.* **1998**, *34*, 375–384. [CrossRef]
24. Rui, X.; Jiang, G. Some advances and comments on the theory and computational methods of basin confluence. *Water Resour. Hydropower Technol.* **1997**, *9*, 43–45.
25. Lai, P.; Xia, C. Variable-velocity geomorphic unit lines. *J. Hefei Univ. Technol. Nat. Sci. Ed.* **1989**, *2*, 102–112.
26. Yan, B.; Duan, M.; Jiang, H.; Li, Z.; Yang, W.; Zhang, J. Study on catchment confluence model based on fractional instantaneous unit line. *Peoples Changjiang* **2020**, *51*, 84–88.
27. Yi, B.; Chen, L. Time-varying distributed unit lines considering dynamic sink paths. *Adv. Water Sci.* **2022**, *33*, 944–954.
28. Guo, J. General Unit Hydrograph from Chow's Linear Theory of Hydrologic Systems and Its Applications. *J. Hydrol. Eng.* **2022**, *27*, 04022020. [CrossRef]
29. Guo, J. Application of General Unit Hydrograph Model for Baseflow Separation from Rainfall and Streamflow Data. *J. Hydrol. Eng.* **2022**, *27*, 04022027. [CrossRef]

30. Guo, J. General and Analytic Unit Hydrograph and Its Applications. *J. Hydrol. Eng* **2022**, *27*, 04021046. [CrossRef]
31. Guo, J. Design Hydrographs in Small Watersheds from General Unit Hydrograph Model and NRCS-NOAA Rainfall Distributions. *J. Hydrol. Eng.* **2023**, *28*, 5942. [CrossRef]
32. Chen, R.; Kang, E.; Yang, J.; Zhang, J. A review of hydrological modeling studies. *Chin. Desert* **2003**, *18*, 15–23.
33. Hou, D.; Zhang, X.; Wu, Q.; Li, Z.; Wang, L.; Zhang, L. Construction and application of a flash flood forecasting model based on runoff coefficients. *Hydropower Energy Sci.* **2022**, *40*, 83–86.
34. Ren, B.; Liang, J.; Yan, B.; Lei, X.; Fu, W.; Ni, X.; Guo, J. Parameter Optimization of Double-Excess Runoff Generation Model. *Pol. J. Environ. Stud.* **2018**, *27*, 809–817. [CrossRef] [PubMed]
35. Duan, Q.; Sorooshian, S.; Gupta, V.K. Optimal use of the SCE-UA global optimization method for calibrating watershed models. *J. Hydrol.* **1994**, *158*, 265–284. [CrossRef]
36. Duan, Q.Y.; Sorooshian, S.; Gupta, V. Effective and Efficient Global Optimization for Conceptual Rainfall-Runoff Models. *Water Resour. Res.* **1992**, *28*, 1015–1031. [CrossRef]
37. Jeon, J.-H.; Park, C.-G.; Engel, B.A. Comparison of Performance between Genetic Algorithm and SCE-UA for Calibration of SCS-CN Surface Runoff Simulation. *Water* **2014**, *6*, 3433–3456. [CrossRef]
38. Kang, T.; Lee, S. Modification of the SCE-UA to Include Constraints by Embedding an Adaptive Penalty Function and Application: Application Approach. *Water Resour. Manag.* **2014**, *28*, 2145–2159. [CrossRef]
39. Huang, L.; Wang, L.; Zhang, Y.; Xing, L.; Hao, Q.; Xiao, Y.; Yang, L.; Zhu, H. Identification of Groundwater Pollution Sources by a SCE-UA Algorithm-Based Simulation/Optimization Model. *Water* **2018**, *10*, 193. [CrossRef]
40. Li, X.; Cheng, C.; Wu, X.; Lin, J. Research on fuzzy multi-objective SCE-UA parameter preference method for hydrological models. *China Eng. Sci.* **2007**, *3*, 52–57.
41. Boufadel, M.C. Unit hydrographs derived from the Nash model. *J. Am. Water Resour. Assoc.* **1998**, *34*, 167–177. [CrossRef]
42. Xing, Z.; Wang, L.; Wang, X.; Fu, Q.; Ji, Y.; Li, H.; Liu, Y. The Study on Equifinality of Hydrological Model Parameters and Runoff Simulation Based on the Improved Simulation-optimization Algorithm. *J. Basic Sci. Eng.* **2020**, *28*, 1091–1107.
43. Yue, S. A bivariate gamma distribution for use in multivariate flood frequency analysis. *Hydrol. Process.* **2001**, *15*, 1033–1045. [CrossRef]
44. Sun, Y.; Rui, X. Study of Nash unit hydrograph based on random theory. *Water Resour. Power* **2005**, *23*, 18–20.
45. Li, Y.; Wang, G.; Liu, C.; Lin, S.; Guan, M.; Zhao, X. Improving Runoff Simulation and Forecasting with Segmenting Delay of Baseflow from Fast Surface Flow in Montane High-Vegetation-Covered Catchments. *Water* **2021**, *13*, 196. [CrossRef]
46. Turkyilmazoglu, M. Accelerating the convergence of Adomian decomposition method (ADM). *J. Comput. Sci.* **2019**, *31*, 54–59. [CrossRef]

MDPI AG
Grosspeteranlage 5
4052 Basel
Switzerland
Tel.: +41 61 683 77 34

Water Editorial Office
E-mail: water@mdpi.com
www.mdpi.com/journal/water